T0214134

Lecture Notes in Computer Science   12606

Founding Editors

Gerhard Goos
Karlsruhe Institute of Technology, Karlsruhe, Germany
Juris Hartmanis
Cornell University, Ithaca, NY, USA

Editorial Board Members

Elisa Bertino
Purdue University, West Lafayette, IN, USA
Wen Gao
Peking University, Beijing, China
Bernhard Steffen
TU Dortmund University, Dortmund, Germany
Gerhard Woeginger
RWTH Aachen, Aachen, Germany
Moti Yung
Columbia University, New York, NY, USA

More information about this subseries at http://www.springer.com/series/7407

Yong Zhang · Yicheng Xu ·
Hui Tian (Eds.)

Parallel and Distributed Computing, Applications and Technologies

21st International Conference, PDCAT 2020
Shenzhen, China, December 28–30, 2020
Proceedings

 Springer

Editors
Yong Zhang
Shenzhen Institutes of Advanced
Technology
Shenzhen, China

Yicheng Xu ⓘ
Shenzhen Institutes of Advanced
Technology
Shenzhen, China

Hui Tian
Griffith University
Gold Coast, QLD, Australia

ISSN 0302-9743 ISSN 1611-3349 (electronic)
Lecture Notes in Computer Science
ISBN 978-3-030-69243-8 ISBN 978-3-030-69244-5 (eBook)
https://doi.org/10.1007/978-3-030-69244-5

LNCS Sublibrary: SL1 – Theoretical Computer Science and General Issues

© Springer Nature Switzerland AG 2021
This work is subject to copyright. All rights are reserved by the Publisher, whether the whole or part of the material is concerned, specifically the rights of translation, reprinting, reuse of illustrations, recitation, broadcasting, reproduction on microfilms or in any other physical way, and transmission or information storage and retrieval, electronic adaptation, computer software, or by similar or dissimilar methodology now known or hereafter developed.
The use of general descriptive names, registered names, trademarks, service marks, etc. in this publication does not imply, even in the absence of a specific statement, that such names are exempt from the relevant protective laws and regulations and therefore free for general use.
The publisher, the authors and the editors are safe to assume that the advice and information in this book are believed to be true and accurate at the date of publication. Neither the publisher nor the authors or the editors give a warranty, expressed or implied, with respect to the material contained herein or for any errors or omissions that may have been made. The publisher remains neutral with regard to jurisdictional claims in published maps and institutional affiliations.

This Springer imprint is published by the registered company Springer Nature Switzerland AG
The registered company address is: Gewerbestrasse 11, 6330 Cham, Switzerland

Preface

The International Conference on Parallel and Distributed Computing, Applications, and Technologies (PDCAT) is a major forum for scientists, engineers, and practitioners throughout the world to present their latest research, results, ideas, developments, and applications in all areas of parallel and distributed computing. Beginning in Hong Kong in 2000, PDCAT 2020 was held in Shenzhen, China, after 20 years of a successful journey through various countries/regions including Taiwan, Japan, China, Singapore, Australia, New Zealand, and Korea. For the 21st event we invited new and unpublished papers.

The conference papers included in the proceedings cover the following topics: PDCAT of Networking and Architectures, Software Systems and Technologies, Algorithms and Applications, and Security and Privacy. 34 papers were selected from 109 submissions. Accepted and presented papers highlight new trends and challenges of parallel and distributed computing, applications, and technologies. We hope readers will find these contributions useful and inspiring for their future research.

Our special thanks go to the Program Chairs, and all the Program Committee members and Reviewers for their valuable efforts in the review process that helped us to guarantee the highest quality of the selected papers for the conference.

December 2020

Yong Zhang
Yicheng Xu
Hui Tian

Organization

Honorary Chair

Guoliang Chen   Nanjing University of Posts and Telecommunications, China

General Chairs

Jianping Fan   Shenzhen Institute of Advanced Technology, Chinese Academy of Sciences, China
Shengzhong Feng   National Supercomputing Center in Shenzhen, China
Hong Shen   Sun Yat-sen University, China
Chengzhong Xu   University of Macau, China

Program Chairs

Yong Zhang   Shenzhen Institute of Advanced Technology, Chinese Academy of Sciences, China
Li Ning   Shenzhen Institute of Advanced Technology, Chinese Academy of Sciences, China
Hui Tian   Griffith University, Australia
Francis Lau   The University of Hong Kong, China
Haiying Shen   University of Virginia, USA

Organizing Chair

Yanjie Wei   Shenzhen Institute of Advanced Technology, Chinese Academy of Sciences, China

Advisory Chair

Francis Chin   The University of Hong Kong, China

Publications Chairs

Vincent Chau   Shenzhen Institute of Advanced Technology, Chinese Academy of Sciences, China
Yingpeng Sang   Sun Yat-sen University, China

Registration and Finance Chair

Yicheng Xu   Shenzhen Institute of Advanced Technology,
                                        Chinese Academy of Sciences, China

Local Arrangement Committee

Yuyang Li   Shenzhen Institute of Advanced Technology,
                                        Chinese Academy of Sciences, China
Fumin Qi   National Supercomputing Center in Shenzhen, China

Program Committee

Costin Bădică   University of Craiova, Romania
Guangdong Bai   The University of Queensland, Australia
Vincent Chau   Shenzhen Institute of Advanced Technology,
                                        Chinese Academy of Sciences, China
Yawen Chen   University of Otago, New Zealand
Yichuan Dong   National Supercomputing Center in Shenzhen, China
Chunru Dong   Hebei University, China
Zisen Fang   National Supercomputing Center in Shenzhen, China
Guichen Gao   Shenzhen Institute of Advanced Technology,
                                        Chinese Academy of Sciences, China
Huaxi Gu   Xidian University, China
Longkun Guo   Qilu University of Technology, China
Ajay Gupta   Western Michigan University, USA
Xinxin Han   Shenzhen Institute of Advanced Technology,
                                        Chinese Academy of Sciences, China
Lu Han   Academy of Mathematics and Systems Science,
                                        Chinese Academy of Sciences, China
Ping He   Heibei University of Economics and Business, China
Qiang Hua   Hebei University, China
Zhiyi Huang   University of Otago, New Zealand
Haiping Huang   Nanjing University of Posts and Telecommunications,
                                        China
Mirjana Ivanović   University of Novi Sad, Serbia
Graham Kirby   University of St Andrews, UK
Kai-Cheung Leung   University of Auckland, New Zealand
Kenli Li   Hunan University, China
Yamin Li   Hosei University, Japan
Shuangjuan Li   South China Agricultural University, China
Min Li   Shandong Normal University, China
Shangsong Liang   Sun Yat-sen University, China
Zhongzhi Luan   Beihang University, China
Marin Lujak   IMT Lille Douai, France

Li Ning	Shenzhen Institute of Advanced Technology, Chinese Academy of Sciences, China
Cheng Qiao	University College Cork, Ireland
Yingpeng Sang	Sun Yat-sen University, China
Neetesh Saxena	Cardiff University, UK
Guang Tan	Sun Yat-sen University, China
Haisheng Tan	University of Science and Technology of China, China
Jingjing Tan	Weifang University, China
Lei Wang	Peking University, China
Yinling Wang	Dalian University of Technology, China
Xin Wang	Fudan University, China
Zijun Wu	Hefei University, China
Jigang Wu	Guangdong University of Technology, China
Weigang Wu	Sun Yat-sen University, China
Jun Wu	Beijing Jiaotong University
Chenchen Wu	Tianjin University of Technology, China
Di Wu	Sun Yat-sen University, China
Yicheng Xu	Shenzhen Institute of Advanced Technology, Chinese Academy of Sciences, China
Qiang Xu	Wenzhou University Oujiang College, China
Ruiqi Yang	University of Chinese Academy of Sciences, China
Xinfeng Ye	University of Auckland, New Zealand
Dongxiao Yu	Shandong University, China
Jingjing Yu	Beijing Jiaotong University, China
Filip Zavoral	Charles University, Czech Republic
Feng Zhang	Heibei University, China
Zhenning Zhang	Beijing University of Technology, China
Haibo Zhang	University of Otago, New Zealand
Xiaoyan Zhang	Nanjing Normal University, China
Zonghua Zhang	Huawei France, France
Yong Zhang	Shenzhen Institute of Advanced Technology, Chinese Academy of Sciences, China
Yu Zhang	University of Science and Technology of China, China
Dongmei Zhang	Shandong Jianzhu University, China
Cheng Zhong	Guangxi University, China
Chunyue Zhou	Beijing Jiaotong University, China
Rong Zhou	Shenzhen Institute of Advanced Technology, Chinese Academy of Sciences, China
Yifei Zou	The University of Hong Kong, China

Contents

Blood Leukocyte Object Detection According to Model Parameter-Transfer and Deformable Convolution

Kaizhi Chen[1]($\boxtimes$), Wencheng Wei[1], Shangping Zhong[1], and Longkun Guo[2]

[1] College of Mathematics and Computer Science,
Fuzhou University, Fuzhou 350100, Fujian, China
{ckz,spzhong}@fzu.edu.cn, 18506904188@163.com
[2] Shandong Key Laboratory of Computer Networks,
School of Computer Science and Technology, Shandong Computer Science Center (National Supercomputer Center in Jinan), Qilu University of Technology (Shandong Academy of Sciences), Jinan 250353, People's Republic of China
Longkun.guo@gmail.com

Abstract. Currently, leukocyte detection has the problem of scarcity of labeled samples, so a focal dataset must be expanded by merging multiple datasets. At the same time, given the difference in the dyeing methods, dyeing time, and collection techniques, some datasets have the problem of different homology distributions. Moreover, the effect of direct training after dataset merging is not satisfactory. The morphology of the leukocyte types is also variable and stain contamination occurs, thereby leading to the misjudgment of using traditional convolutional networks. Therefore, in this paper, the model parameter-transfer method is used to alleviate the problem of less leukocyte labeled data in the training model and deformable convolution is introduced into the main network of target detection to improve the accuracy of the object detection model. First, numerous leukocyte datasets are used to train the blood leukocyte binary classification detection network, and the model parameters of the blood leukocyte binary classification detection network are transferred to the blood leukocyte multi classification detection network through the transfer of model parameters. This method can make better use of datasets of the same origin and different distributions so as to solve the problem of scarcity in blood leukocyte data sets. Finally, the multi classification detection network is trained quickly and the accurate blood leukocyte detection results are obtained through fine tuning. The experimental results show that compare our method with the traditional Faster RCNN object detection algorithm, $mAP_{0.5}$ is 0.056 higher, $mAP_{0.7}$ is 0.119 higher, with higher recall by 4%, and better accuracy by 5%. Thus, the method proposed in this paper can achieve highly accurate leukocyte detection.

Keywords: Blood leukocyte object detection · Model parameter-transfer · Deformable convolution · Transfer learning

© Springer Nature Switzerland AG 2021
Y. Zhang et al. (Eds.): PDCAT 2020, LNCS 12606, pp. 1–16, 2021.
https://doi.org/10.1007/978-3-030-69244-5_1

1 Introduction

The leukocyte [1] is an important part of human immune cells. In clinic, the determination of the number, type proportion, morphological analysis, and cell development and maturity in blood [2] has vital clinical significance for the judgment of patients' symptoms. We can ascertain a patients' physical conditions by preliminary analysis of these leukocyte parameters.

In the past, the detection and recognition methods [3] of blood leukocytes generally adopt the following steps: segmentation of the blood leukocytes image, artificial design of blood leukocytes features, and shallow classifier training [4]. Therefore, in the traditional detection and recognition methods of blood leukocytes, deviation in any of the above steps would affect the result of blood leukocytes classification. The traditional methods cannot segment and classify blood leukocytes accurately because of the differences in staining, the pollution of staining agents, the incomplete features of image acquisition equipment, and artificial design. N. Humaimi et al. [5] proposed a segmentation method according to cell color and which uses Lab color space to segment blood leukocytes and red blood cells quickly, but the algorithm is inefficient and prone to errors. Huang et al. [6] used the Otsu algorithm of dazin method to segment and remove the nucleus from blood leukocytes, used principal component analysis (PCA) [7] to select and process the features of the nucleus, and finally employed K-means clustering algorithm to classify the nucleus; however, the monocyte in the blood leukocytes and part of the nucleus in the lymphocyte were similar, thereby leading to misjudgment. Wang et al. [8] extracted blood leukocytes from blood micrographs on the basis of image color information and gradient vector flow, realized and extracted high saturation features, and extracted white nuclei by morphology. Finally, the color, morphology, and texture features of leukocytes were classified by a support vector machine. Pan et al. [9] proposed a fast and simple framework for blood leukocytes image segmentation and which uses the extreme learning machine [10] to learn the samples through the simulation visual system and designs the feature training and classification model for the blood leukocytes image manually. Tabrizi et al. [11] combined a feature selection algorithm and neural network classifier to classify blood leukocytes. The first method is to select leukocyte image features by PCA and then use the LVQ neural network [12] to classify. The second method entails selecting features of a leukocyte by SFS and Fisher judgment criteria and finally using SVM to classify leukocytes.

In recent years, object detection [14] on the basis of deep learning [13] has rapidly improved in terms of recognition accuracy and recognition speed and has excellent performance in other tasks. Shirazi et al. [15] used Wiener filter and curvelet transform to enhance image and remove noise for eliminating false edge to achieve blood leukocyte segmentation, employed the combination of threshold processing and mathematical morphology to perform blood leukocyte segmentation and boundary detection, and finally utilized a BP neural network to conduct five classifications of blood leukocytes. In line with the deep learning theory, Qin et al. [16] constructed a blood leukocytes classifier according to the deep residual neural network. Zhao et al. [17] removed the blood leukocytes region by binary processing of the blood cell image and then applied the morphological method to automatically detect the blood leukocytes classification combined with the self-designed CNN model. However, given staining pollution and noise interference, the region cutting effect is not satisfactory.

Using object detection according to deep learning to complete the task of blood leukocytes recognition can simplify the steps of blood leukocytes detection, replace the difficult leukocyte segmentation with the automatic search for the location of the leukocyte region, and realize end-to-end automatic recognition of blood leukocytes. However, traditional deep learning has the main assumption that training, test, and future data must be in the same feature space and have the same data distribution. When using deep learning to build an object detection model, numerous data annotations are needed. For a medical image, image data is easy to collect but requires the professional knowledge of doctors to label the image. Doctors, however, cannot spend much time to label accurately. In this case, if we can transfer the knowledge we have learned, then we will avoid the tedious and consuming data marking work and considerably improve the generalization performance of the model. In recent years, transfer learning [18] has become a new learning framework to solve this problem. It aims to fine-tune the trained model through the existing data so as to reduce the training time and improve model performance. Under the dyed blood micrograph, leukocyte morphology is varied and changed greatly, and training the deep learning model is difficult because of the light, dye, and dyeing time. The deformable convolutional network proposed by Dai et al. in 2017 [20] can increase the original fixed receptive field into a deformable receptive field by adding an offset field without adding additional datasets and additional training costs. This approach improves the deep model's ability to extract spatial deformation features, such that the deep learning model has significantly improved the effect of complex tasks.

The main contributions of this article are as follows:

1) This study introduces deformable convolution and deformable region of interest pooling on the basis of ResNet-101 layers to improve the feature extraction ability of the object detection model;
2) In this research, the model parameter-transfer is used to build the model of blood leukocyte binary classification detection. Then transfer model parameters to blood leukocyte multi-class object detection model to reduce the impact of less data volume on model performance.
3) Traditional blood leukocyte detection uses segmentation followed by classification. This article adopts an end-to-end deep object detection method to avoid the limitation of poor classification and unsatisfactory classification because of the segmentation effect.

2 Related Work

2.1 Deformable Convolution

In stained blood microscopic images, leukocytes have various shapes and are easily affected by staining. Therefore, making the traditional convolutional neural network adapt well to the stained leukocyte image after data enhancement of the blood leuko-cyte image is difficult. Given that traditional convolutional neural networks use fixed-geometry convolution kernels when modeling images, this method restricts traditional convolutional neural networks to only perform fixed-position feature extraction on input images, an approach which results in the model's poor ability to extract features from the target object in the image. In the pooling process, similar operations are adopted to reduce the resolution of the extracted feature space by a certain percentage, thereby achieving a higher feature loss rate and further causing a slower decline in the loss function, a low degree of fit, and low detection accuracy and poor performance for the trained model.

Deformable convolution [20] adds a 2D offset to the position of the grid sampling on the basis of the traditional convolutional layer and allows free deformation of the convolutional layer sampling to replace the traditional sampling fixed geometric network grid mode. As shown in Fig. 1, (a) is a traditional sampling grid, and (b), (c), and (d) depict those based on the traditional sampling grid to introduce offsets for expansion.

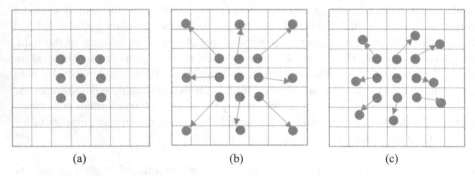

(a)   (b)   (c)

Fig. 1. Traditional convolution and deformable convolution sampling

2.2 Region of Interest Pooling

In the early object detection task, the region-based sliding window method [21] and selective search [22] are used to obtain the region proposals. This technique of obtaining the region proposal will generate numerous region proposals and lead to the disadvantages of slow processing speed and inability to achieve the end-to-end training model. Region of interest pooling (RoI pooling) solves these problems and significantly improves the speed of the object detection model during the training and testing phases and improves model detection accuracy.

ROI pooling can transform the rectangular region of any size input into a small feature map of fixed size. Given the feature map x, a ROI of w × h and the vertex of the upper left corner p_0, RoI pooling divides RoI into k × k (k is a hyperparameter) bins and output a k × k feature map y. The resulting Formula (1) is as follows:

$$y(i, j) = \sum_{p\in\text{bin}(i,j)} x(p_0 + p)/n_{ij},\qquad(1)$$

where n_{ij} is the number of pixels in the chunk.

On the basis of RoI pooling, deformable RoI pooling adds an offset field { } in the spatial bins. We refer to Formula (1) to obtain Formula (2):

$$y(i, j) = \sum_{p \in bin(i,j)} x(p_0 + p + \Delta p_{ij})/n_{ij}, \tag{2}$$

Usually Δp_{ij} is a decimal, and the feature map is normalized by the full convolution layer after RoI pooling to generate an offset field $\Delta \hat{p}_{ij}$ of size k × k. Given the size of the feature map and the inconsistent RoI, they cannot be used directly. Therefore, by using an empirical gain value $\gamma = 0.1$ to process with the width w and height h of the feature map, we can obtain $\Delta p_{ij} = \gamma \Delta \hat{p}_{ij} \cdot (w, h)$.

2.3 Parameter-Based Transfer Learning

Parameter-based transfer learning [19] is a type of transfer learning. In this way, the source and target domains share model parameters. The introduction of transfer learning is due to the fact that current deep learning requires massive data as a basis to acquire a satisfactory model, but insufficient data is an inevitable problem in some fields. Transfer learning seeks to resolve the issue of unsustainable data. It applies the knowledge learned in a certain task or field to similar counterparts to improve the learning of a target task or field.

In transfer learning, a domain $\mathcal{D}$ is composed of a feature space χ and an edge probability distribution $P(X)$, where $X = \{x_1, x_2, \ldots, x_n\} \subset \chi$. For example, in the classification of blood leukocytes, χ is the feature space, X is the set of leukocytes, x_i is the i-th leukocyte, and $P(X)$ is the classification of a leukocyte. Given a domain $\mathcal{D} = \{\chi, P(X)\}$, the task is composed of the label space Y and the prediction function, that is, the conditional probability distribution P(X|Y),where P(X|Y) is obtained from the learning and training of feature vectors and label sets, and the type of sample x_i is judged by P(X|Y). According to the above definition, given a source domain $\mathcal{D}_S$, the corresponding task T_S. Then, another target domain $\mathcal{D}_T$ is given and the corresponding task T_T.

In our task $\mathcal{D}_S \neq \mathcal{D}_T$ or $T_S \neq T_T$, through the introduction of model parameter-transfer in transfer learning, the parameters of the $\mathcal{D}_S$ trained model are transferred to $\mathcal{D}_T$ as the initial parameters, by learning the knowledge information of $\mathcal{D}_S$ and T_S to ascertain the prediction function P(X|Y) of $\mathcal{D}_T$. Applying this method can make better use of existing datasets with different distributions and improve the performance of our object detection model effectively.

3 Formulation

3.1 Main Idea

The Faster RCNN proposed by He et al. [27] is used as the backbone network of our blood leukocyte detection model in this paper. The ResNet-101 layer is used in combination as the feature collection network of the Faster RCNN, and according to the proposal by Dai et al. [20], the deformable convolutional network is utilized to reconstruct the convolutional layer and pooling layer of the ResNet-101 layer to improve the ability of the object detection model to extract leukocyte deformation.

In training the model, we adopt a two-stage method. In the first stage, we train the object detection network of the binary classification of blood leukocyte and then transfer the learned features and parameters to the object detection network of the multi-classification of blood leukocytes for fine-tuning.

3.2 Deformable Convolution Network Structure

Traditional convolutional networks use fixed-structure convolution kernels for sampling, and their receptive field also involves a fixed geometric structure. After multiple convolutional sampling in the network, the receptive field expands but remains a fixed structure. The scope may not match the target shape in the task. In deformable convolution, the sampling of the convolution kernel becomes flexible by adding an offset to the convolution layer. Thus, deformable convolution has a more flexible receptive field, and its sampling range is the same as the target in the task. The shape is more consistent, thereby further improving the efficiency of convolution kernel sampling. The difference between traditional and deformable convolution samplings is shown in Fig. 2.

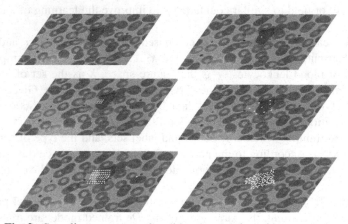

Fig. 2. Sampling processes of traditional and deformable convolutions.

The traditional convolution layer mainly consists two steps: 1) sampling through the grid on the feature map, for which the sampling position is decided by $R = \{(-1,0), (0,-1), ..., (1,1), (0,1)\}$ and 2) input of the collected data to the downsampling layer for pooling operation.

The matrix quantity of the offset matrix is $\{\Delta p_n | n = 1, 2, \ldots, N\}$, where $N = |R|$. In the traditional convolutional network, we perform a convolution operation on each point p_0 collected on the feature map as shown in Formula (3).

$$y(p_0) = \sum_{p_n \in R} \omega(p_n) \cdot x(p_0 + p_n), \tag{3}$$

Formula (3) in deformable convolution becomes Formula (4) as follows.

$$y(p_0) = \sum_{p_n \in R} \omega(p_n) \cdot x(p_0 + p_n + \Delta p_n), \tag{4}$$

The offset Δp_n is usually a fraction, and the pixel value is an integer. At this time, the idea of bilinear interpolation [24] is introduced to solve this problem. The offset matrix is obtained from the offset obtained by learning each point on the input feature map in the convolution layer, and its size equivalent to that of the input feature map plane, but the depth is twice that of the feature map.

As the sampling grid in deformable convolution is no longer a fixed geometric structure, its sampling position can be freely transformed and it constantly tends to the object during the model training process, so that more and more comprehensive information about target features can be obtained. This adaptive sampling method is very effective for blood leukocyte detection and classification.

3.3 Deformable Position-Sensitive RoI Pooling

As the number of model layers continues to increase, the translation and rotation invariance of the model becomes stronger. This property has important significance for the robustness of the classification of the model. In the object detection model, the position of the object requires good perception ability. In the deep model, however, excessive translation and rotation invariance will reduce the model's ability to perceive the object position, thereby affecting the performance of the model. Taking position-sensitive RoI pooling (PS RoI pooling) [23] in the RoI pooling stage can effectively solve this problem. To better obtain the position and type of leukocytes in the blood image, the same method as deformable convolution is adopted to construct deformable PS RoI pooling and use the bilinear interpolation method to obtain pixel coordinates for mean pooling. Fig. 3 shows the structure of the deformable PS RoI pooling.

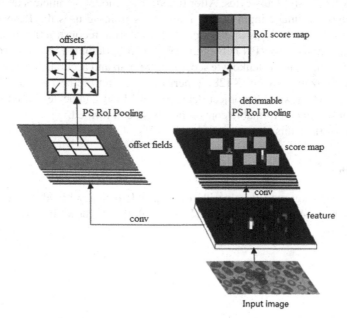

Fig. 3. Structure of deformable PS ROI pooling.

3.4 Model Framework for Blood Leukocyte Object Detection on the Basis of Model Parameter-Transfer and Deformable Convolution

The construction of the blood leukocyte detection model is divided into two stages. The first stage involves the training of the blood leukocyte binary classification (leukocyte, background) detection model. The model structure of the blood leukocyte binary classification detection is consistent with that of the blood leukocyte multiple classification detection. Moreover, the feature extraction network is reconstructed by deformable convolution.

For the first stage, we reconstruct the convolutional layers of res3b, res4b, res5a, res5b, and res5c in the Conv3 to Conv5 stages of ResNet into deformable convolutional networks. After generating suggested regions in the RPN network, we use RoI pooling to classify blood leukocytes and bounding box regression. This stage is trained as a binary classification model, in which the categories are leukocytes and background, and finally the accurate positioning of leukocytes is obtained by NMS [28].

In our datasets, an imbalance exists in the number of leukocyte types. In the first stage of training, an additional BCCD[1] dataset was added because the BCCD dataset and the dataset trained in the multi-classification stage belong to the isomorphic blood leukocyte dataset. The introduction of the BCCD dataset in the binary classification training can endow the model with better feature extraction ability, and the transfer parameters and features up to the classification model can improve the detection rate and classification accuracy of leukocytes.

After the first stage of training for leukocytes, the model parameters and features are transferred to the multi-classification model. That is, the model trained in the first stage is used as the two-stage pre-training model to improve how the model detects leukocytes and its ability to classify leukocytes. After transferring model parameters and features, the blood leukocyte image input during multi-class training uses the ResNet feature extraction network on the basis of deformable convolution reconstruction to obtain the feature map and enters the RPN network to obtain RoI combined with deformable PS RoI pooling for attaining the leukocyte score feature map and accurate location; Finally, multi-class optimization by NMS [28] is performed to get the best result. Thus far, a complete model framework for object detection of blood leukocytes based on model parameter-transfer and deformable convolution has been formed.

The framework of this algorithm is shown in Fig. 4.

3.5 Algorithm

Under the above background, the algorithm of object detection of blood leukocytes based on model parameter-transfer and deformable convolution is shown below:

---

[1] https://github.com/Shenggan/BCCD_Dataset.

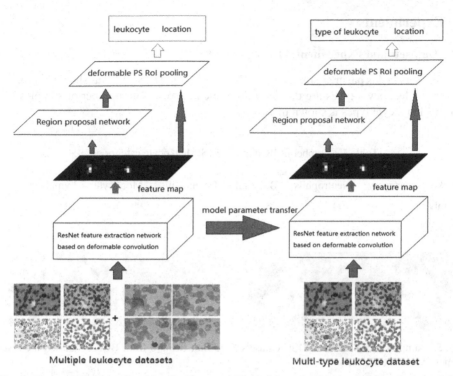

Fig. 4. Framework of the algorithm in this paper (The left branch is binary classification, the right branch is multi classification).

Algorithm 1 Object detection of blood leukocyte on the basis of model parameter-transfer and deformable convolution

---

1. Prepare the leukocyte detection dataset, record as $D = \{d_1, d_2, ..., d_N\}$, D is a collection of five types of leukocyte datasets, where d_i is i-th leukocyte dataset, $d_i = \{x_{i1}, x_{i2}, ..., x_{in}\}$, x_{ij} is the j-th sample of calss i leukocytes.

2. Count the number of samples of each type of leukocyte dataset $\|d_i\|$, divide D into $D_S = \{d_{a1}, d_{a2}, ..., d_{ak}\}$ and $D_T = \{d_{b1}, d_{b2}, ..., d_{bm}\}$, make $\|D_S\| = \|d_{a1}\| + \|d_{a2}\| + \cdots + \|d_{ak}\| > N_1$, $\|D_S\| = \|d_{b1}\| + \|d_{b2}\| + \cdots + \|d_{bk}\| > N_1$, $N_1 \gg N_2 > 1000$。

3. Train the deformable convolution reconstruction leukocyte binary classification Faster RCNN detection model. The training set is D_S and the additional BCCD dataset, the number of rounds is 20, and the pre-training model M_s is obtained.

4. Use M_s as the pre-training model of the leukocyte multi-classification Faster RCNN detection model reconstructed using deformable convolution and deformable PS RoI Pooling. The training set is D_s, the test set is D_T, and the number of rounds is 20. Finally, a multi-class detection model for leukocytes is achieved.

---

4   Experiments

4.1   Datasets and Experimental Environment

The datasets in this paper are mainly from the BCISC[2] and LISC [29] as shown in Fig. 5, where the leukocyte types are divided into five categories. The number of samples of various leukocyte types is shown in Table 1.

Table 1. Number of BCICS and LISC leukocyte categories

Leukocyte category	Neutrophils	Basophil	Eosinophils	Monocyte	Lymphocyte
Number	100	98	89	99	104

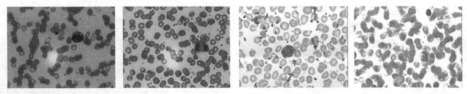

Fig. 5. Samples of BCICS and LISC datasets (The first two are from BCICS dataset, the rest comes from LISC dataset).

The additional BCCD dataset in the first state is shown in Fig. 6, and the number of various leukocyte samples is shown in Table 2.

Table 2. Number of BCICS and LISC leukocyte categories.

Leukocyte category	Neutrophils	Basophil	Eosinophils	Monocyte	Lymphocyte
Number	218	4	94	23	40

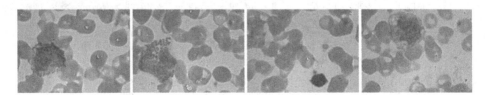

Fig. 6. Samples of BCCD dataset.

We use traditional data enhancement methods such as stretching, flipping, translation, and adding noise to enhance the dataset. BCISC and LISC generate 7350 leukocyte

---

[2] https://github.com/fpklipic/BCISC.

images. Each kind of sample randomly selects 70% of the pictures as the training set and the remaining 30% as the test set. BCCD generates a total of 3000 leukocyte pictures. As BCCD only participates in one-stage binary classification training, all pictures after data enhancement are added to the training.

All experiments are completed on the PyTorch open source platform, the development language is Python version 3.6.1. Moreover, two NVIDIA GTX 1080Ti graphics cards are employed.

4.2 Experimental Results and Analysis

We use the conventional Faster RCNN network and our proposed blood leukocyte detection method based on deformable convolution and model parameter-transfer as a control to prove the effectiveness of our algorithm. The evaluation results adopt $mAP_{0.5}$ and $mAP_{0.7}$, as well as recall and precision.

Recall refers to the proportion of correct results among all detected results. This indicator is used to evaluate the ability of the model to detect the target.

Precision refers to the proportion of correct positive and negative examples among all detected positive and negative examples. This indicator is utilized to evaluate the model's ability to classify targets.

As for mAP, AP is the area of the precision-recall curve. That is, the average accuracy is used to comprehensively evaluate the recall rate and the accuracy rate, for which mAP is the average AP value. The average AP value is obtained for the test set of multiple categories and is a measure of the performance of the model in object detection The most important indicator of the model performance.

In order to test the performance of our method for blood leukocyte target detection, we studied the decrease of the loss function and the map value of the model. We compared our proposed method with a version of it without model parameter-transfer and deformable convolution. As shown in Fig. 7, As the number of training iterations increases, our method achieves the highest mAP value (see Fig. 7 A), and in the loss function decline, our method not only achieves the optimal results in the classification loss function (see Fig. 7 C) and bbox regression loss function (see Fig. 7 D), but also reduces the training loss value to the minimum (see Fig. 7 B).

We compared our proposed method with a version of it without model parameter-transfer and deformable convolution. The results are shown in Table 3. Experimental data in Table 3 indicate that when our method incorporates deformable convolution, the experimental results are significantly better than those without deformable convolution. Moreover, the comparison group without model parameter-transfer reveals that such a transfer also exerted a satisfactory impact on our experiment. Therefore, the addition of model parameter-transfer methods and deformable convolutions have improved our experimental results. Specifically, the addition of deformable convolution has a greater impact on the performance of our object detection model and significantly improved its performance.

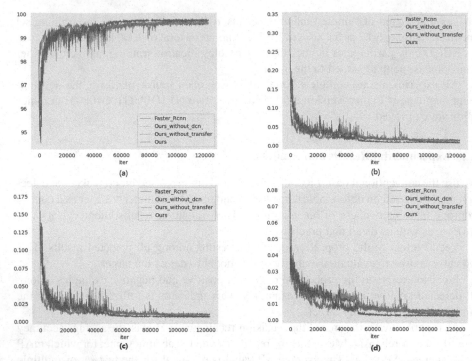

Fig. 7. Convergence studies on mAP and training loss with respect to the number of iterations (a) By changing the structure of our algorithm, the mAP value changes with the increasing number of iterations; (b) By changing the structure of our algorithm, the change of convergence loss with the increasing number of iterations; (c) By changing the structure of our algorithm, the change of the loss value of classification with the increasing number of iterations; (d) By changing the structure of our algorithm, the change of bbox regression loss value with the increasing number of iterations.

Table 3. Comparison the difference of the structure of the proposed method in this paper.

Contrast algorithm	$mAP_{0.5}$	$mAP_{0.7}$	Recall	Precision
Ours (without transfer)	0.967	0.936	0.961	0.913
Ours (without deformable convolution)	0.931	0.866	0.959	0.881
Ours	**0.976**	**0.966**	**0.972**	**0.919**

We compare the algorithm proposed in this paper with the current mainstream object detection algorithm. The experimental results are shown in Table 4. The mainstream object detection algorithms include Faster R-CNN [27], R-FCN [23], SSD [30], and YOLO-V3 [32]. Our method is superior to the current mainstream object detection algorithm in terms of mAP value, recall rate, and accuracy rate.

To further verify the effectiveness of this method, we also made a comparative experiment on the BCCD dataset. The data volume of basophils in the BCCD dataset is

Table 4. Comparison structure between the method in this paper and other object detection algorithms.

Contrast algorithm	$mAP_{0.5}$	$mAP_{0.7}$	Recall	Precision
Faster R-CNN	0.910	0.847	0.932	0.869
R-FCN	0.913	0.828	0.916	0.891
SSD	0.835	0.792	0.893	0.787
YOLO-V3	0.862	0.859	0.916	0.803
Ours	**0.976**	**0.966**	**0.972**	**0.919**

too low, there being only four pieces. Thus, we excluded the basophils and conducted experiments on the other four categories. The experimental results are shown in Tables 5 and 6.

Table 5. Comparison the difference of the structure of the proposed method in this paper.

Contrast algorithm	$mAP_{0.5}$	$mAP_{0.7}$	Recall	Precision
Ours (without transfer)	0.944	0.913	0.923	0.906
Ours (without deformable convolution)	0.924	0.873	0.934	0.884
Ours	**0.963**	**0.942**	**0.965**	**0.923**

Table 6. Comparison structure between the method in this paper and other object detection algorithms.

Contrast algorithm	$mAP_{0.5}$	$mAP_{0.7}$	Recall	Precision
Faster R-CNN	0.907	0.826	0.916	0.879
R-FCN	0.916	0.831	0.923	0.891
SSD	0.882	0.843	0.913	0.846
YOLO-V3	0.894	0.855	0.917	0.858
Ours	**0.963**	**0.942**	**0.965**	**0.923**

As shown in Fig. 8, finally our model can locate the whole blood leukocyte of the input blood micrographs, and classify the located white blood cells to obtain the category confidence. A small number of incomplete or unclear blood leukocyte classification will lead to misjudgement (see Fig. 8, top right).

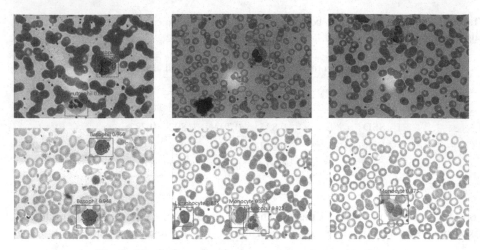

Fig. 8. The final results of our proposed work over test data.

5 Conclusion and Discussion

In object detection, many datasets are often needed for training so as to obtain a model with high accuracy and strong robustness. However, in our blood leukocyte detection task, datasets are difficult to obtain and many problems occur in the collection of leukocyte images, such as inconsistent use of staining agents and different microscope observations. To solve these problems, this paper adopts model parameter-transfer to maximize leukocyte datasets. The BCISC and LICS datasets are used in the first-stage leukocyte binary detection training and an additional BCCD dataset is added to improve the model's ability to extract leukocyte features. Subsequently, the model parameters and features obtained from the first-stage training are transferred to the second-stage training. Furthermore, deformable convolution and deformable PS ROI pooling are combined to train the leukocyte multi classification detection. Experimental results show that our method can effectively solve the problems of scarce data and unbalanced data distribution.

Through the method of this paper, we successfully add a BCCD dataset to the first stage of training. However, we only use binary classification to extract the additional dataset. In future work, we hope to use datasets of the same origin and different distributions for multi classification training before transfer.

Acknowledgments. This work is supported by the National Natural Science Foundation of China (NSFC) under Grant 61972187, the Scientific Research Project of Science and Education Park Development Center of Fuzhou University, Jinjiang under Grant 2019-JJFDKY-53 and the Tianjin University-Fuzhou University Joint Fund under Grant TF2020-6.

References

1. Nakanishi, N., et al.: White blood-cell count and the risk of impaired fasting glucose or Type II diabetes in middle-aged Japanese men. Diabetologia **45**(1), 42–48 (2002). https://doi.org/10.1007/s125-002-8243-1

2. 李海波. 血常规检验中各项指标的意义. 世界最新医学信息文摘 **16**(42), 110–111 (2016)
3. AL-Dulaimi, K., et al.: Classification of blood leukocytes types from microscope images: techniques and challenges. In: Microscopy Science: Last Approaches on Educational Programs and Applied Research, vol. 8. Formatex Research Center (2018)
4. Rawat, J., et al.: Review of leukocyte classification techniques for microscopic blood images. In: 2015 2nd International Conference on Computing for Sustainable Global Development (INDIACom). IEEE (2015)
5. Mahmood, N.H., Lim, P.C., Mazalan, S.M., Azhar, M., Razak, A.: Blood cells extraction using color based segmentation technique. Int. J. Life Sci. Biotechnol. Pharma Res. **2**(2), 2250–3137 (2013)
6. Huang, D.-C., Hung, K.-D., Chan, Y.-K.: A computer assisted method for leukocyte nucleus segmentation and recognition in blood smear images. J. Syst. Softw. **85**(9), 2104–2118 (2012)
7. Price, A.L., et al.: Principal components analysis corrects for stratification in genome-wide association studies. Nature Genet. **38**(8), 904–909 (2006)
8. Wang, W., Su, P.: Blood cell image segmentation on color and GVF Snake for Leukocyte classification on SVM. Opt. Precis. Eng. **12**, 26 (2012)
9. Pan, C., et al.: Leukocyte image segmentation by visual attention and extreme learning machine. Neural Comput. Appl. **21**(6), 1217–1227 (2012)
10. Huang, G.-B., Zhu, Q.-Y., Siew, C.-K.: Extreme learning machine: theory and applications. Neurocomputing **70**(1–3), 489–501 (2006)
11. Tabrizi, P.R., Rezatofighi, S.H., Yazdanpanah, M.J.: Using PCA and LVQ neural network for automatic recognition of five types of blood leukocytes. In: 2010 Annual International Conference of the IEEE Engineering in Medicine and Biology. IEEE (2010)
12. Liu, J., et al.: Nonwoven uniformity identification using wavelet texture analysis and LVQ neural network. Expert Syst. Appl. **37**(3), 2241–2246 (2010)
13. Lecun, Y., Bengio, Y., Hinton, G.: Deep learning. Nature **521**(7553), 436 (2015)
14. Liu, L., et al.: Deep learning for generic object detection: a survey. arXiv preprint arXiv: 1809.02165 (2018)
15. Shirazi, S.H., et al.: Efficient leukocyte segmentation and recognition in peripheral blood image . Technol. Health Care **24**(3), 335–347 (2016)
16. Qin, F., et al.: Fine-grained leukocyte classification with deep residual learning for microscopic images Comput. Methods Programs Biomed. **162**, 243–252 (2018)
17. Zhao, J., et al.: Automatic detection and classification of leukocytes using convolutional neural networks. Med. Biol. Eng. Comput. **55**(8), 1287–1301 (2017)
18. Torrey, L., Shavlik, J.: Transfer learning. In: Handbook of Research on Machine Learning Applications and Trends: Algorithms, Methods, and Techniques, pp. 242–264. IGI Global (2010)
19. Sinno, J.P., Yang, Q.: A survey on transfer learning. IEEE Educational Activities Department (2010)
20. Dai, J., et al.: Deformable convolutional networks. In: Proceedings of the IEEE International Conference on Computer Vision (2017)
21. Girshick, R., et al.: Rich feature hierarchies for accurate object detection and semantic segmentation. In: Proceedings of the IEEE Conference on Computer Vision and Pattern Recognition (2014)
22. Uijlings, J.R.R., et al.: Selective search for object recognition. Int. J. Comput. Vis. **104**(2), 154–171 (2013). https://doi.org/10.1007/s11263-013-0620-5
23. Dai, J., et al.: R-FCN: object detection via region-based fully convolutional networks. In: Advances in Neural Information Processing Systems (2016)
24. Gribbon, K.T., Bailey, D.G.: A novel approach to real-time bilinear interpolation. In: Proceedings DELTA 2004. Second IEEE International Workshop on Electronic Design, Test and Applications. IEEE (2004)

25. He, K., Zhang, X., Ren, S., et al.: Deep Residual Learning for Image Recognition (2015)
26. Simonyan, K., Zisserman, A.: Very deep convolutional networks for large-scale image recognition. arXiv preprint arXiv:1409.1556 (2014)
27. Ren, S., He, K., Girshick, R., et al.: Faster R-CNN: towards real-time object detection with region proposal networks. IEEE Trans. Pattern Anal. Mach. Intell. **39**(6), 1137–1149 (2017)
28. Neubeck, A., Gool, L.J.V.: Efficient non-maximum suppression. In: International Conference on Pattern Recognition. IEEE Computer Society (2006)
29. Rezatofighi, S.H., Soltanian-Zadeh, H.: Comput. Med. Imaging Graph. **35**, 333 (2011)
30. Liu, W., et al.: SSD: single shot multibox detector. In: Leibe, B., Matas, J., Sebe, N., Welling, M. (eds.) Computer Vision – ECCV 2016: 14th European Conference, Amsterdam, The Netherlands, October 11–14, 2016, Proceedings, Part I, pp. 21–37. Springer, Cham (2016). https://doi.org/10.1007/978-3-319-46448-0_2

Heterogeneous Software Effort Estimation via Cascaded Adversarial Auto-Encoder

Fumin Qi[1,4](✉), Xiao-Yuan Jing[2], Xiaoke Zhu[3], Xiaodong Jia[2], Li Cheng[2], Yichuan Dong[4], Ziseng Fang[4], Fei Ma[5], and Shengzhong Feng[4](✉)

[1] College of Computer Science and Software Engineering, Shenzhen University, Shenzhen, China
qfm120@163.com
[2] School of Computer, Wuhan University, Wuhan, China
[3] Henan Key Laboratory of Big Data Analysis and Processing, Henan University, Kaifeng, China
[4] National Supercomputing Center in Shenzhen, Guangzhou, Guangdong, China
[5] School of Computer Science, Qufu Normal University, Rizhao, China

Abstract. In Software Effort Estimation (SEE) practice, the data drought problem has been plaguing researchers and practitioners. Leveraging heterogeneous SEE data collected by other companies is a feasible solution to relieve the data drought problem. However, how to make full use of the heterogeneous effort data to conduct SEE, which is called as Heterogeneous Software Effort Estimation (HSEE), has not been well studied. In this paper, we propose a HSEE model, called Dynamic Heterogeneous Software Effort Estimation (i.e., DHSEE), which leverages the adversarial auto-encoder and convolutional neural network techniques. Meanwhile, we have investigated the scenario of conducting HSEE with dynamically increasing effort data. Experiments on ten public datasets indicate that our approach can significantly outperform the state-of-the-art HSEE method and other competing methods on both static and dynamic SEE scenarios.

Keywords: Heterogeneous effort estimation · Adversarial auto-encoder · Deep learning · Data drought issue

1 Introduction

Software Effort Estimation (SEE) is an important activity for the development of software projects [1,2]. In SEE, project managers often need to learn effort estimation models from historical projects and predict the required efforts for developing new projects. Existing SEE methods can be categorized into two groups, including Within-Company SEE (WCSEE) methods and Cross-Company SEE (CCSEE) methods [3,4]. For the WCSEE methods, researchers conduct SEE by using the within-company data (i.e., learning local lessons from within-company data). Cross-Company Software Effort Estimation (CCSEE) methods try to

Supported by National supercomputing center in Shenzhen.

© Springer Nature Switzerland AG 2021
Y. Zhang et al. (Eds.): PDCAT 2020, LNCS 12606, pp. 17–29, 2021.
https://doi.org/10.1007/978-3-030-69244-5_2

leverage the external companies' data sets to conduct SEE (i.e., learning global lessons from cross-company data).

The commonly used cost-drivers of SEE are shown in Table 1 [5–7]. Researchers use the cost-drivers to collect the effort data and build their predictive models.

Table 1. Commonly used cost drivers or **Size** features

	Data sets	Cost-drivers or **size** features	
Function Point analysis	Albrecht	Input, Output, Inquiry, File, FPAdj, RawFP, AdjFP, Effort	
	China	AdjFP, Input, Output, Enquiry, File, Interface, Added, Changed, Deleted, PDR AFP, PDR UFP, NPDR AFP, NPDU UFP, Resource, Dev.Type, Duration, N.Effort, Effort	
Cobol	Kitchenham	Client.code, Project.type, Actual.start.date, Actual.duration, Actual.effort, Adjusted.function points, Estimated.completion.date, First estimate, First.estimate.method	
	Kemerer	Language, Hardware, Duration, KLOC, AdjFP, RawFP, Effort	
	Miyazaki94	ID, KSLOC, SCRN, FORM, FILE, ESCRN, EFORM, EFILE, MM	
COCOMO	Nasa93, Coc81	Product features	Required software reliability (RELY), Data base size (DATA), Product complexity (CPLX), Require reusability (RUSE), Documentation match to life-cycle needs (DOCU)
		Platform features	Execution time constraint (Time), Platform volatility (PVOL)
		Personnel features	Analyst capability (ACAP), Programmer capability (PCAP), Application experience (AEXP), Platform Experience (PEXP), Language and tool experience (LTEX), Personnel continuity (PCON)
		Project features	Use of software tools (TOOL), Multisite development (SITE), Required development schedule (SCED)
Other	Maxwell	Number of different development languages used (Nlan), Customer participation (T01), Development environment, Adequacy (T02), Staff availability (T03), Standards use (T04), Methods use (T05), Tools use (T06), Softwares logical complexity (T07), Requirements volatility (T09), Quality requirements (T10), Efficiency requirements (T11), Installation requirements (T12), Staff analysis skills (T12), Staff application knowledge (T13), Staff tool skills (T14), Staff team skills (T15), Duration, time, Application size (number of function points)	

1.1 Motivation

The heterogeneous problem is a major 'barrier' for making full use of the cross-company data. In reality, even if we can get sufficient data to conduct SEE, these data are often heterogeneous. Intuitively, if the heterogeneous SEE data can be used to build the predictive model, much time and cost will be saved. It means that people do not need to consume much time and cost to find homogeneous data to conduct SEE.

1.2   Contribution

The major contributions of our study are summarized as follows

1. We propose an end-to-end HSEE approach based on deep learning techniques, which can simultaneously solve the heterogeneity between different data sets and enhance the representation ability of effort data. Specifically, we propose the Cascaded Adversarial Auto-Encoder (CAAE) framework to deal with the heterogeneity problem, and employ CNN to enhance the representation ability of effort data.
2. We investigate the scenario of dynamic HSEE in this paper as well as proposed a suitable solution. To the best of our knowledge, this is the first work on dynamic HSEE.
3. We conduct extensive experiments to evaluate the effectiveness of the heterogeneous effort data in both static and dynamic HSEE scenarios.

1.3   Organization

The organization of the paper is as follows: Sect. 2 reviews the related work. Section 3 describes the proposed approach. The experimental design is illustrated in Sect. 4. Experimental results and conclusion are reported in Sect. 5.

2   Related Work

2.1   Homogeneous Effort Estimation

Homogeneous SEE mainly focus on WCSEE scenario where data sets are collected by using identical cost-drivers. Jorgensen et al. [1] performed a system review on SEE and identified 304 SEE papers in 76 journals. There exist four main homogeneous SEE methods, including regression-based methods [8], Analogy-based methods [9], function point analysis based methods [4,10], expert judgment based methods [11]. Wen et al. [12] found that there also exist three main machine-learning methods to conduct SEE, including (1) Case-Based Reasoning (CBR) based method [13]; (2) Artificial Neural Networks based methods [14]; and (3) Decision Trees (DT) methods [15]. Recently, a new baseline homogeneous SEE method, i.e., Linear Programming for Effort Estimation (LP4EE), has been proposed by Federica et al. [16], which uses a strong optimization technique (i.e., Linear Programming) to learn the coefficient values of linear objective function.

There exist many CCSEE methods such as transfer learning based method [17], Dynamic cross-company Mapped Model Learning (Dycom) [18]. In these methods, researchers try to make full use of cross-company homogeneous data by designing different strategies.

2.2   Heterogeneous Related Work in Software Effort Estimation

In heterogeneous SEE, the data sets are collected by using different cost-drivers. To the best of our knowledge, there exists only one heterogeneous SEE method. Tong et al. [19] proposed a Mixture of Canonical correlation analysis and Restricted boltzmann machines (MCR) method for effort estimation.

Although MCR has achieved interesting results in HSEE, it's still exist two main limitations, including (1) UMR+CCA is sensitive to the order of combination of different data sets when we can get multiple heterogeneous data sets; (2) MCR only can be used for static HSEE scenario.

3   The Proposed Approach

3.1   Overview of Our Approach

Our proposed approach, named DHSEE (stands for Dynamic Heterogeneous Software Effort Estimation), consists of two main procedures: generation procedure and feature enhancement procedure. In the generation procedure, we propose a Cascaded Adversarial Auto-Encoder (CAAE) method to transform different heterogeneous effort data sets into a common space. In the feature enhancement procedure, we introduce the Convolutional Neural Network (CNN) technique into HSEE, which is used to further learn the deep representation of lower layer data and improve the performance of HSEE. Finally, we use a fully-connected layer to generate the effort value. Figure 1 shows the overview of our approach.

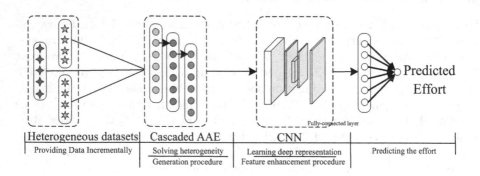

Fig. 1. Overview of our approach

Q: Why AAE works on SEE data?

A: By adopting proper strategies, we can train a deep learning model on SEE data. **Strategy 1: Controlling the scale of the model.** Compared with the tasks in NLP, Speech, and CV, the SEE task is quite small, and the dimension of the data that need to be modeled is much lower. The scale of the data required to train this model is much smaller than that of the data required to train a

CV or NLP model. Therefore, by controlling the model scale, we can reduce the amount of data that required to train the model.

Strategy 2: Data augmentation by fusing multiple SEE datasets via cascaded AAE. We harness AAE to fuse multiple SEE data sets, which alleviates the data drought problem to some extent. Since multiple data sets can be obtained in HSEE, we can combine them into a larger data set and get more than one thousand examples.

3.2 Cascaded Adversarial Auto-Encoder

Modified Adversarial Auto-Encoder for Eliminating the Heterogeneity. Adversarial AutoEncoder (AAE) is a data enhancement technology was presented by Makhzani [20], whose effectiveness has been widely validated in many existing works [21]. The AAE matches the aggregated posterior of the hidden code vector of the auto-encoder with an arbitrary prior distribution. The solution [20] to this 'antagonism' can be expressed as follows:

$$\min_{G} \max_{D} E_{x \sim p_{data}} \left[\log D(x) \right] + E_{z \sim p(z)} \left[1 - \log D(G(z)) \right] \tag{1}$$

In Eq. (1), $\log D(x)$ denotes the probability of assigning the correct label to both training examples and samples from G, AAE trains G as well as minimizes $1 - \log D(G(z))$ simultaneously. The generator G provides data to D and adjusts the parameters of the encoder network according to the feedback of the discriminator D. If the D can not classify whether the newly arrived sample is provided by G or not, which means that the training of the AAE is completed. Then we can use the trained network to align two heterogeneous data sets.

Let X_1 and X_2 be the two heterogeneous effort data sets, and Z be the latent code vectors (hidden units) of an auto-encoder with a deep encoder and decoder. In HSEE, we aim to impose the shared latent information among X_2 into the $p(X_1)$, and meanwhile preserve the distinctive information of X_2. In the procedure, we first use the same number of neuron layers with specific width units to align X_1 and X_2. Then, we encode X_1 and learn the prior $p(X_1)$. To preserve the distinctive information of X_2, we *add a new constraint* and use the accuracy of $p(X_1)$ for X_2 to conduct the validation. Therefore, we re-formula the Eq. (1) as follows:

$$\min_{G} \max_{D} E_{X_1 \sim p_{data}} \left[\log D\left(X_1\right) \right] \\ + E_{X_2 \sim p(X_2)} \left[\log \left(1 - D\left(G\left(X_2\right)\right) \right) \right] + \min_{p(Z)} error \tag{2}$$

where, $\min_{p(Z)} error = \begin{cases} Q_0 - Q_1, \ if \ Q_0 - Q_1 > 0 \\ 0, \ otherwise \end{cases}$, Q_0 denotes the accuracy of the original X_2, Q_1 represents the accuracy of using the predictor to predict the X_2 where the predictor is trained by using generated samples from prior $p(X_1)$. $\min_{p(Z)} error$ represents the gap between Q_0 and Q_1, which is used to ensure that the generated samples can keep the information of X_1 and meanwhile preserve the information of X_2. Here, we named it as *Heterogeneous AAE (HAAE)*.

Cascaded Adversarial Auto-Encoder for Dealing with Dynamical HSEE. In practice, the collection of the effort data sets is an ongoing process, we call the SEE under this situation as dynamical HSEE (DHSEE). To make the proposed approach can be used for DHSEE, we design a cascaded strategy based on HAAE and name it Cascaded Adversarial Auto-Encoder (CAAE). Specifically, when we have multiple data sets, each newly arrived data set can be added into the network incrementally. We only need to adjust the parameters of current HAAE network at each iteration. Since not all the heterogeneous data sets are useful to HSEE, we design a constraint for HAAE. If the learned latent feature is useful to improve the performance of HSEE, we use the learned feature as one of the input of next HAAE, otherwise we will discard the arrived data and restore our approach into previous state. The CAAE framework is illustrated in Fig. 2.

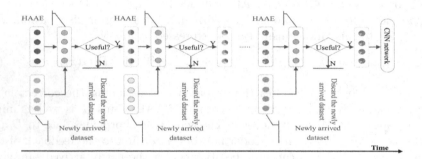

Fig. 2. Architecture of the CAAE

In Fig. 2, different colors in the circle denote different information contained in the heterogeneous data sets.

3.3   Feature Enhancement and Effort Estimation Procedures

Data representation capability is a key element for training a good predictive model. Previous stages only use for data alignmentation, here, the convolutional and pooling layers are used to further improve the representation capability of SEE.

We provide a simple example to illustrate how we use CNN to learn deep representation of effort data. In the traditional COCOMO-II model, Boehm et al. used Eq. 3 to predict the required effort of a new project.

$$effort = \alpha \prod_i EM_i * KLOC^{\beta+0.01\sum_i SF_j} \tag{3}$$

In Eq. 3, EM, KLOC and SF are effort multipliers, size driver and scale factors respectively and α, β are the local calibration parameters (with default values of 2.94 and 0.91). Software professionals usually calibrate the parameters

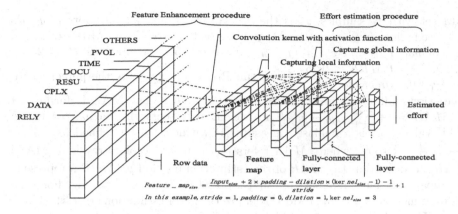

Fig. 3. An example of using CNN to learn deep representation of SEE data

α and β before using it to conduct SEE. If we transform it into Eq. 4, we can see that Eq. 4 is actually a regression problem.

$$ln(effort) = \sum_i ln(\alpha EM_i) + (\beta + 0.1 \sum_i SF_i) * ln(KLOC) \qquad (4)$$

Assuming that we have a data set A described by the COCOMO II with 22 metrics, including PVOL, TIME, DOCU, RESU, CPLX, DATA, RELY and etc. The procedure of using CNN to re-construct the representation of the SEE data can be illustrated in Fig. 3.

4 Experimental Design

4.1 Datasets

In experiments, eight close-source data sets and two open-source data sets are used to evaluate the proposed approach. The eight close-source data sets are introduced in Table 1, which can be accessed at PROMISE Repository[1]. The open-source data consists of two data sets Open-1 and Open-2 [4], which can be accessed publicly[2].

In experiments, since our paper mainly focuses on HSEE, some data sets with identical metrics have been combined into the same data set. In particular, 'Coc81' and 'Nasa93' form 'CocNas', 'Open-1' and 'Open-2' form 'Opens'. Finally, eight heterogeneous data sets are used in the experiments.

4.2 Evaluation Measures

Three measures are adopted in this paper, including Median Absolute Error (MdAE), Pred(25) and Standardized Accuracy (SA) [1,13,16]. Assume that a

---

[1] https://zenodo.org/communities/seacraft/search?keywords=effort.
[2] https://sites.google.com/site/whuxyjfmq/home/see_osp_dataset.

data set with N samples, x_i is the ith sample, whose actual effort is y_i, the predicted effort is $\bar{y}_i$, then the Magnitude of Relative Error (MRE) of x_i can be calculated by $MRE_i = |y_i - \bar{y}_i| / y_i$. Pred(25) is defined as $Pred(25) = \frac{100}{N} \sum_{i=1}^{N} \begin{cases} 1, & if\ MRE_i \leq 0.25 \\ 0, & otherwise \end{cases}$. Here, the Pred(25) represents the percentage of predictions falling within 25% of the actual values. The MAE is formulated as $MAE = \frac{1}{N} \sum_{i=1}^{N} |y_i - \bar{y}_i|$. The MdAE represents the median of the N MAE values. SA is a unbiased measure [16], which can be defined as $SA_{pi} = 1 - \frac{MAR_{pi}}{MAR_{p0}} \times 100\%$, where, MAR refers to the mean absolute residual (MAR), MAR_{pi} is defined as the MAR of the estimation method p_i, MAR_{p0} represents the mean of a large number (in our case 1000) of random guessing. For these adopted measures, the lower MdAE denote the lower estimation error, the higher Pred(25) and SA represent the higher prediction accuracy.

4.3   Compared Methods

Three compared methods are introduced in this paper including LP4EE, MCR and HDP-LP4EE. *LP4EE* [16] is a recently proposed SEE method for WCSEE scenario by Federica et al; *MCR* [19] is a HSEE targeted method by Tong et al; *HDP-LP4EE* [22] is a hybrid method consists of data alignment part (HDP) and LP4EE. For method LP4EE, we perform experiments with the code provided by the original authors. For the methods of HDP- and MCR without public implementation, we re-implement these methods and carefully tune the parameters.

4.4   Experimental Settings

Settings for the proposed approach: In the experiments, CNN uses a vector with 3 dimensions (i.e., $1 * 3$ vector) and 2 strides, as the convolution kernel to obtain the feature map. In the effort estimation stage, we create 15 neurons for the full connection layer (here, 15 is an empirical value, a larger number of neurons will increase the complexity of weight optimization). Here, we set the iteration number as 5000 and define the batch sizes of the China, Kitchenham, Albrecht, Maxwell, Cocnas, Opens, Kemerer and miyazaki94 as 50, 15, 5, 15, 15, 15, 10 and 15 respectively.

We conduct 20 runs for each experiment. The data partition rule is set as 30% to 70% (i.e., 30% for testing, 70% for training). Before conducting an experiment, the eight data sets are randomly divided into two parts according to the rule.

We implement the DHSEE by using Chiner. The source code of the proposed approach and the experimental data are publicly available[3].

---

[3] https://github.com/CAAE-HSEE/DHSEE.

5   Experimental Results and Conclusions

Can the proposed approach outperform the existing methods in static HSEE?
In this experiment, we select one data set as the target data set at each time, and use the other seven heterogeneous data sets as the source data sets. In the procedure, we can get seven estimation results for each target data set respectively. Then, the median value is adopted as the result of the target data set. The experimental results of static HSEE with three measures are reported in Table 2.

From Table 2, we can see that our approach significantly better than the compared HSEE methods (i.e., heterogeneous SEE methods). Specifically, our approach improves the Pred(25) values by 0.2% ∼ 32.2%, 8.1% ∼ 46.6%, 3.1% ∼ 76%, 1.8% ∼ 29.7%, 21.1% ∼ 87.9%, 66% ∼ 77.7%, 3% ∼ 57.2% and 1.7% ∼ 26.4%, and decreases MdAE values by 1028.9 ∼ 1.8, 2120.6 ∼ 287.4, 2080.1 ∼ 393.5, 809.3 ∼ 61.3, 4472.9 ∼ 461.3, 1268.8 ∼ 118.6, 1596.4 ∼ 2.5 and 1724.2 ∼ 0.5 on Albrecht, Kitchenham, China, Kemerer, Maxwell, CocNas, Opens and Miyazaki94 respectively. At the sametime, our approach can get better SA values.

The reasons why the proposed approach can achieve better performance than the compared methods are summarized as follows:

(1) We introduce ΛΛE, which encodes the two data distributions into a common space and keeps the distinctive information of each data set. The encoded information can be used to improve the performance of each data set.
(2) In the proposed approach, the HAEE model is followed by a CNN network which can be used to learn deep representation for the lower layer data and improve discrimination capability.
(3) For the MCR, although it tries to use RBM to refine the feature of the transformed data, the alignment order of UMR for padding the data sets significantly decreases the stability of the predictive accuracy.

Table 2. Experimental results of the compared methods for the static HSEE scenario

	HDP-LP4EE			MCR			DHSEE		
	Pred(25)	MdAE	SA	Pred(25)	MdAE	SA	Pred(25)	MdAE	SA
Albrecht	10.0	1050.5	29.6	10.7	185.6	28.7	**42.0**	**21.6**	**52.1**
Kitchenham	12.5	1486.6	39.6	18.0	2660.5	36.5	**59.1**	**539.9**	**90.9**
China	28.4	1622.9	49.2	18.6	2102.9	27.1	**94.7**	**22.8**	**92.2**
Kemerer	15.0	240.7	49.3	15.5	741.4	27.0	**40.0**	**78**	**56.5**
Maxwell	7.7	1616.8	27.5	22.1	1726.4	16.0	**94.7**	**607.8**	**77.9**
CocNas	18.2	1286.9	58.9	13.8	774.1	34.3	**91.5**	**18.5**	**82.1**
Opens	12.5	1616.9	39.7	23.1	158.2	24.1	**69.7**	**20.5**	**83.8**
Miyazaki94	37.5	129.5	49.8	30.6	1751.6	18.4	**51.7**	**25.4**	**86.0**

Table 3. Experimental results of the compared methods for the dynamic HSEE scenario

	HDP-LP4EE			MCR			DHSEE		
	Pred(25)	MdAE	SA	Pred(25)	MdAE	SA	Pred(25)	MdAE	SA
Albrecht	54.5	983.2	69.4	6.6	2205.0	24.7	**95.7**	**18.9**	**92.1**
Kitchenham	36.4	1321.2	29.8	2.9	1108.0	12.1	**85.7**	**403.5**	**90.9**
China	27.8	1542.2	39	15.4	1752.3	21.4	**95.6**	**21.4**	**92.2**
Kemerer	18.2	622.9	29.2	5.1	2208.4	24.8	**93.6**	**55.0**	**96.5**
Maxwell	18.2	1150.3	29.4	3.1	792.3	8.9	**89.4**	**305.6**	**97.9**
CocNas	27.8	1152.6	39.2	4.3	910.3	11.9	**88.9**	**22.5**	**92.1**
Opens	27.3	1373.5	39.2	9.0	809.0	28.8	94.8	**11.4**	**93.8**
Miyazaki94	18.3	148.5	29.7	0.0	1508.1	14.8	**81.3**	**25.5**	**96.0**

Can the proposed approach outperform the existing methods in dynamic HSEE? In dynamic HSEE scenario, we aim to add the newly arrived data into the model to improve the existing model. In the experiment, to simulate dynamic HSEE scenario, we add the source data set into the model incrementally according to the ascending order of the collection date. The experimental results are reported in Table 3 as well as Figs. 4(a)–4(h).

From Table 3 and Figs. 4(a)–4(h), we can see that the accuracy of our approach is better than those of the compared methods in most dynamic HSEE cases. For example, our approach can improve the Pred(25) value by 14.1% (i.e., MCR method for Maxwell data set, 14.1% = 89.4% − 75.3%) ∼ 90.3% (i.e., MCR method for Albrecht data set, 90.3% = 95.7% − 5.4%) over the HDP-LP4EE and MCR methods on all the adopted data sets, and meanwhile decrease the MdAE value.

The reasons why the proposed approach can achieve better performance than the compared methods are that:

(1) DHSEE designs a CAAE for the dynamic HSEE scenario. In the procedure, DHSEE can get more useful data to conduct HSEE.
(2) DHSEE incorporates a data filtering strategy into the CAAE, which prevents the performance compromised by irrelevant data.

These experimental results mean that: (1) although the cost-drivers or **size** metrics of the effort data sets can be described in different forms, there exist some inherent relationships between the collected effort data; (2) The implication is that the selection of source data sets is an important aspect for conducting HSEE.

Fig. 4. Pred(25) measure in dynamic HSEE scenario (when new data set is added incrementally)

Acknowledgement. The authors would like to thank Dr. Federica Sarro of Department of Computer Science, University College London, UK for providing code of the *LP4EE* code. This research was supported by the National Key R&D Program of China (Grant No. 2018YFB0204403) and the Project of Chinese Postdoctoral Science Foundation No. 2019M652624.

References

1. Jorgensen, M., Shepperd, M.: A systematic review of software development cost estimation studies. IEEE Trans. Softw. Eng. **33**(1), 33–53 (2007). https://doi.org/10.1109/TSE.2007.256943
2. Idri, A., Abnane, I., Abran, A.: Evaluating Pred (p) and standardized accuracy criteria in software development effort estimation. J. Softw. Evol. Process **30**(4), e1925 (2018). https://doi.org/10.1002/smr.1925
3. Minku, L.L.: A novel online supervised hyperparameter tuning procedure applied to cross-company software effort estimation. Empirical Softw. Eng. **24**(5), 3153–3204 (2019). https://doi.org/10.1007/s10664-019-09686-w
4. Qi, F., Jing, X.-Y., Zhu, X., Xie, X., Xu, B., Ying, S.: Grid information services for distributed resource sharing. Inf. Softw. Technol. **92**, 145–157 (2017). https://doi.org/10.1016/j.infsof.2017.07.015
5. Boehm, B.W., Madachy, R., Steece, B.: Software cost estimation with Cocomo II with Cdrom, pp. 540–541. Prentice Hall PTR (2000). book/10.5555/557000
6. Symons, C.R.: Function point analysis: difficulties and improvements. IEEE Trans. Software Eng. **14**(1), 2–11 (1998). https://doi.org/10.1109/32.4618
7. Mohagheghi, P., Anda, B., Conradi, R.: Effort estimation of use cases for incremental large-scale software development. In: 27th International Conference on Software Engineering, New York, pp. 303–311. IEEE (2005). https://doi.org/10.1109/ICSE.2005.1553573
8. Idri, A., Abnane, I., Abran, A.: Support vector regression-based imputation in analogy-based software development effort estimation. J. Softw. Evol. Process **30**(12), e2114 (2018). https://doi.org/10.1002/smr.1925
9. Benala, T.R., Mall, R.: DABE: differential evolution in analogy-based software development effort estimation. Swarm Evol. Comput. **38**, 158–172 (2018). https://doi.org/10.1016/j.swevo.2017.07.009
10. Silhavy, R., Silhavy, P., Prokopova, Z.: Analysis and selection of a regression model for the use case points method using a stepwise approach. J. Syst. Softw. **125**, 1–14 (2017). https://doi.org/10.1016/j.jss.2016.11.029
11. Altaleb, A., Gravell, A.: An empirical investigation of effort estimation in mobile apps using agile development process. J. Softw. **14**(8), 356–369 (2019). https://doi.org/10.17706/jsw.14.8.356-369
12. Wen, J., Li, S., Lin, Z., Hu, Y., Huang, C.: Systematic literature review of machine learning based software development effort estimation models. Inf. Softw. Technol. **54**(1), 41–59 (2012). https://doi.org/10.1016/j.infsof.2011.09.002
13. Kocaguneli, E., Menzies, T., Bener, A., Keung, J.W.: Exploiting the essential assumptions of analogy-based effort estimation. IEEE Trans. Software Eng. **38**(2), 425–438 (2012). https://doi.org/10.1109/tse.2011.27
14. Heiat, A.: Comparison of artificial neural network and regression models for estimating software development effort. Inf. Softw. Technol. **44**(15), 911–922 (2002). https://doi.org/10.1016/s0950-5849(02)00128-3
15. Jørgensen, M., Indahl, U., Sjøberg, D.: Software effort estimation by analogy and "regression toward the mean". J. Syst. Softw. **68**(3), 253–256 (2003). https://doi.org/10.1016/s0164-1212(03)00066-9
16. Sarro, F., Petrozziello, A.: Linear programming as a baseline for software effort estimation. ACM Trans. Softw. Eng. Methodol. **27**(3), 12:1–12:28 (2018). https://doi.org/10.1145/3234940

17. Kocaguneli, E., Menzies, T., Mendes, E.: Transfer learning in effort estimation. Empirical Softw. Eng. **20**(3), 813–843 (2015). https://doi.org/10.1007/s10664-014-9300-5

18. Minku, L.L., Yao, X.: How to make best use of cross-company data in software effort estimation? In: 36th International Conference on Software Engineering, Hyderabad, pp. 446–456. IEEE (2014). https://doi.org/10.1145/2568225.2568228

19. Tong, S., He, Q., Chen, Y., Yang, Y., Shen, B.: Heterogeneous cross-company effort estimation through transfer learning. In: 23rd Asia-Pacific Software Engineering Conference, Hamilton, pp. 169–176. IEEE (2016). https://doi.org/10.1109/APSEC.2016.033

20. Makhzani, A., Shlens, J., Jaitly, N., Goodfellow, I., Frey, B.: Adversarial autoencoders (2015). https://arxiv.org/abs/1511.05644

21. Creswell, A., Pouplin, A., Bharath, A.A.: Denoising adversarial autoencoders: classifying skin lesions using limited labelled training data. IET Comput. Vision **12**(8), 1105–1111 (2018). https://doi.org/10.1049/iet-cvi.2018.5243

22. Nam, J., Fu, W., Kim, S., Menzies, T., Tan, L.: Heterogeneous defect prediction. IEEE Trans. Softw. Eng. **44**(9), 874–896 (2017). https://doi.org/10.1109/TSE.2017.2720603

A Novel Distributed Reinforcement Learning Method for Classical Chinese Poetry Generation

Liangliang Ma, Hong Shen[✉], and Shangsong Liang

Sun Yat-sen University, Guangzhou, China
`shenh3@mail.sysu.edu.cn`

Abstract. Poetry generation has been a classic natural language generation task recently. But so far the methods for this topic mainly imitate and reproduce the poems on the training data set, which indicates that they either have not much connotation or overfit too much like plagiarism of the existing poems. To solve this problem, unlike previous work, instead of tuning the trade-off between connotation and innovation, we propose a distributed reinforcement learning framework, which consists of two stages of training, to generate creative and meaningful poetry. At the first stage we train a model in parallel on a large poetry corpus at word level to master how poets write poems. At the second stage we train the model with a distributed architecture to learn how connotation is developed in human literary art works at sentence level and force the model to imitate itself when it composes some 'good poems' to further improve performance. Experiments on generating classical Chinese poetry demonstrate that the proposed model is able to achieve better performance and the high efficiency of training compared to the state-of-the-art.

Keywords: Reinforcement learning · Natural Language Generation · Distribution

1 Introduction

Poetry is one of the most beautiful art in human history and still very commonly seen and used in people's daily lives across the world. Especially, classical Chinese poetry is widely used in every traditional Chinese festival and many kinds of celebrations.

Poetry generation is a Natural-Language Generation (NLG) task that aims to generate new poems automatically. This technique can be used for education, advertisement and entertainment. The application purpose of poetry generation demands the generated poem to be different from existing poems composed by human and look like a human work. In other words, the generated poetry should be creative and have reasonable connotation. However, although there are some previous works for poetry generation [7,15,16,20], none of them meets the huge demand of the market.

© Springer Nature Switzerland AG 2021
Y. Zhang et al. (Eds.): PDCAT 2020, LNCS 12606, pp. 30–42, 2021.
https://doi.org/10.1007/978-3-030-69244-5_3

Automatic poetry generation has always been a challenging work in the field of artificial intelligence. In recent years, deep learning has proven to be effective in this task and many related works have been published. For instance, in Recurrent Neural Network (RNN) based methods [7], the whole content of poems is put forward to a language model as training corpus. After training, the model can take some initial content as input, sample the next word according to the probability distribution of the output and repeat the process to produce a complete poem.

However, these memory neural network based methods are different in algorithms but most of them are superficial imitation of poems from training data on different degree. Different from the previous works, our goal is to generate poetry which has connotation and is different from existing poems. To this end, we propose a method to solve this task, the framework of which is shown in Fig. 1. It consists of two stages of learning. At the first

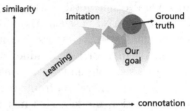

Fig. 1. Sketch map for poetry generation.

stage, it makes the model learn to imitate the real poems. At the second stage, it makes the model learn to compose new poems with much reasonable meaning.

In real life and writing theories, the way we learn to compose a poem is also a two-stage procedure. At first, we learn the poetry attributes by reading poetry masterpieces, and then we compose poems ourselves and get scores by teachers, which is very much like reinforcement learning agent that learns how to maximize its return. Thus, it is very natural for our two-stage method to train a model that can generate poems like a real person. At the first stage we utilize Long-Short Term Memory (LSTM) [2] to learn the poetry corpus. The LSTM network is trained in word level to generate poem like real poet does. At the second stage the model is trained with reinforcement learning (RL) [14] in distributed architecture Ape-x [3] and applies self-imitation learning (SIL) [8] method to calculate priorities for experience replay. The rewarder of RL framework is functioned by another model, which is trained in sentence level to score how good the connotation is developed for generated poem. Through our method, we can generate innovative and meaningful poetry. We conduct two experiments to test the generation quality and the efficiency experiment verifies the improvement brought by parallel structure.

Our contributions can be summarized as: 1) A novel distributed reinforcement learning method for poetry generation is proposed. 2) A two-stage training to generate poetry with connotation and innovation is proposed for our reinforcement learning method. 3) We systematically conducted experiments on our distribution architecture and the experimental results demonstrate the effectiveness of our proposed method.

2   Related Work

In this section, we only discuss the most related works in two lines: poetry generation and reinforcement learning.

2.1   Poetry Generation

In addition to RNN-based [7] methods introduced in the previous section, there are several other good works for poetry generation. A Generative Adversarial Networks (GAN) based method [4,20] uses RNN as its generative model to get new poems and CNN as its discriminative model to judge whether the poem is written by man or by machine. A planning-based method [15] tries to imitate the process of planning a writing outline before writing like human do. The whole poetry generation framework consists of two components: planning model and generation model. This method takes connotation of poetry into account rather than only imitating the training data. Another encoder-decoder based method [16] tries to simulate the process of repeated modification of human poetry and a polishing mechanism is added to improve the quality of poetry generation through repeated iterations. This method is also planning based, which can keep the meaning of the generated poem consistent with the title, but the meaning of sentences is relatively flat and convergence. Like the last two methods, our proposed one does not content with similarity with existing poems but focus more on the connotation in the generated poem.

2.2   Reinforcement Learning

Reinforcement learning [14] is an area of Machine Learning and the study of agents and how they learn by trial and error. For poetry generation, there is only one previous study [18] making use of reinforcement learning as we know. It simultaneously trains two generators which learn from the rewarder and from each other. This method yields good results but relies too much on poetry knowledge, which is hard to generalize to other NLG system. As far as we know, there is no distributed RL work that tackles poetry generation task like our proposed method.

In this work, we build our model based on Ape-x architecture [3], self-imitation learning (SIL) [8] and Proximal Policy Optimization (PPO) [11] algorithm. Ape-x architecture only uses one learner and one replay buffer, but multiple actors to generate data in distributed way. Through different actors, we can get replays with different priority, which greatly improves the ability of explore and prevents over fitting. The SIL algorithm is used to calculate priorities for experience, which makes the model learn to imitate state-action pairs in the replay buffer only when the return in the past episode is greater than the agent's value estimate. Experiments of SIL prove that exploiting past good experiences can indirectly drive deep exploration, which meets our needs that exploiting past experiences can benefit development of connotation and deeper exploration reduces the similarity with real Poetry. PPO is an off-policy reinforcement learning algorithm, which means the trained policy is different from the policy that interacting with the environment. The feature can increase the learning speed and make it easier to implement in parallel way (Ape-x architecture in this work).

3   The Proposed Method

Our goal is to generate poetry with connotation and innovation. To this end, we separate the learning process into two stages, which is very similar to real student's experience of learning how to compose poetry at school.

3.1   Stage One

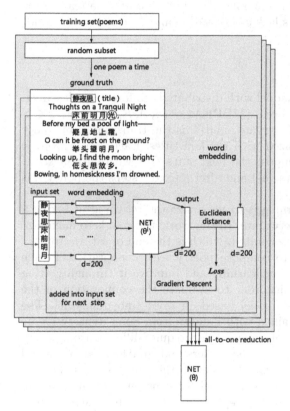

Fig. 2. Model flow chart at stage 1. The input set adds a word at each step.

The first stage is to learn with existing poems at word level in order to remember what a poem looks like. It corresponds to people reading a lot of masterpiece and learning writing rules. The purpose of this stage is imitation, which is usually the first step of human learning everything.

To realise the first stage: (See Fig. 2), we first map every word of every poem in the training set to a float vector with a pre-trained word embedding model. Here we utilize Tencent AI Lab Embedding Corpus [12]. We maintain a set which is initialized with the title of a ground poem from training data, as the input of our model. The output is a vector, which represents the first word of generated poem. This vector and the embedded vector of the first word in the ground poem are used to calculate the loss. Then the first word of the ground poem is added into the input set for the model to generate the next word. The training process keeps going until one whole ground poem is added to the input set. Then the set is cleared, and another ground poem is ready to be learned.

The objective of this stage is

$$\mathcal{L}_1 = \|v_i - v_*\|_2, \tag{1}$$

where v_i is the generated vector, v_* is the embedding vector of word in ground truth poem, and $\|\cdot\|_2$ is the 2-norm. The neural network in this stage is LSTM, for it can learn longer poems, and to some extent, it can improve the semantic coherence before and after.

In order to speed up training, we perform Stochastic Gradient Descent (SGD) in parallel. We equally divide the training set in random order and send those subsets to machines. In each machine the model parameter is trained and together we make an all-to-one reduction for the parameter. The convergence of parallelized stochastic gradient descent is ensured [22].

After the first stage, the machine should be able to imitate the real poems well. But this is not enough, for its lack of innovation and its connotation is not ensured.

3.2   Stage Two

At the second stage, we train the model with distributed reinforcement learning, which is on a Markov Decision Process (MDP) $\mathcal{M} = (S, A, R, T)$, where $s \in S$ is state, $a \in A$ is action, $r \in R$ is reward, and T is the transition probability matrix. In this work, s_0 is initialized with random titles. Each step a word is chosen by policy $\pi(a_t|s_t)$ and compose the next state s_{t+1} with current state. The rewarder provides rewards $r_t(s_t, a_t)$ and after a fixed c steps the policy π is updated according to $r_0...r_c$.

The goal of reinforcement learning is to maximize the return after taking a sequence of actions, starting in every state. In this work, the goal is to maximize the total rewards after a poem is generated, which is the end of a trajectory.

The second stage is to learn to compose poetry with connotation. People create their own works after primary learning and improve it through getting evaluated for their works. Thus, in order to evaluate the connotation of the generated poem, we train another model based on word embeddings [6]. The evaluating model (rewarder) is trained with iambics from Song dynasty for its characteristics of writing and rules of using words are quite different from poetry and its literary connotation is easy to be understood and learned by neural networks. We make the rewarder trained basically the same way as the first stage except that it is trained at sentence level to focus on inner meaning of the poem and avoid paying too much attention to words. The sentences of a training iambic is evenly divided into k groups, where k is decided according to the number of sentences of poetry we want to generate. Then we map every group to a float vector with pre-trained word embedding model. The rewarder is trained by taking embedding vectors of groups as input and trying to output the vector of the next group. The loss of rewarder is

$$\mathcal{L}_{rewarder} = \left\| v_j - \frac{1}{M} \sum_{q=1}^{M} v_q \right\|_2, \tag{2}$$

where v_j is the generated vector, M is the number of words in the next group, and v_q is the embedding vector of word in the next group. The rewarder remembers how the connotation is developed in thousands of iambics which means it can give a higher reward when the generated poem has the similar way to develop connotation.

Because the generating model generate one word at a time, the rewarder evaluates the poem at sentence level and in real life the poetry is composed sentence by sentence, we provide the reward function as follows:

$$r = \begin{cases} 0, & \text{if } w \text{ is not the end of a sentence} \\ -\left\| \frac{1}{N} \sum_{i=1}^{N} v_i - f_e(s_t) \right\|_2, & \text{if } w \text{ is the end of a sentence} \end{cases} \tag{3}$$

where N is the number of words in a sentence, v_i is the generated vector which composes the state as word vector of the generated poem, and f_e represents the evaluating model (rewarder) which takes vectors of the state as input.

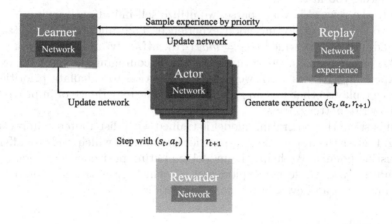

Fig. 3. Our reinforcement learning framework for poetry generation.

Now we have a well-trained generating model which can imitate the ground poems and a rewarder. We use custom Ape-x architecture to continue training the generating model (see Fig. 3). Multiple actors, each with its own instance of the network, take its own actions and add the experience to the shared priority replay buffer. We set a shared rewarder to take place the one environment instances of each actor in original Ape-x architecture, for the environment is same and stable about agent's actions, which helps to save memory as well. The parallel of actors can highly increases the exploration of RL algorithm and the shared experience replay can ensure the learner to reach global optimization.

The single learner samples experience by priorities from the buffer to update its network. The objective of learner is clip version of PPO [11]:

$$\mathcal{L}^{ppo}(\theta) = \hat{\mathbb{E}}[min(b_t(\theta)\hat{A}_t, clip(b_t(\theta), 1 - \epsilon, 1 + \epsilon)\hat{A}_t)], \tag{4}$$

$$b_t(\theta) = \frac{\pi_\theta(a_t|s_t)}{\pi_{\theta_{old}}(a_t|s_t)}, \tag{5}$$

where θ is the network parameter, θ_{old} is the actor network parameter which is also old version of learner network, and $\hat{\mathbb{E}}$ is the expectation over a batch of samples. The fraction $b_t(\theta)$ is importance sampling probability ratio, which is widely used in off-policy RL algorithms to calculate return (cumulative rewards). $\hat{A}_t$ is an estimator of the advantage function at timestep t, which is defined as state-action value minus state value, or approximated by $(r_{t+1} + v_\theta(s_{t+1}) - v_\theta(s_t))$. ϵ is a hyperparameter and clip function is used to limit importance sampling fraction in $[1 - \epsilon, 1 + \epsilon]$. The objective takes the minimum of the clipped and unclipped importance sampled advantage function, which prevents the network from updating too fast.

Due to the reward is very sparse, we utilize self-imitation learning (SIL) [8] method to enhance exploiting good experience. The function $(R - V_\theta(s))_+$ is the core idea of SIL, in which $(\cdot)_+ = max(\cdot, 0)$. SIL only updates the parameter when R is bigger than current estimated value, encouraging the model to exploit good experience. In this work we used this function to calculate priorities for experience replay so that we can give priority to the experience of improving the ability to write better poems.

At this stage the generating model is trained with distributed reinforcement learning. It gives the model enough space for exploration which leads to difference from existing poems. SIL helps the model exploiting past good experiences and it can indirectly drive deeper exploration [8], which gives the generated poems good connotation and even bring more innovation.

3.3   Algorithm

The Pseudo code of the two stages is showed in Algorithm 1.

4   Experiment

In this section, we describe our experimental setup and analyse the experimental results.

4.1   Data Set

We train our model with a data set of 315 thousand classical Chinese poems from Tang dynasty and Song dynasty, and 21 thousand iambics of Song dynasty.[1] These works are all from famous ancient poets. It ensures that the poems we generate are at a high level.

---

[1] The data sets are publicly available from: https://github.com/chinese-poetry/chinese-poetry.

4.2   Experimental Settings

Our work is aiming at generating poetry with connotation and innovation, so we set the experiment with two parts. The first one measures the similarity degree of generated poetry and human-composed poetry. The second one measures the connotation degree inside the generated poetry. We use Bilingual Evaluation Understudy (BLEU) [9] score as the first part and human evaluation as the second part. We also test the efficiency improvement of utilizing parallel structure.

Algorithm 1. Proposed Poetry Generation Algorithm

---

initialize $\theta \leftarrow initialized\_network()$
training dataset $\mathcal{D}$
#stage one
for each iteration until convergence **do**
    divide $\mathcal{D}$ into c subsets:$d_0, d_1, ..., d_c$
    for $i \in \{0, 1, ..., c\}$ in parallel **do**
        for every word w of every poem in d_i **do**
            $\mathcal{L} \leftarrow compute\_loss(\theta_i, w, equ.1)$
            $\theta_i \leftarrow update\_parameter(\theta_i, \mathcal{L})$
        end for
    end for
    $\theta = \frac{1}{c} \sum_{i=0}^{c} \theta_i$ #all-to-one reduction
end for
#stage two
initialize replay buffer $\mathcal{E} \leftarrow \emptyset$
each Actor parameter $\theta_i \leftarrow \theta$
experience replay network paramer $\theta_e \leftarrow \theta$
pre-trained rewarder f_e
#each Actor continually generate experience and send it to $\mathcal{E}$
for each epoch **do**
    for each step **do**
        $\tau \leftarrow SIL\_priority\_relay.pop()$ #sample experience
        $l_t \leftarrow compute\_loss(\theta, \tau, equ.4)$
        $\theta \leftarrow update\_parameter(\theta, l_t)$
    end for
    each Actor parameter $\theta_i \leftarrow \theta$, $\theta_e \leftarrow \theta$
end for

BLEU is originally designed to automatically judge the machine translation quality and usually use in poetry generation experiments to evaluate imitating or reproducing poetry. In our experiment we randomly choose titles from human works and generate poems by different algorithms. Then we can compare the similarity between generated poems and the true human poems with BLEU. N-gram is set to be 2 (BLEU-2) since most words in classical Chinese poetry consist of one or two Chinese character.

We use human evaluation as the second experiment part because poetry is a kind of human art and it is hard to clearly judge how good it is by machine. But it is a simple job to understand connotation and evaluate artistic degree for professors and students in related fields. Our human judgement consists of five aspects: fluency (Does the poem read fluently?), poeticness (Does the poem have some poetic attributes?), coherence (Is the poem related to the title?), meaning (Does the poem have a certain meaning?), overall (How good is the poem on the whole?). This multi-aspect experiment has also been used by many important works [1,13,15,17–19]. The score of each aspect ranges from 1 to 5 with the higher score the better. We invite 50 experts on Chinese poetry to help our experiment and the rating scores are averaged as the final score.

4.3   Baselines

Our experiment baselines are state-of-the-art Chinese poetry generation algorithms: GPT-based [5], seqGAN [20] and Jiuge [21].

GPT-based method is the first to employ Generative Pre-training (GPT) [10] in developing a poetry generation system. It is far simpler than existing methods based on recurrent neural networks and can generate better poems. This work has released an online mini demonstration program[2] so we can experiment with it. SeqGAN was introduced in Sect. 2 and its source code is released.[3] Jiuge[4] is a poetry generation program that continuously maintained and developed by THUNLP.[5] Many great poetry generation works have published and contributed to this program.

4.4   Experimental Results and Analysis

Table 1. BLEU-2 scores

Models	BLEU-2
GPT-based	0.6670
Jiuge	0.6218
SeqGAN	0.7389
Proposed	**0.5451**

The BLEU-2 and human evaluation experiment results are shown in Table 1 and Table 2. We can see that our proposed method achieves the lowest BLEU-2 scores which indicates that our generated poems is the most different from the data poems. In another word, our method can generate the most innovation. In terms of human evaluation, it shows our method has a quite good overall performance and the best meaning score.

---

[2] For GPT-based method, User may have to register a Wechat account and add "EI体 or 验空间".
[3] seqGAN code is available from: https://github.com/LantaoYu/SeqGAN.
[4] Jiuge is available from: http://118.190.162.99:8080/.
[5] The Natural Language Processing Group at the Department of Computer Science and Technology, Tsinghua University.

Table 2. Human evaluation comparison.

Models	Fluence	Poeticness	Coherence	Meaning	Overall
GPT-based	**4.41**	3.93	**3.85**	3.53	3.64
Jiuge	4.34	**4.10**	3.92	3.74	**3.80**
SeqGAN	4.09	3.96	3.87	3.37	3.01
Proposed	3.97	4.04	3.89	**4.10**	3.77

A concrete example of comparison between generated poem by Jiuge and our method is shown in Fig. 4. It clearly shows that our method has a great advantage in innovation for its low repetition rate. Its connotation is fairly reasonable and poetic.

静夜思 Thoughts on a Tranquil Night(title)

相思静夜阑，
Lovesickness at quiet night,
明月照人寒。
The moon shines on the cold.
回首西湖上，
Looking back on the West Lake,
烟波万里寒。
Smoke waves and everywhere is cold.

(a)

晚催惊昭拨，
The night came, scaring the day,
几龄人離段。
People of all ages left the place.
疑蝉入鹤石，
I suspect the cicada has entered the ancient stone,
又有蚊一宣。
And I hear the Dragon roar.

(b)

Fig. 4. (a) Generated poem by Jiuge method (b) Generated poem by our method. Title is set to be the same as a famous real poem written by Bai Li, a famous poet in China. The Chinese words on green marks and phrases on yellow marks represent their presence in real poem of this title.

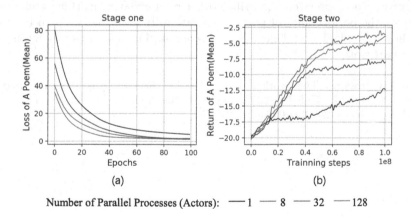

Number of Parallel Processes (Actors): —— 1 —— 8 —— 32 —— 128

Fig. 5. (a) Performance of stage one on different number of processes (b) Performance of stage two on different number of actors.

This result proves our method achieves our goal – having innovation and connotation at the same time.

4.5 Efficiency Test

We test the efficiency improvement by measuring the loss reduction of stage one and average return raise of stage two. The result is shown in Fig. 5. It proves that SGD and Ape-x architecture achieved very good performance on both stages, especially when the number of parallel processes (Actors) increased.

5 Conclusion

Researchers have studied the problem of poetry generation for decades. From randomly combining words, template-based methods to memory neural networks, previous studies mainly try to imitate human works in increasing similarity. To make machine generated poetry have reasonable connotation and innovation, in this work we proposed a novel method for poetry generation, which consists of a distributed reinforcement learning framework and a two-stage learning process. The model learns to compose poetry as past human works at the first stage and learns to compose innovative and meaningful poetry at the second stage. Experiments on generating classical Chinese poetry showed that the proposed method can achieve better results compared to the state-of-the-art, which proved that our method can generate better poetry with connotation and innovation. And the efficiency test proved that our parallel settings are effective. As to future work, we intend to utilize other reinforcement learning techniques to improve the performance of poetry generation. In addition, we plan to study about how to embed brand new literary connotation into artificial intelligent language art works.

Acknowledgements. This work is supported by National Key R & D Program of China Project #2017YFB0203201, Key-Area Research and Development Plan of Guangdong Province 2020B010164003.

A   More Examples of Our Method

Table 3. Generated poems of our method

白石山	White Stone Mountain(title)
倚湖熙古霾	Leaning against the lake, surrounded by ancient haze
青垣群霞晚	Green mountain in sunset clouds
诸客此有致	Every guest is interested here
古茶客属檐	Classic tea serves people under eaves

春雨	Spring rain(title)
春乳浊壽石	Spring rain is like milk, moistening longevity stone
冰消解瓦摧	The ice melted and disintegrated
客均顏有參	Every tourist have a expression in their face
更擬論玄妙	And even write for the wonder and mystery

離歌	Farewell song(title)
唇前何绕然	What words surround your lips
事变未煩落	Things changed and I haven't had time to get upset
感然趁妳在	I am touched while sister is still here
長何叭夜濒	Why I always sigh when night comes

早行	Morning Walk(title)
雲雷声身感	The thunder make the body felt
相扰此經人	Annoying passers-by
由谷见日黄	The yellow of the sunrise came out of the valley
天餅安可食	Can we eat the bread in the sky

References

1. Chen, H., Yi, X., Sun, M., Li, W., Yang, C., Guo, Z.: Sentiment-controllable Chinese poetry generation. In: IJCAI, pp. 4925–4931 (2019)
2. Hochreiter, S., Schmidhuber, J.: Long short-term memory. Neural Comput. **9**(8), 1735–1780 (1997)
3. Horgan, D., et al.: Distributed prioritized experience replay (2018)
4. Liang, S.: Unsupervised semantic generative adversarial networks for expert retrieval. In: WWW (2019)
5. Liao, Y., Wang, Y., Liu, Q., Jiang, X.: Gpt-based generation for classical chinese poetry. arXiv preprint arXiv:1907.00151 (2019)
6. Mikolov, T., Chen, K., Corrado, G., Dean, J.: Efficient estimation of word representations in vector space. Computer ence (2013)
7. Mikolov, T., Karafiát, M., Burget, L., Cernock, J., Khudanpur, S.: Recurrent neural network based language model. In: INTERSPEECH 2010 (2010)

8. Oh, J., Guo, Y., Singh, S., Lee, H.: Self-imitation learning. arXiv preprint arXiv:1806.05635 (2018)
9. Papineni, K., Roukos, S., Ward, T., Zhu, W.J.: Bleu: a method for automatic evaluation of machine translation, October 2002. https://doi.org/10.3115/1073083.1073135
10. Radford, A., Narasimhan, K., Salimans, T., Sutskever, I.: Improving language understanding by generative pre-training (2018)
11. Schulman, J., Wolski, F., Dhariwal, P., Radford, A., Klimov, O.: Proximal policy optimization algorithms (2017)
12. Song, Y., Shi, S., Li, J., Zhang, H.: Directional skip-gram: explicitly distinguishing left and right context for word embeddings. In: Proceedings of ACL 2018 (2018)
13. Sun, M., Yi, X., Li, W.: Stylistic chinese poetry generation via unsupervised style disentanglement, pp. 3960–3969, January 2018. https://doi.org/10.18653/v1/D18-1430
14. Sutton, R., Barto, A.: Reinforcement Learning: An Introduction. MIT Press, United States (1998)
15. Wang, Z., et al.: Chinese poetry generation with planning based neural network. arXiv:1610.09889 (2016)
16. Yan, R., Jiang, H., Lapata, M., Lin, S.D., Lv, X., Li, X.: I, poet: automatic chinese poetry composition through a generative summarization framework under constrained optimization. In: 23rd IJCAI (2013)
17. Yi, X., Li, R., Sun, M.: Chinese poetry generation with a salient-clue mechanism. CoNLL, 241–250 (2018)
18. Yi, X., Sun, M., Li, R., Li, W.: Automatic poetry generation with mutual reinforcement learning. Proc. EMNLP **2018**, 3143–3153 (2018)
19. Yi, X., Sun, M., Li, R., Zonghan, Y.: Chinese poetry generation with a working memory model, September 2018
20. Yu, L., Zhang, W., Wang, J., Yu, Y.: Seqgan: sequence generative adversarial nets with policy gradient. In: AAAI-17 (2017)
21. Zhipeng, G., et al.: Jiuge: a human-machine collaborative chinese classical poetry generation system. In: Proceedings of ACL 2019: System Demonstrations, pp. 25–30 (2019)
22. Zinkevich, M., Weimer, M., Smola, A.J., Li, L.: Parallelized stochastic gradient descent. In: Proceedings of NIPS 2010 (2011)

Memory Access Optimization of High-Order CFD Stencil Computations on GPU

Shengxiang Wang, Zhuoqian Li, and Yonggang Che[✉]

Institute for Quantum Information and State Key Laboratory of High Performance Computing, College of Computer, National University of Defense Technology, Changsha, China
ygche@nudt.edu.cn

Abstract. Stencils computations are a class of computations commonly found in scientific and engineering applications. They have relatively lower arithmetic intensity. Therefore, their performance is greatly affected by memory access. This paper studies the issue of memory access optimization for the key stencil computations of a high-order CFD program on the NVidia GPU. Two methods are used to optimize the performance. First, we use registers to cache the data used by the stencil computations in the kernel. We use the CUDA warp shuffle functions to exchange data between neighboring grid points, and adjust the thread computation granularity to increase the data reuse. Second, we use the shared memory to buffer the grid data used by the stencil computations in the kernel, and utilize loop tiling to reduce redundant accesses to the global memory. Performance evaluation is done on an NVidia Tesla K80 GPU. The results show that compared to the original implementation that only uses the global memory, the optimized implementation that utilizes the registers achieves a maximum speedup of 2.59 and 2.79 relatively for 15M and 60M grids, and the optimized implementation that utilizes the shared memory achieves a maximum speedup of 3.51 and 3.36 relatively for 15M and 60M grids.

Keywords: Stencil computation · NVidia GPU · Warp shuffle · Register caching · Shared memory · Loop tiling

1 Introduction

Stencils are a class of algorithms which update elements in a multi-dimensional grid based on neighboring values using a fixed pattern. They are widely in science and engineering applications and image processing applications [1,2]. Since the values of multiple adjacent grid points are used to update the value of a grid point, stencil computations are usually memory bound.

In recent years, GPUs are widely used as accelerators in high performance computers, including some most powerful supercomputers [3,4]. GPU can provide multi-level storage structure to meet the memory access requirements of

© Springer Nature Switzerland AG 2021
Y. Zhang et al. (Eds.): PDCAT 2020, LNCS 12606, pp. 43–56, 2021.
https://doi.org/10.1007/978-3-030-69244-5_4

kernel. Current GPUs usually have a large but relatively low-bandwidth Off-chip global memory that can store all device array variables and scalar variables. The on-chip storage has a higher bandwidth, but its size is relatively small. Starting from the Kepler GPU, NVidia introduced the shuffle functions in CUDA [5]. These functions allow threads to access the register files of neighboring threads in the same warp. Developers can manage these operations manually to cache important data with register file [6]. Current GPUs also have a high-bandwidth on-chip storage structure, the shared memory. Making use of the shared memory is often the key to accelerate the performance of the kernels [7].

HNSC (High order Navier-Stokes simulator for Compressible flow) is an CFD software that solves the compressible Navier-Stokes equations. The application is based on a stable high-order finite difference method presented in [8]. We have ported it to NVidia GPU in [9]. The key kernels of HNSC involve a large number of stencil computations. By performance analysis we find that when only the global memory is used to store the data on the grid points, the overhead of memory access is very high. Therefore, this paper further studies the method of using GPU registers and shared memory to optimize these stencil computations. Our contributions are as follows:

(1) We use registers to cache data used by the stencil computations in the kernel, in unit of warp. We use the shuffle functions to exchange data between neighboring threads, and adjust the thread computation granularity to increase the data reuse. We analyze the resource cost and performance to get optimal settings.
(2) We use the shared memory to buffer the grid data used by the stencil computations in the kernel, and utilize loop tiling to reduce redundant accesses to the global memory. We analyze the impact of bank access mode, sub-array size, and thread computation granularity on performance. We also analyze the performance difference between different kernels, taking into account of the metrics of shared memory and the GPU hardware occupancy.

In this paper, our tests are performed on a server based on 1 NVidia K80 GPU, which consists of two GK210 chipsets. Table 1 shows its specifications. The rest of the paper is organized as follows: Section 2 introduces the principle of warp shuffle on GPU and the algorithm to use registers to accelerate stencil computations, and analyzes the performance impact of the thread granularity. Section 3 introduces the characteristics of shared memory on the NVidia GPU, and the implementation of stencil computations with loop tiling. Then we test and analyze the kernel performance of different bank modes, different sub-array sizes and different thread computation granularities. Section 4 presents the overall performance results. Section 5 summarizes the work of this paper.

Table 1. The specifications of the NVidia Tesla K80 GPU.

Item	Tesla K80
GPU	2*GK 210
Peak performance (Double precision)	2.91Tflops (GPU Boost)
Device memory	24 GB GDDR5
Bandwidth	480 GB/s
CUDA cores	4992
Shared memory/SM	64 KB + (16/32/48)KB
Register files/SM	128 KB 32-bit
Maximum number of threads/SM	2048 (64 warps)
Maximum number of threads/block	1024

2 Using Registers to Optimize Stencil Computations

2.1 Stencil Computations in HNSC

HNSC adopts a fourth-order Runge-Kutta time marching scheme. Each time step contains 4 stages, and each stage involves several time-consuming stencil computation kernels. The program is based on the sixth-order energy stable central difference method presented in [8] to calculate the derivative of the primitive variables and the flux vectors. In the kernel subroutine FLUX_DERIV that calculates the derivative of several flux vectors, the stencil format is shown in Fig. 1. In Fig. 1, in and out are both four dimensional arrays, and I, j, k, n are the four indices of them. This is a 12-point stencil, where 12 elements of array in are used to calculate one element of array out.

```
do n = 1, nvar
  do k = 0, nz
    do j = 0, ny
      do i = 0, nx
        out(i,j,k,n) = ( a1_24 * (in(i+1,j,k,n) - in(i-1,j,k,n)) + a2_24 * (in(i+2,j,k,n) - in(i-2,j,k,n)) + &
                         a3_24 * (in(i+3,j,k,n) - in(i-3,j,k,n)) + a4_24 * (in(i+4,j,k,n) - in(i-4,j,k,n)) + &
                         a5_24 * (in(i+5,j,k,n) - in(i-5,j,k,n)) + a6_24 * (in(i+6,j,k,n) - in(i-6,j,k,n)) ) / dx
      end do
    end do
  end do
end do
                              Stencil computation format
```

Fig. 1. Stencil computation on x plane.

In fact, this subroutine computes 12-point stencil on 4 arrays, 2 of which are along the x direction and the other two are along the y direction. As HNSC is written in Fortran 90, the arrays are stored by column-major. The stencil computations in the y direction will involve a lot of strided memory accesses. We first split this subroutine into two sub-kernels in the x-direction (FLUX_DERIV_X) and y-direction (FLUX_DERIV_Y), and then optimize the two subroutines respectively. For the subroutine FLUX_DERIV_Y, the input arrays and the output arrays of the stencil computation can be respectively transposed in the previous and next adjacent kernels, so as to transform them into the stencil computations similar to the ones in the x direction. Then we divide the constant coefficient by dx to attain a new coefficient.

The original subroutine only uses global memory access to implement the stencil computations. The threads will redundantly load data in global memory and cannot effectively utilize the hardware resources on GPU. In order to improve the efficiency of memory access to improve the performance of these stencil computations, we can use the GPU on-chip storage. We use the stencil computation in the subroutine FLUX_DERIV_X as the baseline implementation. Since in each time-step, after the subroutine FLUX_DERIV, MPI communications are needed to exchange the data on the halo regions, the register files or shared memory reuse occurs in only one time-step iteration. This requires us to make effective use of GPU's on-chip memory.

2.2 CUDA Warp Shuffle Functions

CUDA warp shuffle functions allow threads to access the register files of neighboring threads in the same warp. Lane ID is used to index threads in a warp. The function format of _shfl_up() is var2 = _shfl_up (var1, delta, width). It computes the source Lane ID by subtracting delta from the caller's thread ID. The value of var1 held by the resulting thread ID is returned. The width argument specifies the execution range of this function. In the same way, the _shfl_down() function adds the delta to caller's thread ID to compute the source Lane ID. Figure 2 shows the execution process of shfl_up().

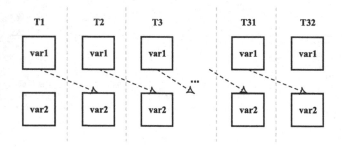

Fig. 2. shfl_up(), var2 = _shfl_up(var1, 1, 32), Var1 is shifted up the warp by 1 lane.

2.3 Using Warp Shuffle Functions in Stencil Computations

The stencil computations in subroutine FLUX_DERIV use the data from 12 grid points (6 on the left and 6 on the right) to calculate the data of current grid point. In the loop body, we can make each thread load the value of 1 or more grid points, and call shfl_up and shfl_down multiple times to make each thread

read the value from its left and right neighbors. This will cache the data in the registers. Since CUDA shuffle functions take warp as the unit, the data block in a warp is relatively small. Generally, the smaller the data block is, the larger is the halo region compared to the whole computation area. Therefore, we must reasonably allocate data block size and the computation granularity of thread to accelerate the performance while saving register files. We implement and test the cases when a thread computes on 1, 2, 3 and 6 grid points in the loop body and get the optimal computation configuration.

2.3.1 Case 1: Each Thread Computes on 1 Grid Point in the Loop Body

In the loop body, each thread reads only the data of one grid point, and then calls the __shfl_up 6 times to read the data of 6 grid points on its left; similarly, calls the __shfl_down 6 times to read the data of 6 grid points on its right. We set the thread block configuration to (32, 4, 1) and map it to the grid by overlapping warps (see Fig. 3).

The thread loads the variable cur on one grid point. Variable l1~l6 and r1~r6 are the data of the left and right neighbors of cur. Only the 20 inner threads can read all the variables needed for stencil computations. In order to compute all the data, warps are cross-mapped on grid in unit of 20.

```
! Define to read register files of neighbor threads.
real(DP) :: l1, l2, l3, l4, l5, l6;      real(DP) :: r1, r2, r3, r4, r5, r6;     real(DP) :: cur
! Index to global memory, warps are cross-mapped on grid.
i = (blockIdx%x - 1) * 20 + threadIdx%x - 7
j = (blockIdx%y - 1) * blockDim%y + threadIdx%y - 1
do n
    do k
        cur = in(i,j,k,n)
        ! read l1~l6, r1~r6 from neighbor threads
        l1 = __shfl_up(cur, 1, 32); l2 = __shfl_up(cur, 2, 32); l3 = __shfl_up(cur, 3, 32)
        l4 = __shfl_up(cur, 4, 32); l5 = __shfl_up(cur, 5, 32); l6 = __shfl_up(cur, 6, 32)
        ! Same operation on the right.
        !sync
        if(threadidx%x < 27 .and. threadidx%x > 6) then
            out(i,j,k,n) = a1 * (r1 - l1) + a2 * (r2 - l2) + a3 * (r3 - l3) + a4 * (r4 - l4) + a5 * (r5 - l5) + a6 * (r6 - l6)
        endif
        !sync
    end do
enddo
                        Stencil by warp shuffle 1
```

Fig. 3. Example of stencil computations on registers (case 1).

2.3.2 Case 2: Each Thread Computes on More Than 1 Grid Points in the Loop Body

In subsection 2.3.1, 37.5% (12/32) of the threads are used to store the data of halo regions. We see that the halo region is very large, which leads to the waste of registers and the thread compute capability. We can increase the number of grid points that each thread reads from global memory in the loop body (cur1 and cur2), to reduce the storage cost of halo regions and improve the thread utilization (see Fig. 4).

```
real(DP) :: l1, l2, l3, l4, l5, l6;        real(DP) :: r1, r2, r3, r4, r5, r6
! Each thread reads 2 mesh points.
real(DP) :: cur1, cur2
! Index to global memory, warps are cross-mapped on grid.
i = (blockIdx%x - 1) * 52 + (threadIdx%x - 1) * 2 - 6
j = (blockIdx%y - 1) * blockDim%y + threadIdx%y - 1
do n
   do k
      cur1 = in(i,j,k,n), cur2 = in(i+1,j,k,n)
      ! read l1~l6, r1~r6 from neighbor threads
      l1 = __shfl_up(cur2, 1, 32); l2 = __shfl_up(cur1, 1, 32); l3 = __shfl_up(cur2, 2, 32)
      l4 = __shfl_up(cur1, 2, 32); l5 = __shfl_up(cur2, 3, 32); l6 = __shfl_up(cur1, 3, 32)
      ! Same operation on the right
      !sync
      if(threadidx%x < 30 .and. threadidx%x > 3) then
         out(i,j,k,n) = a1 * (cur2 - l1) + a2 * (r1 - l2) + a3 * (r2 - l3) + a4 * (r3 - l4) + &
                        a5 * (r4 - l5) + a6 * (r5 - l6)
         out(i+1,j,k,n) = a1 * (r1 - cur1) + a2 * (r2 - l1) + a3 * (r3 - l2) + a4 * (r4 - l3) + &
                          a5 * (r5 - l4) + a6 * (r6 - l5)
      endif
      !sync
   end do
end do
                        Stencil by warp shuffle 2
```

Fig. 4. Example of stencil computations on registers (case 2).

In the loop body, if each thread computes on 2 grid points, each warp can update data of $2 * 32 - 2 * 6 = 52$ grid points. Compared with the previous implementation that calculates 20 grid points in the loop body, the workload of each warp increases by 2.6X. At the same time, the number of threads launched by the kernel is reduced by over 50%, due to the change of thread mapping rules.

We can see that the number of registers used by each thread is larger than the case when a thread computes on 1 grid point in the loop body. In the loop body, when each thread computes on more than two grid points, the implementation of the kernel is similar, and the number of registers required will further increase.

We perform test to investigate the impact of the number of grids calculated by each thread in the loop body. We run one process on a GK210 in this test. We use the grid size of 15M and 60M as the total scale, then change the number of grids calculated by each thread from 1, 2, 3 to 6 in the loop body. Table 2 shows the runtime with respect to the number of grids calculated by each thread. We can see that when the number of grids calculated by each thread in the loop body increases from 1 to 2 and 3, the runtime is reduced. When the number of grids calculated by each thread is 6, the runtime becomes longer. When each thread performs calculations on 3 grid points in the loop body, the shortest runtime is achieved, both the 15M and 60M grids. The maximum speedup is 1.75 for the 15M grids and 1.81 for the 60M grids.

Table 2. Runtime when grids calculated by each thread in the loop body changes.

Kernel	Time (ms) 15M	Time (ms) 60M
1	45.2	179.5
2	25.9	99.4
3	25.9	99.1
6	30.4	121.2

3 Using Shared Memory to Optimize Stencil Computations

Loop tiling [10] is one of the most effective techniques to improve the data local-ity and parallelism of the loops. In the register based optimization, the shuffle functions work in the warp range, which is a relatively small granularity. We know that GPU shared memory is on-chip and has a much higher bandwidth and lower latency than the global memory. We can make use of the shared memory to optimize the memory access performance of the stencil computa-tions. This section introduces the characteristics of the shared memory and our implementation of using shared memory to optimize stencil computations.

3.1 The GPU Shared Memory and Its Characteristics

The global memory on a current GPU has hundreds of GB/s of bandwidth and GB-level memory capacity. Each Streaming Multiprocessors (SM) has an on-chip buffer shared by L1 cache/shared memory with a size of 64 KB, which has higher bandwidth than the global memory. CUDA provides an API to configure the size of the L1 cache and the shared memory. For Tesla GPU, each SM has an extra 64 KB shared memory, which directly increases the GPU hardware occupancy of the kernel with limited shared memory capacity. The number of shared memory bank is 32. When different threads within the same warp access data within the same bank, there are bank conflicts resulting in serial access. To avoid bank conflicts, the best access mode is that 32 threads in the same warp access 32 different banks.

One of the most important metrics to guide the performance optimization on GPU is the hardware occupancy. It refers to the ratio of the number of active warps on the SM to the maximum number of warps. It is related to the number of active thread blocks on the SM. The number of active thread blocks on the SM is limited by several factors, including the shared memory size, the number of registers, and the maximum number of threads on each SM. Each memory access operation consumes hundreds of clock cycles. For memory bound kernels, increasing the hardware occupancy [11] helps to utilize the SM's warp schedule mechanism to hide memory access latency.

3.2 Shared Memory Based Implementation

To make use the shared memory, we tile the loop in the kernel and buffer part of the arrays accessed by the thread in a smaller array defined in the shared memory. This will reduce redundant access to global memory and take advantage of the high bandwidth. In this implementation, the factors that influence the performance include the reading width of bank, the sub-array size, and the computation granularity each thread. We will discuss these factors in detail. The performance evaluation in this paper is done on 1 Tesla K80, with the hardware and software configurations shown in Table 1. In the following test, we set the shared memory capacity of each SM on the Tesla K80 to 112 KB (the maximum capacity), and use 32 threads in a warp to access different banks in parallel.

3.2.1 Stencil Computation with Different Bank Access Modes

Starting from the Kepler GPU, bank has 4-byte and 8-byte access modes [12]. Based on the sub-array size of 1024 and the thread computation granularity of 1, we test the kernel performance when 4-byte and 8-byte access modes are used. The grids sizes used are 15M and 60M. The data are shown in Table 3. We can see that 8-byte access mode achieves 1.28X acceleration ratio compared to the 4-byte access mode. The program uses double-precision floating-point data for the stencil computations. While the bandwidth of the 4-byte access mode is 1.5 times higher than that of the 8-byte access mode, it needs 2 load transactions to read a double-precision variable. On the contrary, the 8-byte access mode only needs one load transaction to read a double-precision variable. Hence, as can be seen in Table 3, the 8-byte access mode outperforms the 4-byte access mode.

Table 3. Metrics of different access mode measured by NVprof.

	4-byte	8-byte
Shared load throughput (GB/s)	793.98	522.26
Shared load transactions per request	1.95	1
Shared efficiency	50.25%	98.43%
Shared replay overhead	0.24	0
The kernel runtime (ms)	119.2	93.1

3.2.2 Stencil Computation with Different Sub-array Sizes

The size of the sub-array is another factor that may influent the performance. We test the kernel performance for different sub-array sizes when using the 8-byte access mode here. We set the second dimension of the thread block to 1 to facilitate thread indexing. By changing the size of the first dimension, we can obtain different mappings. We use one-dimensional array to store shared data. Arrays in the global memory are tiled over the x direction. Figure 5 shows the corresponding code.

```
! Define subarray, compute i~i+n-1
real(DP), shared :: sub(-6 : n-1 +6)
! Index to global memory
i = (blockIdx%x - 1) * blockDim%x + threadIdx%x - 1
j = blockIdx%y - 1
! index to shared memory
ishm = threadIdx%x - 1
do n
  do k
      sub(ishm) = in(i,j,k,n)
      if(ishm < 6) then
          sub(ishm - 6) = in(i-6,j,k,n); sub(ishm + n) = in(i+n,j,k,n)
      call syncthreads()
      out(i,j,k,n) = &
          a1dx * (sub(ishm+1) - sub(ishm-1)) + a6dx * (sub(ishm+6) - sub(ishm-6))
      call syncthreads()
  end do
end do
                          Stencil : different sub-array size
```

Fig. 5. Stencil computations on shared memory with different sub-array size. For brevity, only a small part of the boundary processing is listed.

For different loop tiling sizes, in addition to modifying the size of n, we also need to deal with the boundary differently. For the case that each thread updates a grid point, we test the performance when n respectively is 128, 256, 512, and 1024. The results are shown in Table 4.

This kernel computes the stencil on 2 flux vectors. For example, when n is 128, the required shared memory is $(128 + 6 * 2) * 8 \, \text{Bytes} * 2 = 2240$ Bytes. The shared memory allocation granularity is 256 bytes. So, when n is 128, we allocate 2304 bytes shared memory for the sub-array.

Since the sub-array size in the shared memory is significantly larger than that in the warp, increasing n only slightly improves the utilization of the sub-array $(n/(n + 12))$. According to the data shown in Table 2, for the 60M grids, the shortest runtime is achieved when n is 128 or 256. After that, the runtime of the kernel will increase with the increase of n. For the 15M grids, increasing n above 256 will reduce the performance.

Even though different sub-array sizes lead to different performance, the difference is not so significant. The amount of shared memory and registers required by the kernel will directly affect the GPU occupancy. We denote the thread block size as Blk_{sz}, the size of shared memory used by each thread block as Blk_{shm},

Table 4. The performance of kernels with different sub-array sizes.

n	Shared memory space occupied by sub-array (Bytes)	Time (ms) 15M	Time (ms) 60M
128	2304	19.3	82.4
256	4352	19.1	82.3
512	8448	20.1	85.5
1024	16640	24.5	93.1

and the size of register files used by each thread block as Blk_{shm}. Denote the maximum number of threads, maximum number of thread blocks, shared memory capacity and the number of register files on the SM as SM_{thread}, SM_{blk}, SM_{shm}, SM_{reg}, respectively. Then we can calculate the GPU occupancy based on these parameters, as shown in Eqs. (1) and (2).

$$Active\ blocks\ per\ SM = \min\left(\frac{SM_{reg}}{Blk_{reg}}, \frac{SM_{shm}}{Blk_{shm}}, SM_{blk}, \frac{SM_{thread}}{Blk_{sz}}\right) \quad (1)$$

$$Occupancy = Active\ blocks\ per\ sm * Warp\ per\ block \quad (2)$$

When compiling the program, it shows in PTX that each thread uses 53 registers. Registers are allocated by warp, and the allocation granularity is 256 Bytes. A warp will use $32 * 53 = 1696$ bytes, rounded up to 1792 bytes. Different n corresponds to different Blk_{sz}. When n takes the largest value, 1024, a thread block uses $1024/32 * 1792 = 57344$ registers. On the K80 GPU, each SM 128K registers. This means that each SM can only support 2 active thread blocks due to register file limitation. Similarly, when it comes to the usage of shared memory, each SM can allocate 6 thread blocks. Coupled with the limitations of the hardware device, each SM has a maximum of 2048 active threads and 16 thread blocks. When the thread block is 1024, each SM has 2 active thread blocks. The GPU hardware occupancy is $2 * (1024/32)/64 * 100\% = 100\%$. Similarly, since the computation granularity of each thread (that is, each thread computes one grid point) is fixed, a thread always uses 53 registers. So when n is smaller than 1024, the hardware occupancy is 100%. This explains why different sub-array sizes have the same level of operating efficiency. The reason for the performance drop when n equals to 1024 is that there are only 2 active thread blocks per SM, which is less than the number recommended by NVidia.

3.2.3 Stencil Computation with Different Computation Granularities

The computation granularity of thread also affects the performance of the kernel. Here we evaluate the performance impact of thread computation granularity. Through the previous test and analysis, it can be seen that optimal performance will be obtained when the sub-array size is 128 or 256 with 8-byte access mode.

```
I Define subarray
real(DP), shared :: sub(-6 : n * m - 1 + 6)
I index to global memory, one thread does m computations
i = (blockIdx%x-1) * blockDim%x * m + (threadIdx%x-1) * m
j = blockIdx%y - 1
I index to shared memory
ishm = (threadIdx%x − 1) * m
do n
    do k
        I Each thread loads m mesh points , i ~ i+m-1.
        sub(ishm:ishm+m-1) = in(i:i+m-1,j,k,n)   ···
        I Again, threads in ghost zone loads boundary mesh points m times.
        call syncthreads()
        I Each thread computes m mesh points, i ~ i+m-1.
        out(i:i+m-1,j,k,n) = a1dx * (sub((i:i+m-1) +1) - sub((i:i+m-1) -1)) + a6dx * (sub((i:i+m-1) +6) - sub((i:i+m-1) -6))
        call syncthreads()
    end do
end do
                    Stencil : different computation granularity
```

Fig. 6. Stencil computations on shared memory for different computation granularities.

Here we use the 8-byte access mode and set n to 128. Based on the previous algorithm, threads are mapped across the grid through the changed index to achieve different computation granularity of threads. Figure 6 shows the algorithm.

Based on this algorithm, we adjust the thread computation granularity from 1, 2 to 4 in the performance test. The size of the grids is 15M and 60M. Figure 7 shows the runtime for different thread computation granularities.

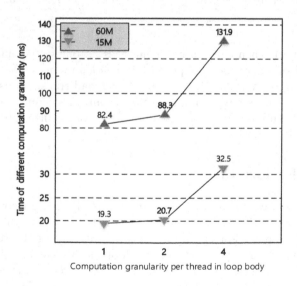

Fig. 7. Performance of stencil computations with different computation granularity.

As can be seen in Fig. 7, when the thread computation granularity increases, the kernel performance decreases. For the 15M grids, the performance loss is not large when the thread computation granularity is increased from 1 to 2. However, there is obvious performance loss when the thread computation granularity in the loop body increases from 2 to 4. For the 60M grids, when the computation granularity increases from 2 to 4, the runtime will significantly increase, too. For memory bound kernels, to take advantage of the high bandwidth and data sharing characteristics of shared memory, the most ideal way is that the 32 threads in a warp access 32 different banks. When each warp accesses the shared memory, different numbers of shared memory units will be loaded due to different computation granularities. When each thread performs computations on more than one grid point in the loop body, bank conflicts will happen. We use the NVprof to measure the metrics such as shared memory bandwidth, sharing efficiency, and conflict reload overhead. The results for the 60M grids are shown in Table 5.

Table 5. Metrics measured by NVprof for different computation granularities.

	1	2	4
Shared load throughput (GB/s)	609.78	332.11	249.28
Shared load transactions per request	1	2	4
Shared store transactions per request	1	2.03	4.34
Shared efficiency	98.47%	84.38%	40.76%
Shared replay overhead	0	0.129	0.324

We can see that when each thread computes on one grid point in the loop body, only one load operation is required for each load request. When the granularity in the loop body is 2 or 4, there are 2 or 4 load transactions due to bank conflicts. Write requests have similar results. From lines 5 and 6 in Table 5, it can be seen that bank conflicts lead to shared memory utilization decrease and additional reload overhead.

4   Overall Performance

This section evaluates the overall performance of the optimized implementations and compares their performance to the original implementation that only uses the global memory. For the optimization that utilizes the registers, each thread loads data on 2 grid points in the loop body, which has the optimal performance. For the optimization that utilizes the shared memory, the optimal configurations are used: 8-byte access mode, 128 size of sub-array, each thread computes on 1 grid point in the loop body. We use 15M and 60M grids in the test. Table 6 shows the runtime for different implementations and the speedup achieved. From Table 6, it can be seen that the two optimized implementations significantly outperform the original implementation. For the 15M and 60M grids, the implementation that utilizes the register respectively achieves a maximum speedup of 2.59 and 2.79, and the implementation that utilizes the shared memory respectively achieves a maximum speedup of 3.51 and 3.36. The results are satisfactory, as the HNSC program is very time-consuming and the stencil computations take a large fraction of the runtime.

Table 6. The runtime for different implementation and the speedup achieved.

Grid size	Kernel implementation	Time (ms)	Speedup
15M	Original	67.1	/
	Register	25.9	2.59
	Shared mem	19.1	3.51
60M	Original	276.5	/
	Register	99.1	2.79
	Shared mem	82.3	3.36

5 Conclusion

This paper has optimized the stencil computations of a high-order CFD program on GPU. Based on the characteristics of the stencil computations and the GPU, two methods are used. The first method utilizes the registers to cache the data used by the stencil computations, uses the shuffle functions to exchange data between neighboring grid points, and adjusts the thread computation granularity to increase the data reuse. The second method uses the shared memory to buffer the grid data used by the stencil computations in the kernel, and utilizes loop tiling to reduce redundant accesses to the global memory. In-depth performance evaluation is conducted for different application configurations on an NVidia K80 GPU. The results show that with the optimal configuration, both methods can significantly improve the performance over the original implementation that only utilizes the global memory. This indicates that for the stencil computations in the high order CFD applications, memory access optimizations have great potential to improve the performance. The optimization methods used in this paper can be extended to other scientific applications and other GPUs.

Acknowledgments. This work was partially supported by the National Key R&D Program under Grant No. 2017YFB0202403, the National Natural Science Foundation of China under grant Nos. 61561146395 and 61772542.

References

1. Tabik, S., Peemen, M., Nicol, G., Corporaal, H.: Demystifying the 16 x 16 thread-block for stencils on the GPU. Concurr. Comput. Pract. Exp. **27**(18), 5557–5573 (2015)

2. Peng, G., Liang, Y., Zhang, Y., Shan, H.: Parallel stencil algorithm based on tessellating. J. Front. Comput. Sci. Technol. **13**(2), 181–194 (2019)
3. Yang, X., Liao, X., et al.: TH-1: China's first petaflop super-computer. Front. Comput. Sci. China **4**(4), 445–455 (2010)
4. Xu, C., Deng, X., et al.: Collaborating CPU and GPU for large-scale high-order CFD simulations with complex grids on the TianHe-1A supercomputer. J. Comput. Phys. **278**, 275–297 (2014)
5. NVIDIA Corp.: CUDA C Programming Guide v11.0, July 2020
6. Falch, T.L., Elster, A.C.: Register caching for stencil computations on GPUs. In: 16th International Symposium on Symbolic and Numeric Algorithms for Scientific Computing, pp. 479–486 (2014)
7. Holewinski, J., Pouchet, L.-N., Sadayappan, P.: High-performance code generation for stencil computations on GPU architectures. In: ICS 2012, pp. 311–320 (2012)
8. Svard, M., Carpenter, M.H., Nordstrom, J.: A stable high-order finite difference scheme for the compressible Navier-Stokes equations, far-field boundary conditions. J. Comput. Phys. **225**(1), 1020–1038 (2007)
9. Wang, S., Wang, W., Che, Y.: GPU acceleration of a high-order CFD program. In: 4th International Conference on High Performance Compilation, Computing and Communications, Guangzhou, China, pp. 123–128 (2020)
10. Lam, M.D., Rothberg, E.E., Wolf, M.E.: The cache performance and optimizations of blocked algorithms. In: ASPLOS 1991, New York, USA, pp. 63–74 (1991)
11. Pikle, N.K., Sathe, S.R., Vyavahare, A.Y.: High performance iterative elemental product strategy in assembly-free fem on GPU with improved occupancy. Computing **100**(12), 1–25 (2018). https://doi.org/10.1007/s00607-018-0613-x
12. NVIDIA Corp, Kepler Tuning Guide v11.0, July 2020

The Dataflow Runtime Environment of DFC

Jing Zhang[1], Jinrong Li[1], Zheng Du[1,2], Jiwu Shu[2], and Qiuming Luo[1(✉)]

[1] NHPCC/Guangdong Key Laboratory of Popular HPC, College of Computer Science and Software Engineering, Shenzhen University, Shenzhen, China
lqm@szu.edu.cn
[2] Department of Computer Science and Technology, Tsinghua University, Beijing, China

Abstract. In this paper, we introduce the DFC dataflow language and its runtime environment. DFC runtime library is in charge of constructing the DAG of the dataflow graph, firing the DFC tasks and the synchronizations between the tasks of successive passes. Basing on an elaborately implemented thread pool and queued Active Data, DFC runtime shows an ideal performance comparing with DSPatch. The experiment of a simple dataflow graph shows that DFC has better performance for the cases that the parallelism beneath the core number, while DSPatch shows a better scalability for the cases of the parallelism exceed the core number. As DFC is still a prototype language, it lacks the ability to construct the DAG dynamically, which leads to low efficiency when coding for a well-structured dataflow graph.

Keywords: Dataflow · DFC · Runtime library

1 Introduction

Facing the challenge of Exascale computing, breakthroughs will be made in four aspects [1]: execution model, memory access mode, parallel algorithm and programming interface. As the execution model is concerned, dataflow execution model dramatically increases both scalability and efficiency to drive future computing across the exascale performance region, even possibly to zetaflops. So, as the number of cores in the host system increases (1~2 orders of magnitude higher than the current one), exascale-level computing will gradually migrate to the data flow execution mode.

Dataflow is a parallel execution model originally developed in the 1970s but explored and enhanced as the basis for non-von Neumann computing architecture and techniques. The dataflow model of computation presents a natural choice for achieving concurrency, synchronization, and speculations. In the basic form, activities in a dataflow model are enabled when they receive all the necessary inputs; no other triggers are needed. Thus, all enabled activities can be executed concurrently if functional units are available. And the only synchronization among computations is the flow of data.

The recent research related to dataflow systems can be classified into three categories: 1) dataflow systems on dedicated customized hardware, 2) macro dataflow (hybrid dataflow/controlflow) systems on off-shelf platform or customized hardware and 3) application-related systems based on underlying dataflow execution software engine.

© Springer Nature Switzerland AG 2021
Y. Zhang et al. (Eds.): PDCAT 2020, LNCS 12606, pp. 57–69, 2021.
https://doi.org/10.1007/978-3-030-69244-5_5

The dedicated hardware can achieve fine-grain dataflow tasks and excellent performance. The Maxeler system contains ×86 CPUs attached to FPGA dataflow engines (DFEs), and speedup the computation by 80/120 times for small/large problem [2]. The Tera-op Reliable Intelligently adaptive Processing System (TRIPS) [3, 4] employs a new instruction set architecture (ISA) called Explicit Data Graph Execution (EDGE). On simple benchmarks, TRIPS outperforms the Intel Core 2 by 10%, and using hand-optimized TRIPS code, TRIPS outperforms the Core 2 by a factor of 3 [3].

Most of the macro dataflow (hybrid dataflow/controlflow) systems used an instruction clustering paradigm; various instructions are clustered together in a thread to be executed in a sequential manner through the use of a conventional program counter. Teraflux [5] is a 4-year project started in 2010 within the Future and Emerging Technologies Large-Scale Projects funded by the European Union. The execution model of Teraflux [6] relies on exchanging dataflow threads in a producer–consumer fashion.

The application-related dataflow softwares include a variety of programs. Labview (Laboratory Virtual Instrument Engineering Workbench) [7] is a system-design platform and development environment for a visual programming. It uses a dataflow model for executing code. Orcc (Open RVC-CAL compiler) [8] can convert the code description in dataflow style into C code to run in a multi-core environment. The deep learning framework, Tensorflow, adopts dataflow engine to perform the forward and backward propagation. The Sunway TaihuLight Supercomputer is equipped with a dataflow execution engine modified from Tensorflow [9].

Besides those dataflow systems, some research focus on lightweight dataflow solution. DSPatch is a C++ library for dataflow framework. StreaMIT is a language for streaming application based on dataflow execution. Programmer of StreaMIT should construct the dataflow graph explicitly in natural textual syntax.

DFC is with a tiny extension of C language to obtain the capacity of describing the dataflow task, without any special underlying hardware [10]. Because it is based on C, it is much easier for programmer. And this make it highly coupled with Unix-like OSs, which means we can build other upper system software layers in DFC environment.

This paper is organized as follows. In Sect. 2, we give an overview of DFC briefly. In Sect. 3, the DFC runtime is introduced. We present a simple evaluation of DFC in Sect. 4. Conclusion is drawn in Sect. 5.

2 Overview of DFC

By extending C language, DataFlow C, short for DFC, provides an efficient method which makes developing a dataflow program with large parallelism easily without sacrificing performance thereby improving programmers' productivity. To set up a dataflow graph, all it requires is that implementing the DF function which is a new type of function defined in DFC. The DF function's body explicitly defines how the node of dataflow graph works, and the edges are derived from the output arguments of the preceding nodes and the input arguments of the following nodes. After implementing all DF functions, the compiler will automatically build the directed acyclic graph of dataflow graph during compilation. Before the execution of DAG of dataflow graph, the number of threads specified by the programmer will be created by DFC runtime library. If all input data for a DFC function are ready, this DF function will be triggered/fired and a thread will be responsible for the computing defined by this DF function. What's more, to pursue higher performance, if more than one group of input data, corresponding to "different passes", are ready, there may be several threads executing the same DF function with all these passes' input data parallelly.

2.1 DF Function

DF function is a unique function in DFC, which is capable to describe dataflow graph. DF function is one of the most important features of DFC and is used to build dataflow graph. The DF function is defined by a statement, as Code 1 shows.

```
1.  void Dataflow_function_name(
2.  arg_input_1, ...,arg_input_K;
3.  arg_output_1, ...,arg_output_L)
4.  {
5.      ...
6.      body
7.      ...
8.  }
```

Code. 1. Declaring a DF function

It can be noted that there exists a semicolon in the argument list of DF function, which is different from C normal function and actually is the only difference between C and DFC in syntax. The semicolon divides DF function's argument list into two argument lists—input argument list and output argument list.

As mentioned before, it is easy to build a dataflow graph by implementing DF functions. DF function describes the most fine-grained execution step of dataflow graph, so all nodes in dataflow graph should be declared as DF functions. By matching the output arguments and input arguments of the neighboring DF functions or nodes, compiler can connect these conjunctional nodes. What's more, DF function will implicitly get data from the input channel and push output data to the output channel without explicit coding.

As a dataflow graph shown by Fig. 1 is under consideration, there needs four DF functions. And the corresponding DFC source code framework can be shown as Fig. 2.

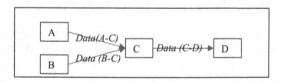

Fig. 1. A demo of dataflow graph

In the demo of Fig. 2, four DF functions are defined, which corresponding to the four nodes (A, B, C and D) in Fig. 1 respectively. During compilation, by matching the input and output parameters of DF function, the compiler can realize the connections between nodes. If the output argument of the preceding node has the same symbol name as the input argument of the succeeding node, it means that there exists a directed edge from the

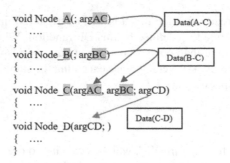

Fig. 2. A demo of DFC source code

preceding node to the succeeding node. Comparing to C++-based dataflow framework, DSPatch, in which not only defining the computing nodes, but also connecting the nodes explicitly are required, obviously it is more convenient to build the irregular structured dataflow graph in DFC.

Additionally, as Fig. 2 shown, it can be seen that a DF function does not need to have both input argument and output argument. Generally, DF function can be divided into Source DF function, Sink DF function and normal DF function. Source DF function has no input arguments, which is the start of dataflow graph, usually responsible for getting data from memory or from network. Sink DF function consumes input data and produce no output data.

2.2 Active Data

Active Data, shorted for AD, is the carrier of data that cascades through the dataflow graph. Intuitively, it is AD that flows from a node to another. In DF runtime environment, the flowing data must be wrapped into AD so that it can be transmitted between nodes, no matter what its data size and data type. And programmers have no worries about getting data from AD or wrapping it up, but just manipulate data as usual. DFC runtime will handle the memory allocating/deallocating, data tagging for various passes, synchronizing for concurrent accessing. DFC has the capability to make sure that the data size and data type of real data is correct while DF function getting data from the input AD channel.

DFC allows that a node can generate new output before its previous output is consumed. The data channel, or the directed edge, is implemented as a FIFO queue that new data should be added to its tail and removed from its head. When a node outputs data, it will give a chance to trigger the succeeding nodes if all its input data are ready. And the data in the FIFO queue will be removed if it is consumed by succeeding nodes.

2.3 Program Framework in DFC

Code 2 shows a demo of program framework in DFC, which corresponds the dataflow graph shown in Fig. 1. After implementing all DF functions, programmer just simply call DF_Run function to execute the dataflow graph. DF_Run requires one int argument,

Round. If the value of this argument is a positive integer, the dataflow graph will be executed for Round times/passes. Otherwise, the execution of the dataflow graph is stopped by calling the DF_Source_Stop function in Source DF function, when the preset conditions are met. All Source DF functions should call the DF_Source_Stop function at the same pass, therefore the dataflow graph will stop executing without generating new data. For each DF function, an int variable named DF_count is declared implicitly, which records current pass of this DF function. So, programmer can directly manipulate the DF_count without declaring but cannot redeclare another variable with the same symbol name in DF function. What's more, in order to adapt to different hardware environment, the appropriate number of threads can be specified by macro command "#define THREADNUM" to achieve proper parallelism.

```
1.  void Node_A(; argAc)
2.  {
3.      ...
4.  }
5.  void Node_b(; argBC)
6.  {
7.      ...
8.  }
9.  void Node_C(argAC, argBC; argCD)
10. {
11.     ...
12. }
13. int main()
14. {
15.     ...
16.     DF_Run(10);
17.     ...
18.     return 0;
19. }
```

Code. 2. A demo of program in DFC

3 DFC Runtime

Hiding the complex implementation details, DFC runtime library provides the environment supporting dataflow program. DFC runtime is implemented elaborately, because it has a great impact on the performance. DFC runtime will be introduced from three aspects as follows: 1) How DFC manages DAG of the dataflow graph; 2) How DFC triggers the dataflow computing; 3) And how DFC distinguishes data of various passes.

3.1 DAG Management

In DFC, for a directive edge, the output channel of the preceding node and the input channel of the succeeding node share a same AD channel. An AD is a global variable and actually is the forms of the edge in the DAG, it means that even though more than one node inputs the same AD, there only exists one copy of this AD in memory. So, it saves memory of storing data.

There are several important data structures in DFC: DF_AD, DF_FN and DF_TFL. In DFC, AD is implemented by a structure named DF_AD, which records the address of the real data. DF_FN manipulates the node of the DAG, and it records information of corresponding DF function, including the entry address of the DF function, the input AD list and output AD list. So, to connect the two neighboring nodes, just add the pointer of the DF_AD to the output list of the preceding DF_FN and also add it to the input list of the succeeding DF_FN. DF_TFL is a table that records information of dataflow graph and configuration. It stores all nodes' addresses, so that the DAG of the dataflow graph can be derived.

3.2 Tasks Trigger and Execution

Different from some other dataflow program framework, in DFC, threads and nodes of dataflow graph are independent, or decoupled. A thread is not work for only a specific node but it can execute task defined by any node. Task is pushed into a queue, and a free thread takes a task from the queue. So when calling the DF_Run function to execute the computation defined by dataflow graph, a thread pool that contains specified number threads will be first created. If the task queue is not empty, one free thread will pick a task from the task queue head and remove it. Unless there is no task ready, the thread will be blocked until new task comes.

After thread pool is created, the main thread will call DF_Loop function to keep adding new task to the task queue until the execution stops. Exactly, the main thread is only in charge of adding tasks defined by Source DF function, the start of the dataflow graph, so that new pass can be triggered. Because dataflow program is data driven, tasks defined by other DF functions—Sink DF function and normal DF function, will be triggered when their preceding nodes output data. To improve dataflow program performance and save memory, it is necessary to have a appropriate strategy to decide when to push task defined by Source DF function into the task queue. In DFC, only when the length of the task queue is less than the number of Source DF function and the Source DF function is not stopped, can this task be added. Due to the particularity of the Source DF function comparing to other DF function, a new structure, DF_SI, is specifically used to describe the Source DF function. Members of DF_SI are DF_FN, status flag 'stop' and remaining running times 'count'. As mentioned before, when calling DF_Run with a positive integer argument, for every DF_SI, its 'stop' will be set to '0' and 'count' will be assigned to this positive argument. Every time a Source DF function is scheduled, its 'count' will decrease by 1. When 'count' equals to 0, its 'stop' will be set to '1' and this Source will not be scheduled any more. Of course, calling DF_Source_Stop in Source DF function will also stop this Source DF function even though its 'count' is greater than 0.

When a node finishes computing, it needs to update its output AD. But as mentioned before, because maybe the same DF function with different passes' input data are being executed, so the current pass' output data should be written after data of previous passes. In AD, queue storing is realized by a resizable array. The queue is shown by Fig. 3. It's obvious that an element of the queue is composed of data and Fanout. Fanout is a counter, implying the times that this data can be accessed in one pass. Once AD is going to be updated, the output data should be written into the data buffer area of the queue slot and the Fanout should be initialized by the number of the node that inputs this data. And every time this data is accessed, Fanout should decreases by one. And when Fanout decreases to 0, it means that no nodes will access this data anymore, so this data should be discarded and the head of the queue should point to the next element.

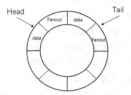

Fig. 3. AD queue

After updating output AD, the preceding nodes should try to trigger their succeeding nodes. As shown in Fig. 4, there is a flag list in DF_FN showing the ready status of the node's input data. The flag is an integer, and each bit of it corresponds to an AD that the node needs. If a node needs n AD inputs, the n lowest bits of its flag are valid. If the input data is ready, the corresponding bit will be set to '1'. Or the bit keeps being '0' until the input data is ready. Only if its flag equals to 2^n-1, all the node's input data are ready. For example, if one node needs 8 AD inputs, then the 8 lowest bits record the status of its inputs. And the binary number '0001 1011' shown in Fig. 4, means that the first to the fifth AD are ready, except for the fourth one. When its flag equals to '1111 1111', all the input data are ready and the task defined by this node can be fired. Because there may exist multi groups of input data corresponding to different passes, a flag list is needed to store all the ready status of input data of various passes.

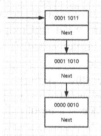

Fig. 4. The ready status list

Before the preceding node finishes working, it need to do the last thing: to modify its succeeding nodes' flags. Go through the ready status list, then set the corresponding bit of the appropriate flag to 1. Or if there exists no flag that its corresponding bit is 0, then create a new flag and set its corresponding bit to 1. And then check the first flag, if it shows that the input data are all ready, the corresponding task defined by this succeeding node will be push into the task queue.

3.3 Distinguish Data of Different Passes

As mentioned before, to boost performance, DFC allows multi graph of different passes are being executed at the same time. It means that maybe multi tasks defined by the same node are fired thereby it may lead to data risk. So it's necessary to take measure to ensure the order of data accessing. In DFC, AD is able to record the pass sequence that the first valid element of the data queue corresponds to by an integer variable named Order. At the K-th pass, the node can get data from the (Order-K)th element relative to the first element in the data queue. Of course, when to output data, the data of current pass have to be inserted to the data queue after the data of previous pass.

4   Evaluation

4.1 Experimental Environment

We have compared the performance of DFC and DSPatch on machines with various number of cores. The first one is based on Intel® Xeon® Gold 6130 CPU with 32 cores and 64 hardware threads. And the operating system on it is Ubuntu with kernel version 4.15.0, and the version of GCC is 7.4.0. The second one is based on Intel® Xeon® E5-2620 v4 with 16 cores and 32 hardware threads. Its operating system is Ubuntu with kernel version5.4.0, and the version of GCC is 7.5.0.

4.2 Brief of DSPatch

DSPatch is a powerful cross-platform dataflow framework based on C++, and has easy-to-use object-oriented API. DSPacth is designed by a concept about "circuit" and "components". The circuit manages the dataflow graph, and component defines the node. Customized component must extends from DSPatch::Component base class and the inherited virtual "Process_()" method must be implemented. A demo of a customized component is shown by Code 3. In this demo, a new "And" component for operating logic "and" is defined. Firstly, call SetInputCount_() and SetOutputCount_() to respectively to set the number of input and output in its constructor. The component's Process function defines the computing of the node. In the Process(), input data is gotten by calling SignalBus:: GetValue() explicitly and SignalBus::SetValue() is called to set output data. After defining all components, they are added to the circuit by calling Circuit::AddComponent(). To connect components, Circuit::ConnectOutToIn() will be called. Code 4 shows a demo of a dataflow program in DSPatch.

```
1.  class And final : public Component
2.  {
3.  public:
4.      And()
5.      {
6.          SetInputCount_( 2 );
7.          SetOutputCount_( 1 );
8.      }
9.  protected:
10.     virtual void Process_( SignalBus const& inputs,
11. SignalBus& outputs ) override
12.     {
13.         auto bool1 = inputs.GetValue<bool>( 0 );
14.         auto bool2 = inputs.GetValue<bool>( 1 );
15.         if ( bool1 && bool2 )
16.         {
17.             outputs.SetValue( 0, *bool1 && *bool2 );
18.         }
19.     }
20. };
```

Code. 3. A demo of a customized component

```
1.  int main()
2.  {
3.      auto circuit = std::make_shared<Circuit>();
4.      auto randBoolGen1 = std::make_shared<RandBool>();
5.      auto randBoolGen2 = std::make_shared<RandBool>();
6.      auto logicAnd = std::make_shared<And>();
7.      auto boolPrinter = std::make_shared<PrintBool>();
8.
9.      circuit->AddComponent( randBoolGen1 );
10.     circuit->AddComponent( randBoolGen2 );
11.     circuit->AddComponent( logicAnd );
12.     circuit->AddComponent( boolPrinter );
13.     circuit->ConnectOutToIn(randBoolGen1,0,logicAnd, 0 );
14.     circuit->ConnectOutToIn(randBoolGen2,0,logicAnd, 1 );
15.     circuit->ConnectOutToIn(logicAnd,0,boolPrinter, 0 );
16.
17.     circuit->Tick();
18. }
```

Code. 4. A demo of DSPatch

There is a little bit difference in tasks trigger between DFC and DSPatch. When calling Circuit::Tick() to run the dataflow graph for a pass, the circuit thread will go through all components and tick them. Maybe the succeeding component will be ticked before the preceding component, so the succeeding component will tick the preceding component. While in DFC, the succeeding node is triggered by the preceding nodes. What's more, the thread is bound with the component in DSPatch. A thread only works for one component. And to enable multi passes concurrence, multi independent groups of threads are needed.

4.3 Experiment Results

We have compared DFC with DSPatch on the same simple dataflow graph shown by Fig. 5. And Table 1 presents the code lines to form the graph of Fig. 5 in DFC and DSPatch respectively. DFC code has 40 lines, which is much shorter than DSPatch code. The program code of such simple dataflow graph in DFC is more concise than it in DSPatch.

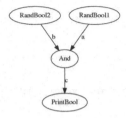

Fig. 5. A simple dataflow graph

Table 1. Comparison DFC with DSPatch

	DFC	DSPatch
Code Lines for Fig. 5	40	92
Code Lines for Fig. 6 (static style)	$31 * 2^{n-1} + 7n + 56$	$43n + 5 * 2^n + 8$
Code Lines for Fig. 6 (static-multi-declaration)		$67 * 2^{n-1} - 14n + 8$
Code Lines for Fig. 6 (construct DAG in iteration)		$61n - 3$
Threads for n parallelism	$n+1$	$5n+1$

We have compared the code to build a graph of two binary trees connected back-to-back as Fig. 6 shown. The first node randomly generate an integer array, and the nodes of the second layer to the (n-1)th layer simply divide the input array averagely into two arrays. The nodes of the n-th layer will bubble sort the input array, and the other nodes will merge the two input orderly arrays.

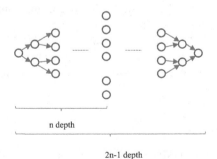

Fig. 6. Binary trees connected back-to-back

Table 3, Fig. 7 (a) and Fig. 7 (b) present the total code line number and the assistant code line number, corresponding to various depth of the tree. And the assistant code is to describe the graph. The number n is the depth of the binary tree on the left just as Fig. 6 shown, and the abscissa is a logarithm of n with a base of two. The ordinate is also logarithmic number. In these cases, the DFC programs are coded in static style, which means all the node are declared individually. And the DSPatch (dynamic style) programs are coded in dynamic style, which means each level of that tree is constructed in a "for/loop" iteration. DSPatch (static-multi-declaration) programs means that all classes of nodes are declared respectively and build the graph without a loop, which is for the case of irregular graph and the nodes are various. DSPatch (static style) programs do not declare classes of nodes repeatedly and only build the graph without any loop. So DSPatch (static-mutli-declaration) programs and DSPatch (static style) share the same number of assistant codes.

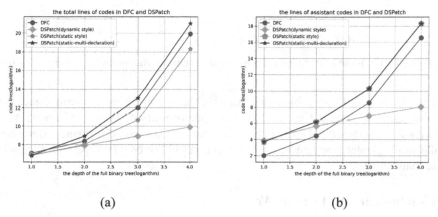

(a)   (b)

Fig. 7. (a) The total number of code in DFC and DSPatch (b) the number of assistant code in DFC and DSPatch

For the well-structured graph, such as the graph of back-to-back-connected binary-tree (as Fig. 6 shown), DSPatch can code that graph in a dynamic style and construct the graph by the code of iteration. If DSPatch programs are coded in nesting loop, the code lines will keep as constant for any depth. While, in DFC, that graph should be code statically and make it have a much more code lines. Because the irregular structured graph cannot be programed by loop/iteration code, so, for the irregular structured graph, or for the small scale graph, DFC is preferred for the easy coding and less code lines.

But DFC runtime needs far less threads to fulfill the same parallelism then DSPatch, as Table 4 shown.

Figure 8(a) and Fig. 8(b) respectively present the time that DFC and DSPatch consumed by the problem of Fig. 5, on the machines with 64 cores and 32 cores, for running 4096 times. The thread number of DFC program and the buffer size of DSPatch is from 2^0 up to 2^9, which corresponds to their parallelism. And the abscissa, representing the parallelism, is a logarithmic coordinate with a base of two. The ordinate on the left shows the time that programs consumed, and ordinate on the right shows the speedup ratio of the programs comparing to the program of serial version.

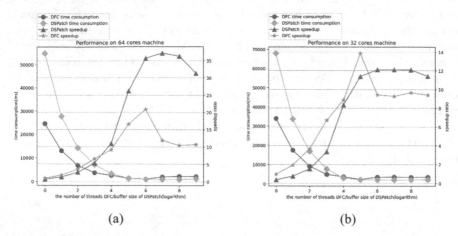

Fig. 8. (a) Time consumption on 64 cores machine (b) time consumption on 32 cores machine

The Fig. 8(a) and Fig. 8(b) shows that DFC have a much better performance for the cases of parallelism below 16 ($2^{\wedge 4}$) in the platform of 64-cores or 32-cores. And, as the parallelism increase, DFC and DSPatch show the similar performance. But when parallelism is greater than number of physical cores (32 or 64), DFC performs slightly worse than DSPatch.

5   Conclusion and Future Works

In this paper, we introduce the DFC dataflow language and its runtime environment. Same as the other dataflow language, DFC is convenient to build a dataflow program easily without concerning the concurrency, synchronization and deadlock, while the traditional parallel programming techniques (such as MPI or OpenMP, etc.) should handle these error prone problems explicitly and carefully.

DSpatch program needs shorter code to describe a given regular, or well-structured dataflow graph, than DFC program does. While for the irregular structured graph, or for the small scale graph, DFC is preferred for the easy coding and less code lines.

DFC runtime library is in charge of constructing the DAG of the dataflow graph, firing the DFC tasks and synchronize between the tasks of successive passes. Basing on an elaborately implemented thread pool and queued Active Data, DFC runtime shows an ideal performance comparing with DSPatch. For the problem shown as Fig. 5, DSPatch needs 5n+1 threads to achieve the parallelism of n, while DFC needs n+1 threads for the parallelism of n. So, DFC consumes less system resource and is more preferable for the large scale dataflow graphs. Even for the parallelism below 16, DFC is outperform DSPatch in matric of execution time.

Although DFC has better performance for the cases that the parallelism beneath the cores numbers, DSPatch shows a better scalability than DFC for such simple graph, as

DSPatch achieves higher speedup for the case of threads number exceed the physical cores. We are enhancing DFC with tracing function to spot the performance bottleneck by recording the timeline of each DFC function's duaration.

By now, as DFC is still a prototype language, it is not capable of constructing the graph dynamically, which make it inefficient for those well-structured dataflow graph. We will extend the syntax to make it capable of constructing the graph dynamically in the next version of implementation.

References

1. Sterling, T., Brodowicz, M., Anderson, M.: High Performance Computing: Modern Systems and Practices, Morgan Kaufmann, pp. 616–618 (2018)
2. Pell, O., Averbukh, V.: Maximum performance computing with dataflow engines. In: Computing in Science & Engineering, vol. 14, no. 4, pp. 98–103 July-August 2012. https://doi.org/10.1109/mcse.2012.78
3. Burger, D., et al.: The TRIPS Team, Scaling to the end of silicon with EDGE architectures. IEEE Comput. **37**(7), 44–55 (2004). https://doi.org/10.1109/MC.2004.65
4. Gebhart, M., et al.: An evaluation of the TRIPS computer system. SIGPLAN Not **44**(3), 1–12 (2009). https://doi.org/10.1145/1508284.1508246
5. Giorgi, R.: Teraflux: exploiting dataflow parallelism in teradevices. In: Proceedings of the 36th Annual IEEE/ACM International Symposium on Microarchitecture, CF 2012, ACM, 2012. New York, NY, USA, 2012, pp. 303–304 (2012). http://doi.acm.org/10.1145/2212908.2212959
6. Portero, A., Yu, Z., Giorgi, R.: Teraflux: exploiting tera-device computing challenges. Procedia CS **7**, 146–147 (2011)
7. Ursutiu, D., Samoila, C., Jinga, V.: Creative developments in LabVIEW student training: (Creativity laboratory — LabVIEW academy). In: 2017 4th Experiment@International Conference (exp.at 2017), Faro, 2017, pp. 309–312 (2017). http://doi.org/10.1109/EXPAT.2017.7984399
8. Chavarrias, M., Pescador, F., Juárez, E., Garrido, M.J.: An automatic tool for the static distribution of actors in RVC-CAL based multicore designs, pp. 1–6. Design of Circuits and Integrated Systems, Madrid (2014)
9. Lin, H., Lin, Z., Diaz, J.M., Li, M., An, H., Gao, G.R.: swFLOW: a dataflow deep learning framework on sunway taihulight supercomputer. In: 2019 IEEE 21st International Conference on High Performance Computing and Communications; IEEE 17th International Conference on Smart City; IEEE 5th International Conference on Data Science and Systems (HPCC/SmartCity/DSS), Zhangjiajie, China, 2019, pp. 2467–2475 (2019)
10. Du, Z., Zhang, J., Sha, S., Luo, Q.: Implementing the matrix multiplication with DFC on kunlun small scale computer. In: 2019 20th International Conference on Parallel and Distributed Computing, Applications and Technologies (PDCAT), Gold Coast, Australia, 2019, pp. 115–120 (2019)

Adaptive Tensor-Train Decomposition for Neural Network Compression

Yanwei Zheng[1], Yang Zhou[2], Zengrui Zhao[1], and Dongxiao Yu[1]

[1] School of Computer Science and Technology, Shandong University,
Qingdao 266237, People's Republic of China
{zhengyw,dxyu}@sdu.edu.cn, zhaozr@mail.sdu.edu.cn
[2] Shenzhen Institutes of Advanced Technology, Chinese Academy of Science,
Shenzhen 518055, People's Republic of China
zhouyangcumt@163.com

Abstract. It could be of great difficulty and cost to directly apply complex deep neural network to mobile devices with limited computing and endurance abilities. This paper aims to solve such problem through improving the compactness of the model and efficiency of computing. On the basis of MobileNet, a mainstream lightweight neural network, we proposed an Adaptive Tensor-Train Decomposition (ATTD) algorithm to solve the cumbersome problem of finding optimal decomposition rank. For its non-obviousness in the forward acceleration of GPU side, our strategy of choosing to use lower decomposition dimensions and moderate decomposition rank, and the using of dynamic programming, have effectively reduced the number of parameters and amount of computation. And then, we have also set up a real-time target network for mobile devices. With the support of sufficient amount of experiment results, the method proposed in this paper can greatly reduce the number of parameters and amount of computation, improving the model's speed in deducing on mobile devices.

Keywords: Tensor decomposition · Parameter compression · Quantization · Mobile target detection

1 Introduction

In recent years, deep convolution neural network has been widely applied to the computer science including image identification, natural language processing and speech recognition [1]. It has made significant breakthrough in solving various tasks. For example, AlexNet [2] achieved an accuracy in classification 8.7% higher than traditional methods in the 2012 ILSVRC [3] competition. With the growing

Y. Zheng and Y. Zhou—Contribute equally to this work and should be considered co-first authors.
This work is partially supported by National Key R&D Program of China with grant No. 2019YFB2102600 and NSFC (No. 61971269, 61832012).

© Springer Nature Switzerland AG 2021
Y. Zhang et al. (Eds.): PDCAT 2020, LNCS 12606, pp. 70–81, 2021.
https://doi.org/10.1007/978-3-030-69244-5_6

application of VGGNet [4], GoogLeNet [5] and ResNet [6], artificial intelligence has caught up with and even surpassed human intelligence in classification of massive image data, which can be seen from the fact that the error rate of top-5 classification for ILSVRC has been as low as 3.5% while that for human eyes is about 5.1%.

In order to improve the performance of neural network models, researchers generally design deeper and more complex networks [7]. Deeper networks will greatly increase the number of parameters and amount of computation, and will thus make higher demands on hardware resources (CPU, GPU memory, and bandwidth). As a result, setting up a deep learning system is quite expensive and the cost has become an obstacle for deep neural networks to deal with assignments with limited computing resource or high real-time requirement.

To deploy large-scale convolutional neural network on edge devices, the problem of limited memory space and computation ability needs to be solved. Studies [8] have shown that there are a large number of redundant structures and parameters in the convolutional neural network, especially in the fully connected (FC) layer. The redundant parameters contribute little to the final result, so the network structure and parameters can be compressed to reduce the model size and speed up the computation.

There are five methods to compress and accelerate deep neural network. (1) Parameter pruning, which finds the redundant neurons and removes them [9]. (2) Parameter sharing, which maps multiple parameters with high accuracy to a single parameter with low accuracy using a certain rule [10]. (3) Low-rank decomposition, which decomposes the large matrix into the product of several approximate kernel matrices to reduce the computation [11]. (4) Designing compact convolutional filters, which reduces the computation and parameters of convolution by redesigning the operation steps or methods of convolution kernel [12]. (5) Knowledge distillation, which transfers knowledge from large network to compact distillation model [13,14].

In this paper, we focus on the low-rank decomposition. In 2013, Denil [15] et al. analyzed the effectiveness of low rank decomposition in solving the redundancy problem of deep neural networks. Jaderberg et al. [16] uses the low rank decomposition technique of tensor to decompose the original convolution kernel into two smaller convolution kernels. Using the classical tensor decomposition algorithm CP [17], the parameter tensor can be decomposed into the sum of several smaller rank one matrices. Using the Tucker decomposition [18], the parameter tensor can be decomposed into the product of a core tensor and several smaller tensors. Using Tensor-Train decomposition [19], the original parameter tensor can be decomposed into the product of multiple matrices, as shown in Fig. 1. Compared with CP and tucker decomposition, Tensor-Train decomposition has a more compact structure, and the representation of matrix multiplication makes the compressed tensor easier to operate.

At present, most of the rank of Tensor-Train decomposition is set by adjusting parameters empirically. The decomposition ranks of each layer need to be determined manually, and there are many layers in a neural network. It is

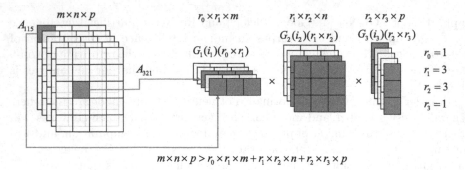

$$m \times n \times p > r_0 \times r_1 \times m + r_1 \times r_2 \times n + r_2 \times r_3 \times p$$

Fig. 1. Tensor-Train Decomposition. A 3D $m \times n \times p$ matrix is approximately decomposed into some small into the product of several small matrices. The sum of calculation amount of the small matrices is less than that of the original matrix.

difficult to achieve low precision loss and high compression ratio at the same time. This paper proposes an Adaptive Tensor-Train Decomposition (ATTD) method. After presetting a precision and the decomposition dimension, singular value decomposition (SVD) is used to directly calculate the optimal decomposition rank. Only one parameter – the precision – needs to be adjusted in the whole process. We summarize our contributions as follows:

- Based on the work of Tensorizing Neural Networks (TNN) [20], we propose an ATTD algorithm which can adaptively compute the optimal decomposition rank of each layer of network. Compared with the manual adjustment method, this method leads to the smaller accuracy loss and the larger compression ratio.
- We use the improved ATTD algorithm to further compress the depthwise separable convolution that is widely used in the current mainstream lightweight networks.
- By combining the compression and acceleration strategy with the quantitative algorithm, a lightweight target detection network based on MobileNet is built on mobile devices. It has nearly doubled the model acceleration effect.

The remainder of this paper is organized as follows. Section 2 discusses the related work about the neural network compression. In Sect. 3, the ATTD model is presented. Section 4 provides the experimental results. Section 5 concludes this paper and outlines the future work.

2   Related Work

This section discusses the compression and acceleration methods in deep neural networks.

Parameter Pruning. Parameter pruning reduces the amount of model parameters by deleting redundant parameters in neural network. For unstructured pruning, [9,21,22] all reduced the amount of network parameters while ensuring a certain accuracy. In order to solve the problem that unstructured pruning would cause a large number of unstructured sparse connectivities, researchers have proposed methods based on structured pruning [23–25], which, while directly compressing the neural network, have effectively accelerated the computing speed of the entire model.

Parameter Sharing. Parameter sharing aims to map multiple parameters with high accuracy to a single parameter with low accuracy using a certain rule. Parameter quantization uses low-precision data type to replace the original 32-bit full-precision floating-point [26–28]. The binarization method uses binary weights to represent the parameters or activation functions of the model [29–31]. Methods such as hash function [15] and structured linear mapping [16] enable the parameter of the FC layer to be shared, which significantly reduces the memory needed for neural network.

Low-Rank Decomposition. Applying matrix or tensor decomposition algorithms to convolutional and FC layers can compress and accelerate deep neural networks [32]. The classical tensor decomposition algorithms such as CP [33] decomposition, Tucker [18] decomposition, and Tensor-Train [19] decomposition all decompose the original parameter tensor with low rank. In [17], the low rank filter with rank 1 is constructed by cross channel and filter redundancy. [34] finds the exact global optimizer of the decomposition, and proposes a method for training low-rank constrained Convolution Neural Networks (CNNs) from scratch. [35] attempts to reduce spatial and channel redundancy directly from the visual input for CNN acceleration.

Designing Compact Convolutional Filters. SqueezeNet [36] replaces the original convolution structure with the Fire Module structure, which can significantly compress the model parameters. Google's MobileNet [37] replaces the original convolution with depthwise separable convolution which is of a smaller amount of computation and parameters; ShuffleNet [38] uses group convolution and channel shuffle to design a new convolution structure. Both have reduced the amount of model parameters and computations.

Knowledge Distillation. The mean idea of knowledge distillation is to distillate certain knowledge from a complex teacher model to a simpler student model. The compression and acceleration strategy based on knowledge distillation [13,14,39,40] can convert knowledge of large-scale networks to small-scale ones, which effectively reduces the amount of computation than the original network.

3 The Proposed Method

3.1 Compress Traditional Network by TT Decomposition

High-order tensors in practice are generally sparse, and direct operations between tensors will waste a lot of computing resources. Decomposition of high-order tensors can effectively reduce the amount of computation. The principle of Tensor-

Train [19] decomposition is to represent each element in a high-dimensional tensor as a matrix multiplication. The Tensor-Train decomposition form of tensor A can be written as:

$$A(i_1, i_2, ..., i_d) = G_1(i_1)G_2(i_2)...G_d(i_d), \tag{1}$$

where $G_k(i_k)$ is a matrix of $r_{k-1} \times r_k$, and the dimension of tensor A is d, so the number of matrices obtained by decomposition is also d. The result of multiplying several matrices represents an element in tensor A. To ensure that the final result is a scalar, we set $r_0 = r_d = 1$.

Figure 1 shows the Tensor-Train decomposition process of a 3D tensor. Any element in tensor A, like A_{321}, can be written in the form of continuous multiplication of 3 matrices. Here, the decomposition rank of the Tensor-Train is set to $(1, 3, 3, 1)$, and the size of each matrix is $r_{k-1} \times r_k$, which of the example mentioned before are 1×3, 3×3, 3×1, respectively. The position of each matrix in G_k is determined by the element's subscript $i_k \in [1, n_k]$, which is 3, 2, and 1, respectively. The original tensor has a total of $5 \times 4 \times 5 = 100$ parameters, while after compression, there are a total of $1 \times 3 \times 5 + 3 \times 3 \times 4 + 3 \times 1 \times 5 = 66$ parameters.

3.2   Adaptive Tensor-Train Decomposition

The decomposition algorithm based on Tensor-Train is able to significantly compress the parameters of the FC layers and convolutional layers, but practically, the Tensor-Train decomposition rank $r_1, ..., r_{k-1}$ of each layer of the network needs to be set manually ($r_0 = r_1 = 1$). If there are n layers to be compressed, $n(k-1)$ decomposition ranks will have to be adjusted. If the k and n are large, there will be lots of parameters that need to be set manually, and it will be difficult to set the optimal decomposition rank to obtain a relatively high compression rate while ensuring a small loss of accuracy.

The larger singular value will determine the main feature of the original matrix. The Tensor-Train decomposition considers the larger singular value and ignores the smaller singular value, that is, only the main feature is considered and the secondary feature is ignored. Thus, Tensor-Train decomposition precision is positively correlated with the singular value. The larger the singular value, the greater the contribution to the precision. Inspired by this, we define the precision as

$$\varepsilon \approx \frac{\sum_{i=1}^{r_k} \sigma_i}{\sum_{i=1}^{n} \sigma_i}, \tag{2}$$

where ε is the ratio of the selected top r_k singular values to the sum of all singular values, σ_i is the singular value of i, and $\sigma_i \leq \sigma_{i-1}$.

According to Eq. 2, using ATTD algorithm with the preset precision ε, the optimal decomposition rank at a current accuracy can be directly computed according to the pre-trained network after the decomposition dimension d is set, as shown in Algorithm 1 .

Algorithm 1. ATTD Algorithm

Input: A, d, ε
Output: $[G_1, ..., G_d]$
1:   $B = A, r_0 = 1, P = \prod_{s=1}^{d} n_s$
2:   **for** $k = 1$ to $d - 1$ **do**
3:      $S_{left} = r_{k-1}n_k, S_{right} = \frac{P}{r_{k-1}n_k}$,
        $B = reshape(B, [S_{left}, S_{right}])$
4:      Apply $\varepsilon - truncated$ SVD on B: $[U_k, \Sigma_k, V_k] = SVD(B)$
5:      $P = \frac{P}{r_{k-1}n_k}r_k$
6:      $G_k = reshape(U_k, [r_{k-1}, n_k, r_k])$
7:      $B = \Sigma_k V_k$
8:   **end for**
9:   $G_d = B$
10: **return** TT-Cores $[G_1, ..., G_d]$.

The whole process only needs to adjust one parameter of ε. Given the value of ε, how many singular values participate in the decomposition is determined. That is, the decomposition ranks of each layer are determined. We do not have to adjust the decomposition ranks of each layer manually.

3.3 Depthwise Separable Convolution Based on ATTD

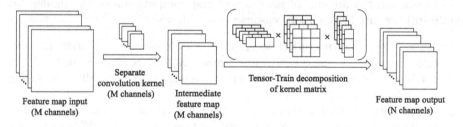

Fig. 2. Tensor-Train Decomposition. A 3D $n_1 \times n_2 \times n_3$ matrix is approximately decomposed into some small into the product of several small matrices. The sum of calculation amount of the small matrices is less than that of the original matrix.

Google proposed depthwise separable convolution in MobileNet, which separates the convolution into two steps: depthwise convolution and pointwise convolution. According to the statistics [37], 95% of the computations and 75% of the parameters in MobileNet come from 1×1 convolution, and the overall distribution of weight parameters roughly conforms to a normal distribution. There are a large number of parameters whose value is around 0, which does not contribute to the network, so 1×1 convolution has a large number of redundant parameters.

Suppose the shape of convolution kernel parameter matrix is $1 \times 1 \times M \times N$, where M and N are the numbers of input and output feature map channels, thus the total parameter of 1×1 convolution is MN. The parameter matrix can be regarded as a FC matrix. We use the adaptive Tensor-Train decomposition algorithm to compress it. The specific steps are:

(i) Transform the convolution kernel matrix of current layer into a tensor A with dimensions $(m_1 n_1, ..., m_d n_d)$, where $\prod_{i=1}^{d} m_i = M$, $\prod_{i=1}^{d} n_i = N$.
(ii) Apply the adaptive Tensor-Train decomposition algorithm to A to obtain the kernel matrix $G_k[m_k, n_k]$.
(iii) Decompose M with the same way in step (ii) to obtain matrix $\mathcal{X}(x, y, m_1, ..., m_d)$, where $\prod_{i=1}^{d} m_i = M$. After tensor operation, we obtain the output feature map $\mathcal{Y}(x, y, n_1, ..., n_d)$, where $\prod_{i=1}^{d} n_i = N$.

The operation process of 1×1 convolution can be shown as:

$$\mathcal{Y}(x, y, n_1, ..., n_d) = \sum_{i=1}^{k} \sum_{j=1}^{k} \sum_{m_1,...,m_d} \mathcal{X}(x, y, m_1, ..., m_d) G_0 G_1[m_1, n_1]...G_d[m_d, n_d] \quad (3)$$

The depth separable convolution after the introduction of Tensor-Train decomposition is shown in Fig. 2.

3.4   Improvement of Inference Speed and Optimization of GPU

After the addition of the adaptive Tensor-Train decomposition module, the actual inference speed is not significantly improved compared to the original model although the amount of parameters and computation of the model has significantly reduced. The main reasons are as follows:

(i) Tensor-Train decomposition decomposes a large parameter matrix into several compact 3D tensor forms, also known as matrix product states. Such small tensor operations cannot effectively use the GPU's parallel computing capabilities for large matrices.
(ii) The adaptive Tensor-Train decomposition algorithm tends to find a decomposition method with a higher decomposition dimension, which will cause increasing small 3D tensors after decomposition, and a larger decomposition rank, which will lead to parameter redundancy.
(iii) The order of computing irregular matrix after decomposition will affect the final computation amount theoretically. The difference in computation amount between the worst and the best order varies from several times to several tens of times.

For problems (i) and (ii), we adopt a strategy of low decomposition dimension and moderate decomposition rank. For problem (iii), after training the model, we use dynamic programming to compute the optimal matrix operation order of each layer and adjust them. The principle of the algorithm is:

(i) Let $A[i : j] = A_i A_{i+1}...A_j$, where $A[i : j]$ denote the product of the i-th matrix to the j-th matrix. For $k(i \leq k < k)$, the computation amount of

$A[i:j]$ is: the sum of there computation amount including the computation amount of $A[i:k]$ and that of $A[k+1:j]$, and the amount of computation multiplied by the two. Let $P[i-1]$ and $P[i]$ be the dimensions of the i-th matrix.

(ii) Let $C[i][j]$ be the amount of computation required for the optimal order of computation of $A[i:j]$, then:

$$C[i][j] = \begin{cases} 0, & if(i == j) \\ min\{C[i][k] + C[k+1][j] + P[i-1]P[k]P[j]\}, & i \le k < j \end{cases} \tag{4}$$

(iii) Update the cost matrix $C[i][j]$ and the marking matrix $S[i][j] = k$ from the bottom up.

(iv) Return the optimal computation method of $C[i][j]$ according to the separation operation provided by $S[i][j]$.

4   Experiments

4.1   Datasets

CIFAR-10 contains 50,000 training pictures and 10,000 test pictures, both in a total of 10 categories. We use this dataset to verify the adaptive Tensor-Train decomposition algorithm's ability of neural network parameter compression.

COCO dataset is currently the mainstream target detection dataset. We built a target detection network based on our proposed method on mobile devices to verify the feasibility and effectiveness on COCO.

4.2   Implementation Details

On CIFAR-10 dataset, we use the SGD optimizer with $momentum = 0.9$. The batchsize is set to 32, and the initial learning rate is 0.1. After 100 iterations, drop the learning rate 10 times every 20 iterations. In order to verify the compression ability of the algorithm on the FC layer and the convolutional layer, we designed 3 structures in this chapter: the network dominated by the FC layer, the network dominated by the convolutional layer, and a classic network composed of both layers. Use adaptive Tensor-Train decomposition algorithm to compress the above 3 models, where the Tensor-Train decomposition ranks are set manually and initialized randomly. Set ε from 0.7 to 0.98, and set $d = 3$.

On the pre-trained MobileNet V1 and V2 [41] models, we use the adaptive Tensor-Train decomposition algorithm to compress the deep separable convolution. The parameter settings remain unchanged.

4.3   Effectiveness of Proposed Method

Effectiveness on Traditional Network. Table 1 shows the results that Tensor-Train decomposition works well on compressing the FC layers where the

Table 1. Results of accuracy and compression ratio of different network structures

Network	Average acc	Compress rate
FULLY(base)	84.8%	1
FULLY(TT)_1	84.6%	20.32
FULLY(TT)_2	83.9%	43.35
FULLY(TT)_3	82.8%	82.74
CONV(base)	90.2%	1
CONV(TT)_1	88.1%	2.28
CONV(TT)_2	86.9%	2.64
CONV(TT)_3	85.3%	3.72
CONV+FULLY(base)	91.21%	1
CONV+FULLY(TT)	89.01%	43.41
CONV+FULLY(ATTD)	88.94%	65.45

Note: Base represents the uncompressed network, FULLY represents the network dominated by FC layers, CONV represents the network dominated by the convolutional layers, CONV+FULLY represents the classic network composed of FC layers and convolutional layers, TT represents the network compressed by Tensor-Train decomposition algorithm, and ATTD represents the network compressed by ATTD algorithm.

effect reaches up to 83 times, and the accuracy loss of the network is about 2%. The compression effect of this algorithm on the convolutional layer is unsatisfactory with only 2 to 4 times, and the accuracy drops quickly, with the accuracy loss reaching about 5%. This is because the parameters of the convolution layer are far less than that of the fully connected layer, so the compression effect is limited. The compression rate of the classic network can reach 43.41 times, and the accuracy loss is reduced by about 1.2%.

The adaptive Tensor-Train decomposition algorithm surpasses the manual adjustment method in accuracy and compression ratio of the model. Within similar accuracy decline (about 1.3%), the compression rate of adaptive Tensor-Train decomposition algorithm can reach about 65 times, greatly exceeding the result of 43 times of the CONV+FULLY model.

Effectiveness on MobileNet. As Table 2 shows, for MobileNet V1, when the value of ε is 0.9, the parameter amount of the model is reduced by about 3 times, the model accuracy is reduced by about 1.6%, and the theoretical computation (Multi-Adds) is nearly halved, but the improvement of inference speed is insignificant.

For MobileNet V2, when the value of ε is 0.9, the parameter amount of the model is reduced by about 4 times, computation amount is reduced by more than half, and the model accuracy is reduced by about 1.8%, but the inference speed is also not significantly improved compared to the original model.

Table 2. Comparison of V1 and V2 before and after compression when $\varepsilon = 0.9$

Network	Acc	Params	MAdds	FPS	ε	Compr
MobileNet V1(base)	94.23%	4.2M	569M	1053		1
MobileNet V1(ATTD)	92.63%	1.3M	305M	1196	0.9	3.28
MobileNet V1(ATTD_Re)	92.51%	0.9M	181M	**1582**	0.9	4.6
MobileNet V2(base)	95.49%	3.4M	300M	1141		1
MobileNet V2(ATTD)	93.78%	0.78M	141M	1365	0.9	4.32
MobileNet V2(ATTD_Re)	93.72%	0.59M	98M	**1694**	0.9	5.8

Note: The accuracy, parameter amount, computation amount, inference speed and compression rate of MobileNet V1 and V2 when value of ε is 0.9.

Table 3. Comparison of V1 and V2 before and after compression when $\varepsilon = 0.9$

Network	FPS
MobileNet V1(base)	1053
MobileNet V1(ATTD_Re)	1582
MobileNet V1(ATTD_Re_Quan)	**2443**
MobileNet V2(base)	1141
MobileNet V2(ATTD_Re)	1694
MobileNet V2(ATTD_Re_Quan)	**2801**

After fine-tuning the ATTD algorithm and improving the operation order of parameter matrices, the inference speed of the two models has been significantly improved, with FPS reaching 1582 and 1694 respectively, and the amount of parameters has also decreased.

Quantitative technology has a significant effect in acceleration on the inference speed of neural networks. As Table 3 shows, quantization technology enables the model to achieve an inference speed acceleration of about 2 times, which is of great significance for mobile and embedded devices.

5 Conclusion

In this paper, the application of the Tensor-Train decomposition algorithm in model compression and acceleration is reviewed and studied. The algorithm is applied to the mainstream lightweight neural network MobileNet, during which some improvements are made in terms of its advantages and disadvantages. As a result, our method has increased the efficiency of the model and reduced the memory needed. In addition, we have also built a target detection network based on this algorithm on mobile devices, through which we verified the feasibility and effectiveness of our algorithm and strategy on mobile CPUs.

References

1. LeCun, Y., Bengio, Y., Hinton, G.: Deep learning. Nature **521**(7553), 436–444 (2015)
2. Krizhevsky, A., Sutskever, I., Hinton, G.E.: Imagenet classification with deep convolutional neural networks. In: NIPS, pp. 1097–1105 (2012)
3. Russakovsky, O., et al.: ImageNet large scale visual recognition challenge. Int. J. Comput. Vision **115**(3), 211–252 (2015). https://doi.org/10.1007/s11263-015-0816-y
4. Simonyan, K., Zisserman, A.: Very deep convolutional networks for large-scale image recognition (2014)
5. Szegedy, C.: Going deeper with convolutions. In: CVPR, pp. 1–9 (2015)
6. He, K., Zhang, X., Ren, S., Sun, J.: Deep residual learning for image recognition. In: CVPR, pp. 770–778 (2016)
7. Sheng, H., et al.: Mining hard samples globally and efficiently for person reidentification. IoT-J **7**(10), 9611–9622 (2020)
8. Han, S., Pool, J., Tran, J., Dally, W.J.: Learning both weights and connections for efficient neural network. In: NIPS, pp. 1135–1143 (2015)
9. Srinivas, S., Babu, R.V.: Data-free parameter pruning for deep neural networks (2015)
10. Cheng, Y., Wang, D., Zhou, P., Zhang, T.: A survey of model compression and acceleration for deep neural networks (2017)
11. Bach, F.R., Jordan, M.I.: Predictive low-rank decomposition for kernel methods. In: ICML, pp. 33–40. Association for Computing Machinery (2005)
12. Prakash, A., Storer, J., Florencio, D., Zhang, C.: Repr: Improved training of convolutional filters (2018)
13. Hinton, G., Vinyals, O., Dean, J.: Distilling the knowledge in a neural network. arXiv preprint arXiv:1503.02531 (2015)
14. Romero, A., Ballas, N., Kahou, S.E., Chassang, A., Gatta, C., Bengio, Y.: Fitnets: Hints for thin deep nets (2014)
15. Chen, W., Wilson, J., Tyree, S., Weinberger, K., Chen, Y.: Compressing neural networks with the hashing trick. In: ICML, pp. 2285–2294 (2015)
16. Cheng, Y., Yu, F.X., Feris, R.S., Kumar, S., Choudhary, A., Chang, S.F.: An exploration of parameter redundancy in deep networks with circulant projections. In: ICCV, pp. 2857–2865 (2015)
17. Jaderberg, M., Vedaldi, A., Zisserman, A.: Speeding up convolutional neural networks with low rank expansions (2014)
18. Zhang, J., Han, Y., Jiang, J.: Tucker decomposition-based tensor learning for human action recognition. Multimedia Syst. **22**(3), 343–353 (2015). https://doi.org/10.1007/s00530-015-0464-7
19. Oseledets, I.V.: Tensor-train decomposition. SIAM J. Sci. Comput. **33**(5), 2295–2317 (2011)
20. Novikov, A., Podoprikhin, D., Osokin, A., Vetrov, D.P.: Tensorizing neural networks. In: NIPS, pp. 442–450 (2015)
21. LeCun, Y., Denker, J.S., Solla, S.A.: Optimal brain damage. In: NIPS, pp. 598–605 (1990)
22. Han, S., Mao, H., Dally, W.J.: Deep compression: Compressing deep neural networks with pruning, trained quantization and huffman coding (2015)
23. Lebedev, V., Lempitsky, V.: Fast convnets using group-wise brain damage. In: CVPR, pp. 2554–2564 (2016)

24. Molchanov, P., Tyree, S., Karras, T., Aila, T., Kautz, J.: Pruning convolutional neural networks for resource efficient inference (2016)
25. Luo, J.H., Wu, J., Lin, W.: Thinet: a filter level pruning method for deep neural network compression. In: ICCV, pp. 5058–5066 (2017)
26. Gupta, S., Agrawal, A., Gopalakrishnan, K., Narayanan, P.: Deep learning with limited numerical precision. In: ICML, pp. 1737–1746 (2015)
27. Ma, Y., Suda, N., Cao, Y., Seo, J.S., Vrudhula, S.: Scalable and modularized RTL compilation of convolutional neural networks onto FPGA. In: International Conference on Field Programmable Logic and Applications, pp. 1–8 (2016)
28. Gysel, P.: Ristretto: hardware-oriented approximation of convolutional neural networks (2016)
29. Courbariaux, M., Bengio, Y., David, J.P.: Binaryconnect: training deep neural networks with binary weights during propagations. In: NIPS, pp. 3123–3131 (2015)
30. Courbariaux, M., Hubara, I., Soudry, D., El-Yaniv, R., Bengio, Y.: Binarized neural networks: Training deep neural networks with weights and activations constrained to +1 or −1 (2016)
31. Rastegari, M., Ordonez, V., Redmon, J., Farhadi, A.: XNOR-Net: ImageNet classification using binary convolutional neural networks. In: Leibe, B., Matas, J., Sebe, N., Welling, M. (eds.) ECCV 2016. LNCS, vol. 9908, pp. 525–542. Springer, Cham (2016). https://doi.org/10.1007/978-3-319-46493-0_32
32. Denil, M., Shakibi, B., Dinh, L., Ranzato, M.A., De Freitas, N.: Predicting parameters in deep learning. In: NIPS, pp. 2148–2156 (2013)
33. Lebedev, V., Ganin, Y., Rakhuba, M., Oseledets, I., Lempitsky, V.: Speeding-up convolutional neural networks using fine-tuned cp-decomposition (2014)
34. Tai, C., Xiao, T., Zhang, Y., Wang, X.: Convolutional neural networks with low-rank regularization (2015)
35. Lin, S., Ji, R., Chen, C., Huang, F.: Espace: accelerating convolutional neural networks via eliminating spatial and channel redundancy. In: AAAI, pp. 1424–1430 (2017)
36. Iandola, F.N., Han, S., Moskewicz, M.W., Ashraf, K., Dally, W.J., Keutzer, K.: Squeezenet: Alexnet-level accuracy with 50x fewer parameters and 0.5 mb model size. arXiv preprint arXiv:1602.07360 (2016)
37. Howard, A.G.: Efficient convolutional neural networks for mobile vision applications, Mobilenets (2017)
38. Hluchyj, M.G., Karol, M.J.: Shuffle net: an application of generalized perfect shuffles to multihop lightwave networks. J. Lightwave Technol. 9(10), 1386–1397 (1991)
39. Korattikara, A., Rathod, V., Murphy, K., Welling, M.: Bayesian dark knowledge. In: NIPS, pp. 3438–3446 (2015)
40. Luo, P., Zhu, Z., Liu, Z., Wang, X., Tang, X.: Face model compression by distilling knowledge from neurons. In: AAAI, pp. 3560–3566 (2016)
41. Sandler, M., Howard, A., Zhu, M., Zhmoginov, A., Chen, L.C.: Mobilenetv 2: inverted residuals and linear bottlenecks. In: CVPR, pp. 4510–4520 (2018)

Development of a UAV Path Planning Approach for Multi-building Inspection with Minimal Cost

Shiwei Lin[✉], Xiaoying Kong, Jack Wang, Ang Liu, Gengfa Fang, and Yunlong Han

Faculty of Engineering and Information Technology,
University of Technology Sydney, Sydney, Australia
`Shiwei.Lin-1@student.uts.edu.au`

Abstract. This paper presents a UAV path planning approach for multi-building inspection, which is a new application for UAV path planning. It generates helix paths for single building inspection first and defines the possible points for collecting inspection data with reasonable time slots. After inspecting one building, the UAV flies to another building with a trajectory based on a cost matrix and a visited vector defined in this algorithm. The planning of the entire inspection path is evaluated considering several factors, such as distance, time, and altitude. The proposed algorithm is applied to historical giant communal homes, Fujian Tulou, consisting of five buildings.

Keywords: UAV · 3D path planning · Multi-building

1 Introduction

Historical buildings are symbols of specific eras. They have geographical and historical value and can vividly express past people live and the aesthetic and philosophy of the architect. Regular inspection of the historical buildings is essential to monitor their condition. Unmanned Aerial Vehicle (UAV) provides the ability of remote inspection by reaching the area that is hard for people to access. It provides the flexibility of building inspection. UAVs can gain different digital imagery with different cameras or sensors, such as ultrasonic sensors, high-resolution cameras, laser scanners, thermal cameras, near-infrared cameras [1, 2].

The collected data can be processed using computer vision and other technologies to detect surface decencies of infrastructure, including spalling, cracking, distortion, rusting, excessive movements, and misalignment [1]. Path planning for UAV-based inspection requires to find a path which is efficient and informative to obtain data from different views [1]. It also requires obstacle avoidance with an effective and optimal path if possible [1, 2]. In this research, path planning for UAV-based inspection of a group of historical buildings, Fujian Tulou. A Tulou is usually a large, enclosed, and fortified earth building, most commonly rectangular or circular in configuration, with very thick load-bearing rammed earth walls between three and five stories high and housing up to 800 people. Figure 1 shows the Tianluokeng Tulou cluster, consisting of five buildings with a square building in the center.

© Springer Nature Switzerland AG 2021
Y. Zhang et al. (Eds.): PDCAT 2020, LNCS 12606, pp. 82–93, 2021.
https://doi.org/10.1007/978-3-030-69244-5_7

Fig. 1. Photo of the Tianluokeng Tulou cluster

2 Related Work

Regular inspection of buildings is required to ensure their condition is safe. It is more important to the historical buildings as heritage [3]. In general practice, the inspection is conducted by taking photos directly for every element to record the dilapidation or damage, which is expensive and time-consuming [2, 3]. With the development of UAV and relevant photogrammetry technologies, it provides great flexibility in many fields, such as urban planning, security, rural environment monitor [4], and recently building inspection [1, 3, 5–8].

The research of building inspection is to check the environmental condition and the appearance of a building and evaluate the condition of building [3]. The UAV survey is aiming to model the structure and recognize the condition [7]. UAV-based photogrammetry allows full documentation of buildings and evaluates the condition of buildings with less human resources in a short time, particularly for the places that are hard to reach [2–5].

UAVs follow a programmed flight path, capture digital images, and obtain survey data for unreachable areas from a terrestrial platform [2, 4]. UAVs are equipped with sensors for 3D data acquisition, obstacle avoidance, and navigation [1, 2, 4]. It achieves low cost and higher flexibility [8]. The survey data can be different from some special sensors, such as images, Forward-Looking Infrared technology, laser scanner, 3D point cloud, multi-attributed point cloud, and separate strips [2, 4, 8, 9].

Close range photogrammetric images or laser scanning data, or the combination of both are generally used for documentation of building inspection [2, 5]. With proper path planning, UAV can create a precise model of buildings with 2D photographs with dedicated software for identifying the defects [5]. The geometry of buildings is measured in the inspection, and required documentation is usually in the form of photogrammetric images [4, 9]. Deep learning algorithms and advanced photogrammetric techniques are employed to record building damages by a true-orthophoto autonomously [10]. A Convolutional Neural Network is fine-tuned for detecting surface crack [11].

There are many algorithms and methods that can be used for UAV path planning, such as rapidly exploring tree, genetic algorithm, particle swarm optimization, and A* search. They can also be used for building inspections. Particle swarm optimization is improved as enhance discrete particle swarm optimization for path planning of UAV inspection for bridges or buildings which have planar surfaces [1]. Besides, the inspection of the Perak Museum plans the path as flying from the bottom to the top of the building and moving to the right side, then moving to the ground [3]. The most common method for path planning in building inspection is the back and forth path.

However, a specific method should be applied for optimizing UAV path planning in different cases. This paper presents a UAV path planning approach for multi-building inspection, which is applied to a group of historical buildings, Fujian Tulou. They are unique enclosed and fortified earth buildings, and the general back and forth path is not an efficient path for inspection of the entire building. In order to inspect the condition of the entire buildings, including both interior and exterior side of the buildings and their roofs, to check if they are under good condition, optimal path planning is required. The building inspection methods proposed in the literature are only for a single building. But in our case, a group of buildings is inspected together, which requires path planning for different buildings and between the buildings.

3 Path Planning Algorithm

The proposed path planning algorithm generates helix paths for a single building inspection and defines the points for collecting inspection data with reasonable time slots. After inspecting one building, the UAV flies to another building with a trajectory based on a cost matrix and a visited vector. The planning of the entire inspection path is evaluated considering several factors, such as distance, time, and altitude.

3.1 Description of the Algorithm

Using $x(t)$, $y(t)$, $z(t)$ to indicate the position of the UAV at the specific time t. The helix path for single building inspection can be defined as:

$$x(t) = r * \sin(vt/2\pi r) + x_g \tag{1}$$

$$y(t) = r * \cos(vt/2\pi r) + y_g \tag{2}$$

$$z(t) = \frac{vh}{2\pi r} * t + z_g + 1 \tag{3}$$

where x_g, y_g, z_g is the center coordinate of the ground floor of a building; v is the UAV flight speed, h is the height of each floor of the building, and r is the radius of the helix path. For interior side inspection, r is the radius of the building minus a few meters, for exterior inspection, r equals the radius of the building plus 2 m.

With an optimal UAV flight speed v, decided by the onboard sensors and inspection condition, the time required for inspection one building with n floors is:

$$Ts = \frac{2\pi rn}{v} \tag{4}$$

After the inspection of one building's interior and exterior side, the UAV flies to another building for inspection. The principal consideration for path planning between buildings is the costs of distance, then attitude and time. The point of tangency of the helix path is selected to reduce the costs by a smooth path and avoid collisions with buildings, which is represented as (x_0, y_0, z_0), and z_0 is considered as the height as the destination building. The last point of the previous path is (x_1, y_1, z_1), The last point of the previous path on the same height as the point of tangency is (x_1, y_1, z_0). The Pythagorean Theorem is used to calculate the point of tangency, where the distance between the center point and the point of tangency is the radius of the destination building plus 5 m to avoid collisions.

$$c = \sqrt{(x_1 - x_c)^2 + (y_1 - y_c)^2}$$
$$a = r + 5 \tag{5}$$
$$b = \sqrt{c^2 - a^2}$$

where (x_c, y_c) is the coordinate of the center of the destination building, and r is the radius of the destination building.

The equations of x_0, y_0 are calculated with (5) as:

$$a^2 = (x_0 - x_c)^2 + (y_0 - y_c)^2 \tag{6}$$

$$b^2 = (x_0 - x_1)^2 + (y_0 - y_1)^2 \tag{7}$$

where (x_c, y_c) is the coordinate of the center of the destination building.

Then make the solution to equations that is close to the destination as (x_0, y_0, z_0). The path is generated by three points, the start coordinate (x_1, y_1, z_1), the destination coordinate located in the next building, and (x_0, y_0, z_0) which calculated by (5–7). The points of the path between the buildings is represented as $(x(t), y(t), z(t))$ at the specific time t. The equations of $x(t), y(t)$ are:

$$x(t) = a_1 t^2 + b_1 t + c_1 \tag{8}$$

$$y(t) = a_2 t^2 + b_2 t + c_2 \tag{9}$$

where a_1, b_1, c_1 are the coefficients of $x(t)$, and a_2, b_2, c_2 are the coefficients of $y(t)$.

Point 1 is the start coordinate (x_1, y_1, z_1). Point 2 is (x_0, y_0, z_0), and Point 3 is the destination coordinate. The maximum time slots for two path segments are t_{max1} and t_{max2} are calculated by the Euclidean distance:

$$Distance1 = \sqrt{(x_2 - x_1)^2 + (y_2 - y_1)^2 + (z_2 - z_1)^2}$$

$$t_{max1} = \frac{Distance1}{v} \tag{10}$$

$$Distance2 = \sqrt{(x_3 - x_2)^2 + (y_3 - y_2)^2 + (z_3 - z_2)^2}$$

$$t_{max2} = \frac{Distance2}{v} \tag{11}$$

where Point 1 is (x_1, y_1, z_1), Point 2 is (x_2, y_2, z_2), Point 3 is (x_3, y_3, z_3), and v is the constant speed.

The coefficients of $x(t)$, $y(t)$ are calculated by the following system (12), (13):

$$\begin{bmatrix} 0 & 0 & 1 \\ t_{max1}^2 & t_{max1} & 1 \\ (t_{max1} + t_{max2})^2 & t_{max1} + t_{max2} & 1 \end{bmatrix} * \begin{bmatrix} a_1 \\ b_1 \\ c_1 \end{bmatrix} = \begin{bmatrix} x_1 \\ x_2 \\ x_3 \end{bmatrix} \tag{12}$$

$$\begin{bmatrix} 0 & 0 & 1 \\ t_{max1}^2 & t_{max1} & 1 \\ (t_{max1} + t_{max2})^2 & t_{max1} + t_{max2} & 1 \end{bmatrix} * \begin{bmatrix} a_2 \\ b_2 \\ c_2 \end{bmatrix} = \begin{bmatrix} y_1 \\ y_2 \\ y_3 \end{bmatrix} \tag{13}$$

where a_1, b_1, c_1 are the coefficients of $x(t)$, and a_2, b_2, c_2 are the coefficients of $y(t)$; t_{max1} is the maximum time slot for traveling from Point 1 to Point 2, t_{max2} is the maximum time slot for traveling from Point 2 to Point 3.

The equations of $z(t)$ is:

$$z(t) = a_3 \left[(x(t) - x_1)^2 + (y(t) - y_1)^2 \right] + b_3 \sqrt{(x(t) - x_1)^2 + (y(t) - y_1)^2} + c_3 + z_1 \tag{14}$$

where a_3, b_3, c_3 are the coefficients of $z(t)$, and (x_1, y_1, z_1) is the start coordinate of the path segment.

The coefficients of $z(t)$ are calculated by the following system (15):

$$\begin{bmatrix} 0 & 0 & 1 \\ (x_2 - x_1)^2 + (y_2 - y_1)^2 & \sqrt{(x_2 - x_1)^2 + (y_2 - y_1)^2} & 1 \\ (x_3 - x_1)^2 + (y_3 - y_1)^2 & \sqrt{(x_3 - x_1)^2 + (y_3 - y_1)^2} & 1 \end{bmatrix} * \begin{bmatrix} a_3 \\ b_3 \\ c_3 \end{bmatrix} = \begin{bmatrix} z_1 - z_1 \\ z_2 - z_1 \\ z_3 - z_1 \end{bmatrix} \tag{15}$$

where a_3, b_3, c_3 are the coefficients of $z(t)$, Point 1 is (x_1, y_1, z_1), Point 2 is (x_2, y_2, z_2), and Point 3 is (x_3, y_3, z_3).

The path of n points is stored as:

$$\text{Points} = \begin{bmatrix} x_1 & y_1 & z_1 \\ x_2 & y_2 & z_2 \\ x_3 & y_3 & z_3 \\ \vdots & \vdots & \vdots \\ x_n & y_n & z_n \end{bmatrix} \tag{16}$$

3.2 Cost Functions

The cost function is to evaluate the paths, and the path with minimal cost is considered as the best path. The weights of cost functions of the inspection and flying between the buildings are different.

The cost function is defined as:

$$f_{Cost} = w_1 * f_{distance} + w_2 * f_{time} + w_3 * f_{altitude} \tag{17}$$

where the sum of w_1, w_2 and w_3 is 1, $f_{distance}$ is the cost function of the distance, f_{time} is the cost function of flight time, $f_{altitude}$ is the change of the altitude. For inspection, w_1 is smaller than w_2 and w_3, because the distance is not the primary consideration. But for traveling between buildings, w_1 is larger than w_2 and w_3.

Distance
The cost function of distance is:

$$f_{distance} = \sum_{t=0}^{n} \sqrt{(x_{tNext} - x_t)^2 + (y_{tNext} - y_t)^2 + (z_{tNext} - z_t)^2} \tag{18}$$

where n is the size of t in the path, and $\sqrt{(x_{tNext} - x_t)^2 + (y_{tNext} - y_t)^2 + (z_{tNext} - z_t)^2}$ is the distance between the positions of each two t.

Altitude
The cost function of altitude is:

$$f_{altitude} = \sum_{t=0}^{n} \sqrt{(z_{tNext} - z_t)^2} \tag{19}$$

where n is the size of t in the path, and $\sqrt{(z_{tNext} - z_t)^2}$ is the change of altitude between the positions of each two t.

Time
The cost function of time is:

$$f_{time} = \sum_{t=0}^{n} \frac{\sqrt{(x_{tNext} - x_t)^2 + (y_{tNext} - y_t)^2 + (z_{tNext} - z_t)^2}}{v_t} \tag{20}$$

where n is the size of t in the path, and $\sqrt{(x_{tNext} - x_t)^2 + (y_{tNext} - y_t)^2 + (z_{tNext} - z_t)^2}$ is the distance between the positions of each two t, and v_t is the current v. v_t is determined by the change of altitude. v_t is different when the UAV is descent, ascent, or flying at the same altitude.

3.3 Implementation of the Algorithm

We implement the algorithm as follows: building the model for the Fujian Tulou, generate the path, then generate the points of collecting inspection data for exterior and interior side inspection. After inspecting one building, the UAV is planned to fly to another building.

The path of inspection is generated by (1), (2), and (3). The line space of t is set to 0.01. (x_g, y_g, z_g), the number of floors, the radius of each building, the height of each floor, the height of the roof are inputted. Once $z(t)$ reach to the specific height based on the inspection type, the iteration terminates and record current t as t_{max}. $z(t)$ is designed to almost reach the roof for exterior inspection and over the height of the building for interior side inspection to cover more areas. Using an empty matrix to store the points of the path as (16).

The points of generating inspection data are calculated from the generated paths, and they are related to the total number of rooms. The total room number is calculated based on the number of floors in the building and the number of rooms on each floor, and it is inputted with the path for generated points of generating inspection data as (4). The points of generating inspection data are concentrated in the middle of the path to generating valid inspection data for inspection. Then connecting the paths of the interior and exterior side inspection through the door and inspecting the entrance hall. The UAV is assumed to be charged after the interior side inspection or the exterior side inspection of one building.

There are five buildings: 1 2 3 4 5, and the order of buildings is determined by the cost matrix (17). Update the visited vector if the building has been visited and update the related costs of the visited building as 999. After defining the order, using (5)–(7) to get (x_0, y_0, z_0), then generate the path between every two buildings by (8)–(15).

$$Cost = \begin{bmatrix} 999 & cost(1,2) & cost(1,3) & cost(1,4) & cost(1,5) \\ cost(2,1) & 999 & cost(2,3) & cost(2,4) & cost(2,5) \\ cost(3,1) & cost(3,2) & 999 & cost(3,4) & cost(3,5) \\ cost(4,1) & cost(4,2) & cost(4,3) & 999 & cost(4,5) \\ cost(5,1) & cost(5,2) & cost(5,3) & cost(5,4) & 999 \end{bmatrix} \quad (21)$$

When processing to another building, the center and radius of the destination building are inputted to calculate a, b, c for the Pythagorean Theorem. Basing on the start coordinate and the destination coordinate and combining with Pythagorean Theorem, x_0, y_0 can be calculated as (5)–(7). The height of this point is supposed to be the same height as the destination point. Also, there will be two solutions for x_0, y_0, using the Euclidean distance to measure which one is closer to the destination point, and get the one that is closer to the destination point as the only solution.

Setting Point 1 as the start point. Point 2 and Point 3 are determined by the distance. If the (x_0, y_0, z_0) is closer to the start point than the distance between the start point and the destination point, then Point 2 is (x_0, y_0, z_0). Otherwise, Point 2 is the destination point. The coordinates of Point 1, Point 2, Point 3, and the destination are inputted for generating a path between two buildings. The constant speed is set to calculate t_{max1} and t_{max2}. The coefficients $\begin{bmatrix} a_1 \\ b_1 \\ c_1 \end{bmatrix}$, $\begin{bmatrix} a_2 \\ b_2 \\ c_2 \end{bmatrix}$ and $\begin{bmatrix} a_3 \\ b_3 \\ c_3 \end{bmatrix}$ are calculated for getting the equations of $x(t), y(t), z(t)$ as (8)–(15). The line space of t is set to 0.01 for $x(t), y(t), z(t)$. If the path almost reaches the destination, the iteration terminates.

4   Simulation Results

The model of the Fujian Tulou is presented in Fig. 2, and the number is marked with each building:

Fig. 2. The Fujian Tulou

To validate the algorithm, we use the simulation for path planning with MATLAB. The center coordinate and the radius of each building are inputted for generating the path of inspection, but the radiuses of building 5 are different between interior and exterior side inspection due to the particularity of the building. The speed of ascent is set to 6 m/s, the speed of descent is 4 m/s, and the speed is 18 m/s. When the UAV in the points of generating inspection data, UAV hovers. Because Building 5 is rectangular in configuration, we have considered making a similar path as the shape of the building, while the cost is higher than the helix path.

Figure 3 shows the flight path for each building, and the points of generating inspection data are marked by "*".

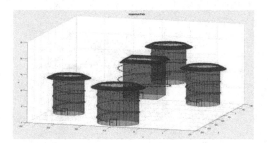

Fig. 3. The flight path for each building

The start building is Building 1. From the cost matrix (17) and the visited vector, the order of traveling the buildings is calculated as 1 2 5 4 3 by Dijkstra's algorithm. The start point for Building 1 is the start of the exterior inspection path, which is close to the top of Building 1.

Figure 4 integrates the flight path as Path 1 for the group of buildings, and (x_0, y_0, z_0) is marked by "o."

Figure 5 presents the 3D flight path for the group of buildings from the top viewpoint.

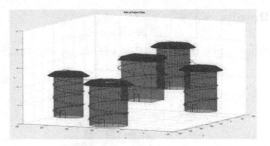

Fig. 4. The 3D flight path (Path 1) from the side viewpoint

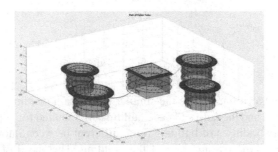

Fig. 5. The 3D flight path (Path 1) from the top viewpoint

Figure 6 shows another 3D flight path as Path 2 for the group of buildings. The path segments between the buildings are different. The destination points of Building 5 and 4 are different, so the direction of the inspection paths has been revised.

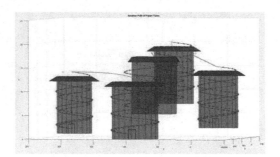

Fig. 6. Another 3D flight path (Path 2) from the side viewpoint

Figure 7 displays another 3D flight path for the group of buildings from the top viewpoint.

The traditional back and forth paths are usually applied to the exterior area of each building to get the inspection data. The back and forth paths are programmed to compare with the proposed methodology. The distance between the inspected building and UAV is equal to or less than 2 m [3]. The path length of the horizontal flight is 2 m. Figure 8

Fig. 7. Another 3D flight path (Path 2) from the top viewpoint

demonstrates the back and forth paths for the exterior area without defining the points of generating the inspection data.

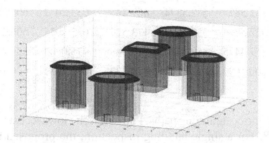

Fig. 8. The back and forth paths (Path 3) for each building

Setting the back and forth path for the vertical flight for Path 4 in Fig. 9.

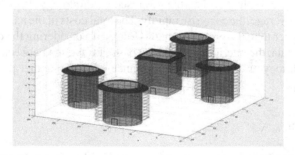

Fig. 9. Another back and forth paths (Path 4) for each building

The costs of the entire path are related to the path of each building, and the paths between every two buildings and the direction of the flight path are important. Table 1 shows the costs for the paths generated by the proposed methodology. Path 1 and Path 2 are generated by the proposed approach. Total cost calculates the costs of inspecting

Table 1. Costs

Path	Each building					The path between				Total cost
	Building 1	Building 2	Building 3	Building 4	Building 5	Building 1 and 2	Building 2 and 5	Building 5 and 4	Building 4 and 3	
Path 1	532.5874	532.5874	469.2459	633.2072	606.2504	55.3632	46.3745	51.9892	43.9798	2971.5849
Path 2	532.5874	532.5874	473.5241	633.2072	606.2504	55.3632	46.3745	72.9840	68.6634	3021.5415

buildings and traveling between buildings. Path 3 and Path 4 are the back and forth paths without defining the inspection points. Table 2 shows the costs of the path of each building.

Table 2. Costs of each building

Path	Building 1	Building 2	Building 3	Building 4	Building 5
Path 1	532.5874	532.5874	469.2459	633.2072	606.2504
Path 2	532.5874	532.5874	473.5241	633.2072	606.2504
Path 3	859.4475	859.4475	746.5488	975.0643	859.4475
Path 4	600.1694	600.1694	542.8375	695.7231	600.1694

From comparing the costs among paths, Path 1 has the lowest costs, and the total cost of Path 1 is 2971.58. For a single building inspection, the proposed methodology has much lower costs than the back and forth path. The points of generating the inspection data are not defined for the back and forth paths, and the back and forth paths are only for the exterior area. The costs of the proposed methodology for the single building include the exterior and interior side inspection path, and the consumed time for hovering and generating the inspection data. The costs of Path 3 and Path 4 only include the exterior path, but they are usually higher than the proposed methodology. If the back and forth path wants more coverage, the costs rise rapidly. The total costs of the back and forth paths must be much higher than the proposed methodology if considering the consumed time of generating the data, the interior path, and the path of traveling between buildings. The proposed methodology ensures higher performance with a smoother path, fewer costs, and more coverage.

5 Conclusion

The path planning approach is developed for multi-building inspection with UAV, with the paths for traveling between buildings. The algorithm uses the helix path for collecting single building inspection data because the helix paths have a smooth turn angle and broader coverage. For traveling between buildings, both cost matrix and a visited vector are employed to calculate the order of processing inspections. The cost function assigns different weights to the cost for a single building and for traveling between buildings, basing on the intention of inspection. The factors considered in the cost function are distance, time, and altitude, as altitude change affects speed and energy consumption.

The proposed path planning has been compared with the traditional back and forth path, and it has much better performance with less cost.

References

1. Phung, M.D., Quach, C.H., Dinh, T.H., Ha, Q.: Enhanced discrete particle swarm optimization path planning for UAV vision-based surface inspection. Autom. Constr. **81**, 25–33 (2017)
2. Mader, D., Blaskow, R., Westfeld, P., Weller, C.: Potential of Uav-based laser scanner and multispectral camera data in building inspection. In: ISPRS - International Archives of the Photogrammetry, Remote Sensing and Spatial Information Sciences, pp. 1135-1142. Copernicus GmbH, Czech Republic (2016)
3. Zainorizuan, M.J., et al.: Visual inspection of historical buildings using micro UAV'. In: MATEC Web of Conferences, vol. 103, EDP Sciences (2017)
4. Roca, D., Lagüela, S., Díaz-Vilariño, L., Armesto, J., Arias, P.: Low-cost aerial unit for outdoor inspection of building façades. Autom. Constr. **36**, 128–135 (2013)
5. Markova, M., Kravchenko, D.: 3D photogrammetry application for building inspection of cultural heritage objects. Bull. Prydniprovs'ka State Acad. Civil Eng. Architect. **1**, 91–96 (2018)
6. Grosso, R., Mecca, U., Moglia, G., Prizzon, F., Rebaudengo, M.: Collecting built environment information using UAVs: time and applicability in building inspection activities. Sustainability **12**(11), 4731 (2020)
7. Buffi, G., Manciola, P., Gambi, A., Montanari, G.: Unmanned aerial vehicle (UAV) and building Information modelling (BIM) technologies in concrete dam management: the case of Ridracoli'. In bo **9**, 36–43 (2018)
8. Vacca, G., Furfaro, G., Dessì, A.: The use of the Uav Images for the building 3D Model Generation. ISPRS – Int. Arch. Photogramm. Remote Sens. Spatial Inf. Sci. **XLII-4/W8**, 217–223 (2018)
9. Pan, N.-H., Tsai, C.-H., Chen, K.-Y., Sung, J.: Enhancement of external wall decoration material for the building in safety inspection method. J. Civil Eng. Manage. **26**, 216–226 (2020)
10. Nex, F., Duarte, D., Steenbeek, A., Kerle, N.: Towards real-time building damage mapping with low-cost UAV solutions. Remote Sens. **11**(3), 287 (2019)
11. Kucuksubasi, F., Sorguc, A.: Transfer learning-based crack detection by autonomous UAVs. In: 35th International Symposium on Automation and Robotics in Construction (ISARC 2018), arXiv.org, (2018)

Construction of Completely Independent Spanning Tree Based on Vertex Degree

Ningning Liu[1], Yujie Zhang[2,3], and Weibei Fan[2,3(✉)]

[1] School of Computer Science and Technology, Soochow University, Suzhou, China
[2] School of Computer, Nanjing University of Posts and Telecommunications, Nanjing, China
20175427005@stu.suda.edu.cn
[3] Jiangsu High Technology Research Key Laboratory for Wireless Sensor Networks, Nanjing 210003, China
{zhangyujie,wbfan}@njupt.edu.cn

Abstract. Interconnection networks have been extensively studied in the field of parallel computer systems. In the interconnection network, completely independent spanning tree (CISTs) plays an important role in the reliable transmission, parallel transmission, and safe distribution of information. Two spanning trees T_1 and T_2 of graph G are completely independent if, for any two distinct vertices u and v of G, the two paths from u to v on T_1 and T_2 are internally disjoint. The spanning trees $T_1, T_2, \ldots, T_k$ of G are completely independent spanning trees if they are pairwise completely independent. In 2015, Hasunuma proof that G has $\lfloor \frac{n(G)}{k} \rfloor$ CISTs if $\delta(G) \geq n(G) - k$, $3 \leq k \leq \frac{n(G)}{2}$ and $n(G) \geq 7$. In this paper, we prove that G has $\lfloor \frac{5n(G)}{12} \rfloor$ CISTs if $\delta(G) \geq n(G) - 2$ and $n(G) \geq 12$, and G has $(t+1)$-CISTs if $\delta(G) \geq n(G) - 3$ and $n(G) = 3t - 2 \geq 23$.

Keywords: Node disjoint paths · ISTs · CISTs · Reliable transmission

1 Introduction

Parallel computing is becoming more and more important in many areas, such as scientific research, education, petroleum, meteorology and so on. Multiprocessor interconnect networks (in brief, interconnection networks) play critical roles in parallel computing systems [7,18]. An interconnection network can be represented by a graph, where a vertex is a processor and an edge is a communication link between the processors.

The graph embedding ability of an interconnection network (host graph) is an indicator of how efficiently parallel algorithms with regular task graphs (guest graphs) can be executed on this network [17]. An ideal interconnection network (host graph) is supposed to possess excellent graph embedding ability.

Considering graph $G = (V, E)$, two spanning trees T_1 and T_2 of G are independent if they satisfy the following conditions: (1) T_1 and T_2 are rooted at the same vertex, say r, and (2) for each vertex $v \neq (r)$ in G, the path P from v to

© Springer Nature Switzerland AG 2021
Y. Zhang et al. (Eds.): PDCAT 2020, LNCS 12606, pp. 94–103, 2021.
https://doi.org/10.1007/978-3-030-69244-5_8

r in T_1 and the path Q from v to r in T_2 are disjoint, i.e. $E(P) \cap E(Q) = \emptyset$ and $V(P) \cap V(Q) = \{v, r\}$. A set of spanning trees of G are independent spanning trees if they are pairwise independent. Independent spanning trees have applications in networks such as reliable communication protocols, one- to-all broadcasting, and secure message distribution. Thus the embedding of independent spanning trees into several classes of networks has been widely investigated [8, 15].

In the computer interconnection network, CISTs have good applications. If there are n CISTs in an interconnection network G, there are n disjoint paths between any two vertices in G. These n paths are used to transmit information between two processors. Assuming that there are at most $n - 1$ failed processors in the network, there is always a fault-free path between the two processors, so as to achieve the purpose of reliable transmission. If all processors are fault-free, the information can be divided into n parts, and each part is transmitted in a path, so as to achieve the purpose of parallel transmission [4]. When used for information distribution, the information transmitted on each path can be encrypted to improve the security of information distribution.

Hasunuma proposed the definition of complete independent trees in 2001 [3]. And he conjectured that there are k CISTs in any $2k$-connected graph. For $k \geq 2$, Paéterfalvi showed that there exists a k-connected graph that does not contain two CISTs [5]. For $n \in \{10, 12, 14, 20, 24, 26, 28, 30\}$, Pai et al. proved that the n-dimensional hypercube does not contain $\frac{n}{2}$ CISTs [12]. Dirac showed that a graph is hamiltonian if its minimal degree is at least half of the order of it [6]. Hong and Liu showed that G has k CISTs if $n(G) \geq 4k - 1$ and $\delta(G) \geq \frac{k-1}{k}n$ [10]. Araki showed that G has 2 CISTs if $\delta(G) \geq \frac{n(G)}{2}$ and $n(G) \geq 7$ [1]. And he showed the following result in 2015.

Theorem 1 [16]. *For* $3 \leq k \leq \frac{n(G)}{2}$, *$G$ has* $\lfloor \frac{n(G)}{k} \rfloor$ *CISTs if* $n(G) \geq 7$ *and* $\delta(G) \geq n(G) - k$.

Fan et al. [9] proved that for a network with a number of vertices greater than or equal to 3, if the sum of degrees of non-adjacent vertex pairs is greater than or equal to n, then there are 2 CISTs in the network. Chang et al. [2] proved that for any graph G containing n vertices, $n \geq 6$, if The minimum vertex degree of graph G is at least $n - 2$, so there are at least $n/3$ CISTs.

For any n-connected graph, it is an NP-hard problem for embedding independent spanning trees with n vertices (edges) rooted at any vertex. Therefore, researchers are committed to the embedding research of independent spanning trees on some special graphs, such as chorded rings [27], Torus [22], de Bruijn and Kautz graphs [19], star graphs [20], Cayley graph [21], etc. Suppose G is a finite plane graph with n connections. Huck et al. [14] proved that the n independent spanning trees rooted at any vertex can be embedded in G, but they did not give a construction algorithm.

Since the n-dimensional hypercube is a product graph, the algorithm in [11] can be used to embed n independent spanning trees in the n-dimensional hypercube, but this construction method does not take into account the performance of independent spanning trees. Yang et al. [26] modified the algorithm in [11] to be able to embed an independent spanning tree with a height of $n + 1$, and proposed a time complexity of $O(NlogN)$ to embed n trees rooted at 0 recursive algorithm of independent spanning tree, where N is the number of vertices of n-dimensional hypercube. Since the hypercube is a Cayley graph [25], it has symmetrical vertices and symmetrical edges, so Tang et al. solved the embedding problem of an independent spanning tree rooted at any vertex in the hypercube. Based on the concept of Hamming Distance Latin Square (Hamming Distance Latin Square), Yang et al. [23] proposed an algorithm for embedding n independent spanning trees rooted at 0 in an n-dimensional hypercube. The complexity is $O(NlogN)$. The difference is that the algorithm can be executed in parallel, and the parallel time complexity is $O(n)$. Werapun et al. [24] further improved Yang's parallel algorithm and gave a more flexible and effective parallel algorithm that can embed n independent spanning trees rooted at any vertex in an n-dimensional hypercube.

The rest of this paper is organized as follows: Sect. 2 gives some definitions and notations. Section 3 proves that G has $\lfloor \frac{5n(G)}{12} \rfloor$-CISTs if $n(G) \geq 12$ and $\delta(G) = n(G) - 2$. Section 4 proves G has $(t + 1)$-CISTs if $n(G) = 3t + 2 \geq 23$ and $\delta(G) = n(G) - 3$. The final section concludes this paper.

2   Preliminaries

Let $G = (V, E)$ be a simple graph, and V' is a non-empty subset of the vertex set V in the graph G. Take V' as the set of vertices, and take all the edges with both ends in V' as the subgraph of the edge set, which is called the subgraph derived from the non-empty vertex subset V' in G, denoted by $G[V']$. For any two vertex points u and v on G, if and only if $(u, v) \in E$, u and v are said to be adjacent. The neighbor set of vertex u is the set of all vertex points adjacent to x on G, denoted as $N(x)$. The number of edges connected to a certain vertex x in graph G is called the degree of this vertex, or degree for short, denoted as $deg(x)$. The minimum degree of graph G is defined as $\delta(G) = \min\{deg(x)|x \in V\}$. When there is a connected edge between the vertex x and y, this edge is recorded as xy. The set of natural numbers is denoted as N in this paper.

Let G be a simple graph, and $V(G)$ and $E(G)$ are the set of vertices and edges of graph G, respectively. For any two vertices x and y in G, let the paths P and Q be the two paths from x to y in the graph G. If $E(P) \cap E(Q) = \emptyset$, then P and Q are called edges disjoint. If P and Q edges do not intersect and $V(P) \cap V(Q) = \{x, y\}$, then P and Q vertices do not intersect. Let $P_1, P_2, ..., P_n$ be n paths from x to y in the graph G, if for any i and j, $1 \leq i < j \leq n$, P_i and P_j are vertices (edges) disjoint, then $P_1, P_2, ..., P_n$ is the disjoint path of n vertices (edges) from x to y in G. If the trees T and T' satisfy the following conditions: (1) The roots of T and T' are the same vertex, such as r, and $V(T) = V(T') = V(G)$; (2) T and T' Can be embedded into G isomorphically; (3) If for any vertex $v \in V(G) - \{r\}$ in G, the path between r and v in T and the path between r and v in T' are edges (vertices) disjoint, then T and T' are Edge (Vertex)-Independent.

In the history of graph theory, there are similar approaches for sufficient conditions of Hamiltonian graphs. Araki proved the existence of k CISTs is in fact equivalent to the existence of a CIST-partition as follows.

Theorem 2 [1]. *A graph G has k completely independent spanning trees if and only if there is a partition of $V(G)$ into $V_1, V_2, \ldots, V_k$ such that (1) For $i \in \{1, 2, \ldots, k\}$, $G[V_i]$ is connected; (2) For any two distinct elements $i, j \in \{1, 2, \ldots, k\}$, $B(V_i, V_j)$ no tree component, that is, $|E(H)| \geq |V(H)|$ for each connected component H of $B(V_i, V_j)$.*

Lemma 1. *Let V_1 and V_2 be any two distinct node subsets of G. Then $B(V_1, V_2)$ has no tree component if (1.) $|V_1| = |V_2| = 2$ and $|E(B(V_1, V_2))| = 4$, or (2.) $|V_1| \geq 2$, $|V_2| \geq 3$ and $|E(B(V_1, V_2))| \geq |V_1| \times |V_2| - 1$*

3   Number of CISTs for $\delta(G) = n(G) - 2$

Let G be a graph with $n(G) \geq 12$ and $\delta(G) = n(G) - 2$. In this section, we propose an algorithm to partition $V(G)$ into a CIST-partition with size $\lfloor \frac{n(G)}{12} \rfloor$.

Algorithm 1. CIST-partition, $CIST(H_{p,q})$

Require: Graph $H_{p,q}$;
Ensure: A CIST-partition V with size $5p + \lfloor \frac{5q}{12} \rfloor$.
1: **for** $i = 1 : p$ **do**
2:      **for** $j = 1 : 3$ **do**
3:          $V_{5i+j-5} \leftarrow \{x_{12i+4j-15}, x_{12i+4j-14}\}$
4:          $V_{5i-1} \leftarrow \{x_{12i-10}, x_{12i-6}, x_{12i-2}\}$
5:          $V_{5i} \leftarrow \{x_{12i-8}, x_{12i-4}, x_{12i}\}$;
6:      **end for**
7: **end for**
8: $a \leftarrow \lfloor \frac{q}{3} \rfloor$
9: **for** $i = 1 : a$ **do**
10:      $V_{5p+i} \leftarrow \{x_{12p+3i-2}, x_{12p+3i-1}, x_{12p+3i}\}$;
11:      **ElseIf** $q - 3a = 1$
12:      $V_{5p+a} \leftarrow V_{5p+a} - \{x_{12p+q}\}$
13:      **ElseIf** $q - 3a = 2$
14:      $V_{5p+a} \leftarrow V_{5p+a} - \{x_{12p+q-2}\}$
15:      $V_{5p+a+1} \leftarrow \{x_{12p+q-2}, x_{12p+q-1}, x_{12p+q}\}$
16: **end for**
17: **return** $V = \cup_{i=1}^{\lfloor \frac{5q}{12} \rfloor} V_i$

For $p \geq 1$ and $0 \leq q \leq 11$, we set $H_{p,q}$ being a graph with $V(H_{p,q}) = \{x_i \mid 1 \leq i \leq 12p+q\}$ and $E(H_{p,q}) = \{\{x_i, x_j\} \mid i \neq j\} - \{\{x_{2k-1}, x_{2k}\} \mid 1 \leq k \leq \lfloor \frac{12p+q}{2} \rfloor\}$. Let $V = \{V_1, V_2, \ldots, V_{5p+\lfloor \frac{5q}{12} \rfloor}\}$ be the output of $CIST(H_{p,q})$. By Lemma 1, $B(V_i, V_j)$ has no tree component for any $i \neq j$. Obviously, $H_{p,q}[V_i]$ is connected for each $i \in \{1, 2, \ldots, 5p + \lfloor \frac{5q}{12} \rfloor\}$. By Theorem 2, V is a set of CISTs-partition of $H_{p,q}$. Let G be a graph with $\delta(G) = n(G) - 2$ and $n(G) = 12p + q$ for some $p \geq 1$ and $0 \leq q \leq 11$. Since $H_{p,q}$ is a subgraph of G, we have the following result.

Theorem 3. If $n(G) \geq 12$ and $\delta(G) = n(G) - 2$, then G has $\lfloor \frac{5n(G)}{12} \rfloor$-CISTs.

Example 1. We set $V_1 = \{x_1, x_3\}$, $V_2 = \{x_5, x_7\}$, $V_3 = \{x_9, x_{11}\}$, $V_4 = \{x_2, x_6, x_{10}\}$, $V_5 = \{x_4, x_8, x_{12}\}$, $V_6 = \{x_{13}, x_{15}\}$, $V_7 = \{x_{17}, x_{19}\}$, $V_8 = \{x_{21}, x_{23}\}$, $V_9 = \{x_{14}, x_{18}, x_{22}\}$, $V_{10} = \{x_{16}, x_{20}, x_{24}\}$, $V_{11} = \{x_{25}, x_{26}, x_{27}\}$, $V_{12} = \{x_{31}, x_{33}\}$, $V_{13} = \{x_{32}, x_{34}, x_{35}\}$ and $V_{12}^* = \{x_{28}, x_{29}, x_{30}\}$. Then we set $A_0 = \{V_i \mid 1 \leq i \leq 10\}$, $A_3 = A_1 \cup \{V_{11}\}$, $A_6 = A_3 \cup \{V_{12}^*\}$, and $A_8 = A_3 \cup \{V_{12}, V_{13}\}$. For each $i \in \{0, 3, 6, 8\}$, $CIST(H_{2,i})$ outputs A_i which is a CIST-partition of $H_{2,i}$ Fig. 1 shows an illustration.

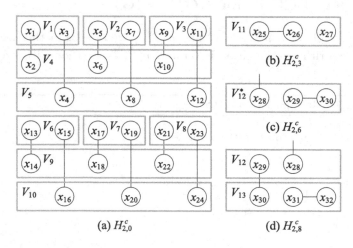

(a) $H_{2,0}^c$ (b) $H_{2,3}^c$ (c) $H_{2,6}^c$ (d) $H_{2,8}^c$

Fig. 1. CIST-partitions of $H_{2,0}$, $H_{2,3}$, and $H_{2,8}$.

Example 2. Let $m = 5p + \lfloor \frac{5q}{12} \rfloor$. Suppose that $H_{p,q}$ has a CIST-partition $V_1, V_2, \ldots, V_{m+1}$. We set $A = \{V_i \mid |V_i| = 2, 1 \leq i \leq m+1\}$. Suppose that $|A| \leq 3p + \lfloor \frac{q}{4} \rfloor$. Then we obtain a contradiction, since $12p + q = n(H_{p,q}) \geq 2 \times |A| + 3 \times (m+1-|A|) = 3m+3 - |A| \geq 12p+q+1$. Thus, $|A| \geq 3p + \lfloor \frac{q}{4} \rfloor + 1$.

Since $2 \times |V(A)| \geq 6p + \lfloor \frac{q}{2} \rfloor + 2 \geq \lfloor \frac{n(G)}{2} \rfloor + 1$, there are two distinct node subsets V_a and V_b in A such that there is a node in V_a and a node in V_b such that there are not adjacent. That is $B(V_a, V_b)$ has a tree component. We obtain a contradiction since $B(V_i, V_j)$ has no tree component. Thus, Theorem 3 is the strictly lower bounded.

4   Number of CISTs for $\delta(G) = n(G) - 3$

In this section, we will discuss that lower bounded the number of CISTs of graph G with $n(G) = 3t + 2 \geq 23$ and $\delta(G) = n(G) - 3$.

For any cycle C_i in G, we set $V(C_i) = \{x_j^i \mid 1 \leq j \leq |V(C_i)|\}$ and $E(C_i) = \{\{x_j^i, x_{j+1}^i\} \mid 1 \leq j \leq |V(C_i)|-1\} \cup \{\{x_1^i, x_{|V(C_i)|}^i\}\}$. Then we set $\mathbb{A}_t = \{G \mid n(G) = 3t+2, \Delta(G) = \delta(G) = 3t-1\}$, $\mathbb{B}_t = \{G \mid n(G) = 3t+2, \Delta(G) = 3t, \delta(G) = 3t-1,$ and $|E(G^c)| = 3t + 1\}$, and $\mathbb{C}_t = \{G \mid n(G) = 3t + 2, \Delta(G) = 3t, \delta(G) = 3t + 1,$ and $|E(G^c)| = 3t\}$ for $t \geq 7$. Obviously, each component in G^c is isomorphic to a cycle if $G \in \mathbb{A}_t$. If G is a graph in $\mathbb{B}_t$, then each component in G^c is either isomorphic a cycle or a path of length 2. Moreover, there are only two distinct nodes x and y in G such that $\deg_G(x) = \deg_G(y) = 3t$. And $\{x, y\} \notin E(G)$. If $G \in \mathbb{C}_t$, then each component in G^c is either isomorphic a cycle or a node. Moreover, there are only one node such that its degree is $3t + 1$. Then we set $\mathbb{G}_t = \mathbb{A}_t \cup \mathbb{B}_t \cup \mathbb{C}_t$ for $t \geq 7$. Note that each graph G in $\mathbb{G}_t$ contains at least 23 nodes.

Before we discuss our main result, we will show that G has $(t + 1)$-CISTs if $G \in \mathbb{G}_t$. The key point of the proof of the following Lemmas is that we try to select a node subset V^* of G where $|V^*| = 14$. Then we separate V^* into 5 disjoint node subsets $V_1, V_2, \ldots, V_5$ with $|V_i| = 3$ for $1 \leq i \leq 4$ and $|V_5| = 2$ such that the following conditions holds: (a) $V_1, V_2, \ldots, V_5$ is a CIST-partition of $G[V^*]$, and (b) $|N_G(V_i) - V^*| \leq 1$ for each $1 \leq i \leq 4$ and $|N_G(V_5) - V^*| = 0$. By Theorem 1, there there is a CIST-partition $V_6, V_7, \ldots, V_{t+1}$ in $G - V^*$. Let m be any element in $\{6, 7, \ldots, t+1\}$. By Lemma 1, condition (b) makes $B(V_m, V_i)$ has no tree component for $1 \leq i \leq 5$. That is if we can find a node subset V^* in a graph $G \in \mathbb{G}_t$ satisfies the conditions above, then G has $(t + 1)$-CISTs.

Lemma 2. Let G be a graph in $\mathbb{G}_t$, and let C_1 be the maximal cycle in G. If $|V(C_1)| \geq 8$, then G has $(t + 1)$-CISTs.

Proof. We have the following cases.
Case 1. $|V(C_1)| \geq 11$. If $|V(C_1)| \geq 14$, then we set $u = x_{12}^1$, $v = x_{13}^1$ and $w = x_{14}^1$. If $12 \leq |V(C)| \leq 13$, then we set $u = x_{12}^1$ and we can choose two nodes v and node w in $G - V(C_1)$ such that $\{v, w\} \in E(G^c)$. If $|V(C_1)| = 11$, then we can choose three nodes u, v and w in $G - V(C_1)$ such that $\{\{u, v\}, \{v, w\}\} \subseteq E(G^c)$. Then we set $V_1 = \{x_1^1, x_2^1, x_3^1\}$, $V_2 = \{x_5^1, x_6^1, w\}$, $V_3 = \{x_7^1, x_8^1, v\}$, $V_4 = \{x_{10}^1, x_{11}^1, u\}$, and $V_5 = \{x_4^1, x_9^1\}$. With the result of Theorem 1, G has $(t + 1)$-CISTs.
Case 2. $8 \leq |V(C_1)| \leq 10$. Since $n(G) - |V(C_1)| - 2 \geq 11$, there is a cycle in $G^c - V(C_1)$. Let C_2 be the maximal cycle in $G^c - V(C_1)$.
Case 2.1. $|V(C_2)| \geq 6$. We set $V_1 = \{x_1^1, x_2^1, x_4^1\}$, $V_2 = \{x_3^1, x_1^2, x_2^2\}$, $V_3 = \{x_6^1, x_7^1, x_6^2\}$, $V_4 = \{x_8^1, x_5^2, x_6^2\}$, and $V_5 = \{x_5^1, x_3^1\}$. See Fig. 2 for an illustration. With the result of Theorem 1, G has $(t + 1)$-CISTs.
Case 2.2. $|V(C_2)| \leq 5$. Since $n(G) - |V(C_1)| - |V(C_2)| - 2 \geq 8$, we can choose the maximal cycle C_3 in $G^c - (V(C_1) \cup V(C_2))$. We set $V_1 = \{x_1^1, x_2^1, x_3^1\}$, $V_2 = \{x_3^1, x_4^1, x_1^2\}$, $V_5 = \{x_2^2, x_2^3\}$. Then we set $V_3 = \{x_5^1, x_6^1, x_3^3\}$ and $V_4 = \{x_3^2, x_4^3, x_5^3\}$

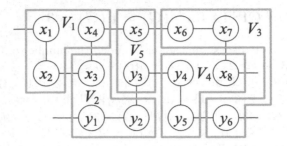

Fig. 2. Partition of a cycle at least 8 nodes and a cycle at least 6 nodes.

if $V(C_3) = 5$; we set $V_3 = \{x_6^1, x_7^1, x_3^3\}$ and $V_4 = \{x_3^2, x_4^3, x_5^1\}$ if $V(C_3) = 4$; and we set $V_3 = \{x_7^1, x_8^1, x_3^3\}$ and $V_4 = \{x_3^2, x_5^1, x_6^1\}$ if $V(C_3) = 4$.

With the result of Theorem 1, G has $(t+1)$-CISTs.

Lemma 3. *If G is a graph in $\mathbb{G}_t$, then G has $(t+1)$-CISTs.*

Proof. Let C_1 be the maximal component in G^c. By Lemma 2, this Lemma holds on $|V(C_1)| \geq 8$. Thus, we consider that $|V(C_1)| \leq 7$. Then we can find the maximal cycle C_2 in $G^c - V(C_1)$, and we can find the maximal cycle C_3 in $G^c - (V(C_1) \cup V(C_2))$.

Case 1. $|V(C_3)| = 3$. Since $|V(G)| \geq 23$, $\max\{|V(C_1)|, |V(C_2)|\} \leq 7$, and $|V(C_3)| = 3$, there are at least two disjoint cycles C_4 and C_5 in $G^c - (V(C_1) \cup V(C_2) \cup V(C_3))$ such that $|V(C_4)| = |V(C_5)| = 3$. We set $V_1 = \{x_1^1, x_1^3, x_1^4\}$, $V_2 = \{x_2^1, x_3^4, x_1^5\}$, $V_3 = \{x_1^2, x_3^2, x_3^3\}$, $V_4 = \{x_2^2, x_3^2, x_3^5\}$, and $V_5 = \{x_2^4, x_2^5\}$. With the result of Theorem 1, G has $(t+1)$-CISTs.

Case 2. $|V(C_3)| = 4$. Obviously, we can find a cycle C_4 in $G^c - (V(C_1) \cup V(C_2) \cup V(C_3))$ where $3 \leq |V(C_4)| \leq 4$. We set $V_1 = \{x_1^1, x_1^2, x_1^3\}$, $V_2 = \{x_3^1, x_3^3, x_3^4\}$, $V_3 = \{x_1^2, x_2^2, x_1^4\}$, and $V_5 = \{x_3^2, x_2^2\}$. Then we set $V_4 = \{x_3^2, x_3^4, x_4^4\}$ if $|V(C_4)| = 4$, and we set $V_4 = \{x_3^2, x_4^2, x_3^4\}$ if $|V(C_4)| = 3$. With the result of Theorem 1, G has $(t+1)$-CISTs.

Case 3. $|V(C_3)| \geq 5$. We set $V_1 = \{x_1^1, x_2^1, x_1^3\}$, $V_2 = \{x_1^2, x_2^2, x_2^3\}$, and $V_5 = \{x_3^1, x_3^3\}$. Then we set $V_3 = \{x_4^1, x_5^1, x_5^3\}$ and $V_4 = \{x_3^2, x_4^2, x_4\}$ if $|V(C_3)| = 5$, $V_3 = \{x_4^1, x_5^1, x_6^3\}$ and $V_4 = \{x_3^2, x_4^3, x_5^3\}$ if $|V(C_3)| = 6$, and $V_3 = \{y_4^1, x_6^3, x_7^3\}$ and $V_4 = \{x_3^2, x_4^3, x_5^3\}$ if $|V(C_3)| = 7$. With the result of Theorem 1, G has $(t+1)$-CISTs.

Let G be a graph in $\mathbb{G}_t$. We set $A(G) = \{x \mid \deg_G(x) = 3t\}$ and $B(G) = \{y \mid \deg_G(y) = 3t + 1\}$. Note that $|A|$ is even.

Lemma 4. *If $n(G) = 3t + 2 \geq 23$, $\Delta(G) \leq 3t$ and $\delta(G) = 3t - 1$, then G has a subgraph in $\mathbb{G}_t$.*

Proof. Obviously, $G \in \mathbb{G}_t$ if $|A(G)| = 0$. Suppose that $A(G) = \{x, y\}$. If $\{x, y\} \notin E(G)$, then $G \in \mathbb{G}_t$. If $\{x, y\} \in E(G)$, then $G - \{\{x, y\}\} \in \mathbb{G}_t$. If $|A(G)| \geq 4$, then we can find a matching M in $G[A(G)]$ where $|M| = |A(G)|/2$. Thus, $G - M \in \mathbb{G}_t$.

Theorem 4. *If $n(G) = 3t + 2 \geq 23$ and $\delta(G) = 3t - 1$, then G has a subgraph in $\mathbb{G}_t$.*

Proof. By Lemma 4, G has a subgraph in $\mathbb{G}_t$ if $|B(G)| = 0$. Thus, we consider that $|B(G)| \geq 1$. Obviously, $G \in \mathbb{G}_t$ if $|B(G)| = 1$ and $|A(G)| = 0$. If $|A(G)| \geq 2$, then we can find a hamiltonian path P in $G[\{x, y\} \cup B(G)]$ joining x to y where x and y are two distinct nodes in $A(G)$. If $|A(G)| = 0$ and $|B(G)| \geq 2$, then we can find a hamiltonian path P in $G[B(G)]$ joining two distinct nodes in $B(G)$. By Lemma 4, $G - E(P)$ has a subgraph in $\mathbb{G}_t$.

By Lemma 3 and Theorem 4, we have the following result.

Theorem 5. *If $n(G) = 3t + 2 \geq 23$ and $\delta(G) = n(G) - 3$, then G has $(t + 1)$-CISTs.*

5   Conclusions

In this paper, we mainly try to find the number of CISTs in a graph as much as possible. Firstly, for a graph with $n \geq 12$ and $\delta(G) = n(G) - 2$, we have proved that G has $\lfloor \frac{5n(G)}{12} \rfloor$ CISTs. Nextly, let G be a graph with $n(G) = 3t + 2 \geq 23$ and $\delta(G) = n(G) - 3$. Then we can find that G has $(t + 1)$ CISTs and we have given the relevant proofs. Finally, we have a conjecture. With the result of Theorem 1 and Theorem 5, we conjecture that G has $\lceil \frac{n}{3} \rceil$ CISTs if $\delta(G) = n - 3$ and the number of nodes of G is large enough. And it is a tight lower bounded.

Acknowledgment. We would like to express our sincerest appreciation to Prof. Jianxi Fan for his constructive suggestions. This work is supported by supported by National Natural Science Foundation of China (Grant No. 61902195), Natural Science Fund for Colleges and Universities in Jiangsu Province (General Program, Grant No. 19KJB520045), and NUPTSF (Grant No. NY219151, NY219131).

References

1. Araki, T.: Dirac's condition for completely independent spanning trees. J. Graph Theor. **77**(3), 171–179 (2014)
2. Chang, H.-Y., Wang, H.-L., Yang, J.-S., Chang, J.-M.: A note on the degree condition of completely independent spanning trees. IEICE Trans. Fundam. Electron. Commun. Comput. Sci. **98**(10), 2191–2193 (2015)
3. Hasunuma, T.: Completely independent spanning trees in the underlying graph of a line digraph. Discrete Math. **234**(1–3), 149–157 (2001)
4. Fan, W., Fan, J., Zhang, Y., Han, Z., Chen, G.: Communication and performance evaluation of 3-ary n-cubes onto network-on-chips. SCI. CHINA Inf. Sci. **63** (2020). https://doi.org/10.1007/s11432-019-2794-9
5. Paéterfalvi, F.: Two counterexamples on completely independent spanning trees. Discrete Math. **312**(4), 808–810 (2012)
6. Dirac, G.A.: Some theorems on abstract graphs. Proc. London Math. Soc. **1**, 69–81 (1952)

7. Fan, W., Fan, J., Lin, C.-K., Wang, Y., Han, Y., Wang, R.: Optimally embedding 3-ary n-cubes into grids. J. Comput. Sci. Technol. **34**(2), 372–387 (2019)
8. Fan, W., He, J., Han, Z., Li, P., Wang, R.: Reconfigurable fault-tolerance mapping of ternary n-cubes onto chips. Concurrency Comput. Pract. Experience **32**(11), 1–12 (2020)
9. Fan, G., Hong, Y., Liu, Q.: Ore's condition for completely independent spanning trees. Discrete Appl. Math. **177**, 95–100 (2014)
10. Hong, X., Liu, Q.: Degree condition for completely independent spanning trees. Inf. Process. Lett. **116**, 644–648 (2016)
11. Obokata, K., Iwasaki, Y., Bao, F., Igarashi, Y.: Independent spanning trees of product graphs and their construction. IEICE Trans. Fundam. Electron. Commun. Comput. Sci. **79**(11), 1894–1903 (1996)
12. Pai, K.-J., Yang, J.-S., Yao, S.-C., Tang, S.-M., Chang, J.-M.: Completely independent spanning trees on some interconnection networks. Ice Trans. Inf. Syst. **97**(9), 2514–2517 (2014)
13. Hsu, L.-H., Lin, C.-K.: Graph Theory and Interconnection Networks, CRC Press (2008)
14. Huck, A.: Independent trees in planar graphs. Graphs and Combinatorics **15**(1), 29–77 (1999)
15. Fan, W., Fan, J., Han, Z., Li, P., Zhang, Y., Wang, R.: Fault-tolerant Hamiltonian cycles and paths embedding into locally exchanged twisted cubes. Front. Comput. Sci. pp. 1–21 (2020). https://doi.org/10.1007/s11704-020-9387-3
16. Hasunuma, T.: Minimum degree conditions and optimal graphs for completely independent spanning trees. In: Lipták, Z., Smyth, W.F. (eds.) IWOCA 2015. LNCS, vol. 9538, pp. 260–273. Springer, Cham (2016). https://doi.org/10.1007/978-3-319-29516-9_22
17. Fan, W., Fan, J., Lin, C.-K., Wang, G., Cheng, B., Wang, R.: An efficient algorithm for embedding exchanged hypercubes into grids. J. Supercomput. **75**(2), 783–807 (2019)
18. Fan, W., Wang, Y., Sun, J., Han, Z., Wang, R.: Fault-tolerant cycle embedding into 3-Ary n-Cubes with structure faults. In: IEEE International Conference on Parallel & Distributed Processing with Applications, Big Data & Cloud Computing, Sustainable Computing & Communications, Social Computing & Networking (ISPA/BDCloud/SocialCom/SustainCom), pp. 451–458 (2019)
19. Lichiardopol, N.: Quasi-centers and radius related to some iterated line digraphs, proofs of several conjectures on de Bruijn and Kautz graphs. Discrete Appl. Math. **202**, 106–110 (2016)
20. Park, J.H., Lim, H.S., Kim, H.C.: Fault-tolerant embedding of starlike trees into restricted hypercube-like graphs. J. Comput. Syst. Sci. **105**(11), 104–115 (2019)
21. Pai, K.-J., Chang, R.-S., Chang, J.-M.: Constructing dual-cists of pancake graphs and performance assessment of protection routings on some cayley networks. J. Supercomput. **3**, 1–25 (2020)
22. Tang, S.-M., Yang, J.-S., Wang, Y.-L., Chang, J.-M.: Independent spanning trees on multidimensional torus networks. IEEE Trans. Comput. **59**(1), 93–102 (2010)
23. Yang, J.-S., Tang, S.-M., Chang, J.-M., Wang, Y.-L.: Parallel construction of optimal independent spanning trees on hypercubes. Parallel Comput. **33**(1), 73–79 (2007)
24. Werapun, J., Intakosum, S., Boonjing, V.: An efficient parallel construction of optimal independent spanning trees on hypercubes. J. Parallel Distrib. Comput. **72**(12), 1713–1724 (2012)

25. Wang, Y., Feng, Y., Zhou, J.: Automorphism group of the varietal hypercube graph. Graphs and Combinatorics **33**, 1131–1137 (2017)
26. Yang, J.-S., Chang, J.-M.: Optimal independent spanning trees on Cartesian product of hybrid graphs. Comput. J. **57**(1), 93–99 (2014)
27. Yang, J.-S., Chang, J.-M., Tang, S.-M., Wang, Y.-L.: Reducing the height of independent spanning trees in chordal rings. IEEE Trans. Parallel Distrib. Syst. **18**, 644–657 (2007)

Distributed Algorithm for Truss Maintenance in Dynamic Graphs

Qi Luo[1], Dongxiao Yu[1(✉)], Hao Sheng[2], Jiguo Yu[3], and Xiuzhen Cheng[1]

[1] School of Computer Science and Technology, Shandong University,
Qingdao, People's Republic of China
luoqi2018@mail.sdu.edu.cn, {dxyu,xzcheng}@sdu.edu.cn
[2] School of Computer Science and Engineering, Beihang University,
Beijing, People's Republic of China
shenghao@buaa.edu.cn
[3] School of Computer Science and Technology, Qilu University of Technology,
Jinan, People's Republic of China
jiguoyu@sina.com

Abstract. Cohesive subgraphs are applied in various fields. Mining cohesive components such as k-truss have attracted a lot of effort to improve time efficiency in large-scale graphs. The k-truss is a subgraph where each edge is contained in at least $k - 2$ triangles and the problem of truss decomposition is computing the k-trusses of a graph for all k. However, most graphs in real scenarios are usually changing over time. The previous studies take the static graphs as input, and the truss maintenance in dynamic graphs receives little attention. This paper focuses on distributed algorithms for truss maintenance. We present a distributed model underlying the real distributed processing model Pregel. Based on the model, we propose truss decomposition and truss maintenance algorithms. To confirm the effectiveness and efficiency of the proposed algorithms, we conduct extensive experiments over both real-world and synthetic graphs.

Keywords: Distributed algorithm · Graph analytics · k-truss · Dynamic graph

1 Introduction

Graphs have been widely used to model social networks, communication networks, and information networks that emerge in various applications [3,20].

This work was supported in part by the National Key Research and Development Program of China under Grant 2019YFB2102600 and in part by NSFC under Grant 61971269, Grant 61832012 Grant 61672321, and Grant 61771289 (Corresponding author: Dongxiao Yu). The Science and Technology Development Fund, Macau SAR (File no.0001/2018/AFJ), the Fundamental Research Funds for the Central Universities and the Open Fund of the State Key Laboratory of Software Development Environment (No. SKLSDE2019ZX-04).

© Springer Nature Switzerland AG 2021
Y. Zhang et al. (Eds.): PDCAT 2020, LNCS 12606, pp. 104–115, 2021.
https://doi.org/10.1007/978-3-030-69244-5_9

As a fundamental task in graph analysis, detecting cohesive components from graphs has attracted much attention from research and industry communities. As finding the densest subgraph clique is an NP-hard problem [7], some relaxation versions of clique such quasi-clique [1], k-core [8,11,13,18,22,23] and k-truss [5] have been widely used in graph analysis. Motivated by the social structure that if two persons are strongly tied, they should share enough common friends, the concept of k-truss is proposed, which requires each edge in a k-truss is contained in at least $k - 2$ triangles. Due to the tight feature reflecting the closeness of two nodes well, k-truss plays an important role in many applications such as community search [2,9] and social contagion [16].

The notion related to k-truss is *trussness*, which is defined as the maximum k such that $e \in G$ in a k-truss but not in a $(k + 1)$-truss. The problem of *truss decomposition* refers to computing the trussness of each edge in graphs and it is solvable in $O(|E|^{1.5})$ time by the edge peeling algorithm [5]. More recent research [17,21] has occurred in the field of sequential algorithms for truss decomposition. As the size of graphs grows larger and graphs are continuously evolving with time in the real world, the restrictions of single computing core and memory make the algorithms increasingly inefficient and difficult to achieve desired results. Some distributed and parallel techniques [4] have been extensively applied to the study of truss decomposition for Massive graphs. But for dynamic graphs, especially when edges and nodes can be dynamically inserted or removed over time, there are few studies [2,9,12,24,25] to date that have focused on the maintenance of trussness. For a large dynamic graph, the problem of truss maintenance is about how to efficiently update and avoid recomputing the trussness of edges when edges are inserted or removed, considering there might be billions of nodes and edges in graphs. So, Two challenges needed to overcome in truss maintenance for large dynamic graphs, one is to determine which edge trussness will change and another is how much will it change.

To overcome the challenge for truss maintenance, we address the drawbacks of the exiting solutions for truss decomposition and propose new distributed truss maintenance algorithms in dynamic graphs. Our work is motivated by the node-centric model such as Pregel [14] which restricts algorithms to operate the local graph in view of a single node. Based on this, we extend the node-centric model to efficiently implement our distributed truss maintenance algorithms. In our algorithm, each node modifies its local information by receiving messages from neighbor nodes, and sending new messages to its neighbor nodes. Individual nodes may leave the computation when they reach the final convergence state. The algorithms described in this paper run in a synchronous environment and can be directly translated into the Pregel model. In the following, we summarize the main contributions of this paper.

1. We design a distributed model underlying Pregel which is a large-scale graph processing system proposed by Google. Based on this model, we design the storage structure of nodes and the communication mode between nodes, and propose a distributed truss decomposition algorithm.

2. We propose a distributed truss maintenance algorithm under one edge insertion/deletion.
3. We carry out some experiments on synthetic graphs and real-world graphs, the results show that our algorithms can achieve good performance.

Paper Organization. In the rest of this paper, Sect. 2 surveys the work related to cohesive subgraphs and k-truss. After describing preliminaries about basic concepts and problem definitions in Sect. 3, we outline our distributed truss decomposition algorithm in Sect. 4 and distributed truss maintenance algorithm in Sect. 5, respectively. In Sect. 6, we experiment with our algorithms by adopting both synthetic and real-world datasets. Finally, Sect. 7 concludes this paper and discusses some future work about the field.

2   Related Work

Truss Decomposition. A lot of works has been done on truss decomposition since k-truss was first presented in [5]. Cohen [5] proposed a sequential algorithm to find the maximal trussness in graphs. Besides, Cohen [6] offered a way to handle large graphs on a single machine that cannot hold the entire graph as well as enables streaming graph for truss decomposition.

Wang et al. [21] improved the existing in-memory algorithm for computing k-truss in networks of moderate size, and then two I/O-efficient algorithms were proposed to handle massive networks that cannot fit in main memory. Sariyüce et al. [17] proposed efficient algorithms to construct the hierarchy of dense subgraphs like k-truss, to avoid traversal during the peeling process, and provided local algorithms that are highly parallel which can provide fast approximations, to explore time and quality trade-offs for truss decomposition. As evidenced by the recent graph challenges, existing algorithms and implementations for truss decomposition are insufficient for the scale of modern datasets. There are some parallel and distributed algorithms for truss decomposition [4,19].

Truss Maintenance. Compared to truss decomposition, there are only several works on truss maintenance to the best of our knowledge. Zhou et al. [25] have investigated the properties of truss change, and proposed algorithms for truss maintenance for one edge deletion or insertion. [2,9] focused on the truss community querying problem based on truss maintenance. [9] investigated the k-truss community search problem in a dynamic graph setting with frequent insertions and deletions of nodes and edges. [2] introduced a truss-preserving index structure called EquiTruss which can be efficiently updated in a dynamic fashion. Zhang et al. [24] applied the standards of boundedness and presented both a bounded removal algorithm and a near bounded insertion algorithm for truss maintainable problems.

3   Preliminaries

Let $G = (V, E)$ represent an undirected and unweighted simple graph where V is the node set and E is the edge set respectively. We denote $n = |V|$ as the number of nodes, $m = |E|$ as the number of edges of G. *Node* and *vertex* are the same concepts in this paper. We apply $N(u)$ to denote the neighbor node set of node u and $EN(e)$ to denote the set of edges that have a common endpoint with edge e. Following, we will introduce some necessary notations and definitions used throughout this paper.

For an edge $e = (u, v)$, the *support* of e is defined as the number of triangles containing e, denoted as $sup(e) = |\{\triangle_{uvw} : \triangle_{uvw} \in G\}|$. In other words, $sup_G(e) = N_G(u) \cap N_G(v)$. The minimum support of all edges in a subgraph $H \subseteq G$ is denoted by $sup_G(H)$. According to the concept of *support*, the k-truss is defined as follows.

Definition 1 (k-truss). *A k-truss is a non-trivial, induced subgraph that each edge is contained in at least $k - 2$ triangles.*

A maximal k-truss is a k-truss that is not a subgraph of another k-truss. The trussness of an edge is the value of k such that e is a maximal k-truss but not in a $k + 1$-truss.

Definition 2 (Triangle connected). *Given two edges e_1, e_2 in G, if $e_1 \in \triangle^{(1)}$, $e_2 \in \triangle^{(t)}$ and there is a series of triangles $\triangle^{(1)}, \triangle^{(2)}, \ldots, \triangle^{(t)}$, where $t \geq 1$ and every two adjacent triangles share an edge, then e_1 and e_2 is* triangle connected.

A k-triangle is a triangle where the smallest trussness of the three edges equals k, denote as $\triangle^k$.

Definition 3 (k-Triangle connected). *If two edges e_1, e_2 are triangle connected and $t(e_1) = t(e_2) = k$. Beyond that, each triangle in the series of triangles between e_1 and e_2 is a k-triangle, then we denote e_1 and e_2 is k-triangle connected.*

Distributed Model. The distributed system we consider is a network consisting of computable nodes. Each node $u \in V$ in G represents a host, which is able to pass messages with neighbor hosts. The rounds in the system are synchronous, i.e., each host receives messages, computing, and send messages one time during any round. In this system, hosts are allowed to leave and join.

4   Distributed Truss Decomposition

We were inspired by the existing algorithm, distributed k-truss decomposition that is based on the bulk synchronous parallel model [4], which reduces the amount of communication and the iteration number in the graph-parallel abstractions based on the line graphs. However, when a graph is turned into a line graph, a more complex network is created. Hence, we aim to just use the original graph for distributed truss decomposition.

Our algorithm is based on a theoretical result *Locality Property* proposed in [4], which implies that if an edge e with trussness $t(e)$, there are at least $2(k-2)$ edges form total $(k-2)$ triangles with e and each edge in these triangles with trussness at least k. In our algorithm, each node is responsible for the trussness of its adjacent edges and maintains an *adjMap* (the key is the adjacent id, and the value is adjacency list) to stow adjacency lists of neighbor nodes. Each node maintains a *trussMap* (the key is the edge, and the value is trussness), which is used to store the estimated trussness of adjacent edges. In subsequence, we denote the estimate trussness of e as $t(\hat{e})$.

The pseudo-code of the distributed truss decomposition algorithm is shown in Algorithm 1. In the *Initialization* phase, u and its neighbors exchange adjacency list. Each node computes the *support* of its adjacent edges as the initial value of the estimated trussness and sends the *trussMap* to neighbor nodes. In the *Execution phase*, u receives *trussMap* from its neighbor nodes and updates the estimated trussness of adjacent edges according to the *Locality Property* (the boolean variable *Changed* as the update flag). If *Changed* equals *True*, u sends only the updated portion of its adjacent edges to the neighbor nodes. Repeating the above steps until the termination condition is reached. The function *ComputeTruss* follows the Locality property [4] of k-truss.

Algorithm 1: Distributed truss decomposition process: node u

Input : A Graph $\mathcal{G}(V, E)$
Output: Trussness of each edge

1 **Initialization**
2 | send $adjMap[u]$ to $v \in N(u)$;
3 | receive $adjMap[v]$ for $v \in N(u)$;
4 | **foreach** $v \in N(u)$ **do**
5 | | $trussMap[(u,v)] \leftarrow |N(u)| \cap |N(v)| + 2$;
6 | send trussMap to $v \in N(u)$;
7 | changed $\leftarrow$ *False* ;

8 **Execution**
9 | **repeat**
10 | | receive trussMap ;
11 | | counter $\leftarrow \emptyset$;
12 | | **foreach** $v \in N(u)$ **do**
13 | | | **for** $w \in N(u) \cap N(v)$ **do**
14 | | | | $k_{min} \leftarrow min(trussMap[(u,w)], trussMap[(v,w)])$ counter.add(k_{min}) ;
15 | | | | $t \leftarrow$ ComputeTruss($counter, |N(u) \cap N(v)|$);
16 | | | | **if** $t < trussMap[(u,v)]$ **then** // update
17 | | | | | $trussMap[(u,v)] \leftarrow t$;
18 | | | | | changed $\leftarrow$ *True* ;

19 | | **if** changed **then** // only send the changed part
20 | | | send $trussMap[(u,*)]$ to $v \in N(u)$;

21 | **until** *terminal condition*;

Termination Mechanism. In the above algorithm, there are several alternatives for the *termination condition* given in [15]. We choose the *barrier synchronization* mechanism. That is, none of the nodes update the *trussMap* during the previous round then the system will be terminated.

Figure 1(a) shows the keys of *adjMap* that are maintained by ①, and the values are received from neighbor nodes of ①. Figure 1(b) shows the *trussMap* state of node ① after initialization and termination. The *trussMap* consists of two parts, one is the estimated trussness of the adjacent edges of ①, and the other is the estimated trussness of the adjacent edges of the neighbor nodes.

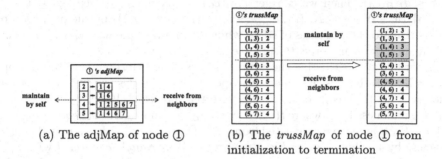

(a) The adjMap of node ①

(b) The *trussMap* of node ① from initialization to termination

Fig. 1. The auxiliary data structure for node ①.

Theorem 1. *The number of rounds for Algorithm 1 will not be larger than* $m - t_{min} + 2$, *where* $m = |E|$, t_{min} *is the minimum trussness of the graph.*

Proof. The estimated value of at least one trussness in each round will be reduced to the true trussness. As long as the system is running, the minimum trussness of edges will be determined in each round. That is, there will be at least one edge reach the trussness in one round, the minimum trussness will be determined. In addition, it takes time for one round of initialization and the last round of judgment to terminate. So the running time will not be large than $m - t_{min} + 2$.

Theorem 2. *The message complexity of Algorithm 1 is bounded by* $2 \sum_{e \in E}(sup(e))$.

Proof. In the worst case, an endpoint of edge e can receive at most $|sup(e) + 2 - t(e)|$ messages from neighbor nodes. The minimum trussness is at least 2 if the graph is connected. From above, the upper bound of the message complexity is $\sum_{e \in E}(sup(e) + sup(e')) = 2 \sum_{e \in E}(sup(e))$.

5   Distributed Truss Maintenance

In this section, we consider truss maintenance in dynamic graphs. We observe that inserting or deleting one edge forms or breaks at most one triangle in a graph. The support of an edge is the number of triangles in which the edge is located. The trussness of an edge depends on the maximum support of all subgraphs it is in. So the trussness may increase 1 or decrease 1 at most after one edge inserting or deleting. For one edge insertion, after inserting e_0 to G, it generates a new graph $G' = G \cup e_0$ (the endpoint of e_0 may be new nodes to G). The trussness of edges $e \in G$ can be affected by two ways: (1) e, e_0 and another edge form a new triangle, which increases the support and trussness; (2) The trussness of some edges of the triangle where e lies gets larger.

For the trussness of inserted edge $e_0 = (u, v)$, we can get a trussness bound for e_0 based on the trussness of the edges before inserting e_0. Assume $t_{LB}(e_0) \leq t(e_0) \leq t_{UB}(e_0)$. Then we can get $t_{LB}(e_0) = arg\max_{k \geq 2}\{|\{w|t_{min}^w \geq k\}| \geq (k-2)\}$ $(w \in N_G(u) \cap N_G(v), t_{min}^w = \min(t(u,w), t(v,w)))$ by the definition of k-truss. Due to the trussness of edges which form a triangle with e_0 will increase by 1 at most, we can derive that $t_{UB}(e_0) - t_{LB}(e_0) \leq 1$.

According to the existing theoretical results [2], our formal description properties relates to the trussness effected scope to which an edge is inserted or deleted as follows: (1) If an edge $e_0 = (u, v)$ is inserted into $G = (V, E)$, $e_1, e_2 \in E$ may increase trussness by 1, where $t(e_1) = k < t_{UB}(e_0)$ and e_2 is k-triangle connected with e_1. (2) If an edge e_0 is deleted from $G = (V, E)$, $e_1, e_2 \in E \setminus e_0$ may decrease trussness by 1, where $t(e_1) = k \leq t(e_0)$ and e_2 is k-triangle connected with e_1.

After inserting an edge $e_0 = (u, v)$, the endpoints of the edge update the *adjMap* and send the new *adjMap* to their neighbors firstly. Secondly, the two nodes both calculate the upper bound of e_0 (i.e., $t_{UB}(e_0)$) by trussness before inserting e_0. e will be marked if edge e forms a triangle with e_0 and $t(e) < t_{UB}(e_0)$, and those edges are also marked if they are k-triangle connected with e. All the marked edges are those whose trussness may increase by 1. Hence, we here increase the value in its trussMap by 1 and start the same process as Algorithm 1 until all edges converge to stable values.

We divide the endpoints of the edge that might change the trussness into two categories. One category of nodes are the endpoints of the inserted edge (denote as *Origin Node Set*), another category of nodes are the endpoints of edges which are *k-triangle connected* with edges whose trussness is lower than the upper bound of new inserting edges in the newly formed triangles (denote as *Passing Node Set*).

Distributed Incremental Truss Maintenance. For edge insertion, each node maintains two auxiliary data structures: 1) **EESet** (effect edge set) is a set of adjacent edges whose trussness may change; 2) **EVSet** (effect node set) is a set of neighbor nodes whose *trussMap* may change. For the promoter node, the *EESet* stores all the edges whose trussness is lower than the upper bound of the inserting edge; the *EVSet* stores another endpoint of edges in *EEset*. Because the trussness of edges in *EEset* may increase by 1, we re-initialize the estimated trussness of these edges by plus 1 to current trussness. Then the same process as Algorithm 1 is executed until the time when there is no update of estimate trussness. The pseudo-code of promoter node and spread node is shown in Algorithm 2 and Algorithm 3 respectively. In these two algorithms, the *Prepare* phase finds adjacent edges and neighbor nodes that satisfy the condition, and the *Propose* phase resets the estimated trussness in *EESet*.

Algorithm 2: The re-initialize stage: promoter node u

    Input : A graph $G(V, E)$, trussMap, insert edge $e_0 = (u, v)$

1 **Prepare**
2     $adjacentList[u].add(v)$;
3     $trussMap[(u, v)] \leftarrow t_{UB}(e_0)$;
4     $EES \leftarrow \emptyset$;
5     $EVS \leftarrow \emptyset$;
6   **send** $adjacentList$ to $v \in N(u)$;
7   **receive** $adjacentList$;
8   **foreach** $w \in N(u) \cap N(v)$ **do**
9   **if** $trussMap[(u, w)] < t_{UB}(e_0)$ **then**
10          $EESet.add((u, w))$;
11          $EVSet.add(w)$;

12 **Propose**
13   **foreach** $e \in EESet$ **do**
14        $trussMap[e] \leftarrow trussMap[e] + 1$;
15   **foreach** $w \in EVSet$ **do**
16   **send** $<id>$ to w ;

The re-initialize stage ends when no node receives EES. The time complexity of the re-initialize stage depends on the diameter of the induced subgraph generated by $EVSet$. Let H represent the induced subgraph generated by $EVSet$, and H is a k-truss. The diameter of a connected k-truss with n nodes is no more than $\lceil \frac{2n-2}{k} \rceil$ [5]. According to related results [10], for a graph $G(V, E)$ and a node set $Q \subset V$, we have $dist_G(G, Q) \leq diam(G) \leq 2dist_G(G, Q)$ where $diam(G) = max_{u,v \in G}\{dist_G(u, v)\}$ is the diameter of graph G and $dist_G(u, v)$ is the length of the shortest path between u and v in G. The time complexity of the re-initialize stage is $diam(H)$. After re-setting estimate trussness for edges by the re-initialize stage, Algorithm 1 terminates just in two rounds of execution phase in most sparse graphs (one round is to ensure that all nodes do not update $trussMap$).

Distributed Decremental Truss Maintenance. The algorithm for deleting an edge is simpler compared to the incremental case. After an edge is deleted, the endpoint of the edge and its neighbor nodes only need to update the *adjacencyMap* and eliminate items in *trussMap* that the corresponding edges do not exist, then continue the execution phase of Algorithm 1 until algorithm termination. In this case, only nodes in $EVSet$ will participate in the calculation and just two rounds will be needed in most sparse graphs (one round is to ensure that all nodes do not update $trussMap$).

6   Experiments

In this section, we evaluate the performances of our algorithms using four real-world graphs and four synthetic graphs. Firstly, we compare our distributed truss decomposition algorithm with the GPTruss algorithm [4], and then illustrate the experimental analysis of our distributed truss maintenance algorithms in terms of running time and the number of exchanged messages. All the experiments are performed under a CentOS 7 operating system running on a machine

---

Algorithm 3: The re-initialize stage: spread node p

Input : A graph $G(V, E)$, trussMap
Output: Trussness of each edge

1 **Prepare**
2   receive $adjacentList$;
3   receive $< id >$ from o ;
4     $\bar{k} \leftarrow trussMap[(p, o)]$;
5   **foreach** $w \in N(p) \cap N(o)$ **do**
6   **if** $min\{trussMap[(p, w)], trussMap[(o, w)]\} == \bar{k}$ **then**
7           $EVS.add(w)$;
8   **if** $trussMap[(p, w)] == \bar{k}$ **then**
9              $EES.add((p, w))$;
10   **if** $trussMap[(o, w)] == \bar{k}$ **then**
11             $EES.add((o, w))$;

12 **Propose**
13   **foreach** $e \in EES$ **do**
14       $trussMap[e] \leftarrow trussMap[e] + 1$;
15   **foreach** $w \in EVS$ **do**
16   send $< id >$ to w ;

---

with an Intel Xeon 3.4 GHz CPU, 120 GB RAM and 1 TB SATA hard disk. All experiments were run 20 times and the average is reported.

Datasets. In order to make comprehensive and convective experiments on our algorithm, we firstly make a brief introduction to our datasets. For synthetic graphs, they are generated by applying different models using the SNAP system and have the same node size. We use the tool of NetworkX[1] to generate graphs by different functions. In detail, **ER** is generated by *Erdos-Renyi* model, **WS** is generated by *Watts-Strogatz* model, **BA** is generated by *Barabasi-Albert* model and **HK** is generated by *Holme and Kim* model. For real-world graphs, all graphs can be downloaded from the SNAP[2]. The statistics of the datasets are shown in Table 1.

Table 1. Statistics of datasets. deg_{avg} is the average degree and $diam$ is the diameter of the graph.

(a) Synthetic graphs.					(b) Real-world graphs.					
Name	$|V|$	$|E|$	$|\triangle|$	deg_{avg}	Name	$|V|$	$|E|$	$|\triangle|$	deg_{avg}	$diam$
ER	131K	335K	17	5.11	ca-HepPh	12K	118K	3358K	19.73	13
WS	131K	393K	49k	6.00	com-DBLP	317K	1049K	2224K	6.62	21
BA	131K	786K	7k	11.99	com-Youtube	1134K	2987K	3056K	5.26	20
HK	131K	786K	357k	11.99	roadNet-CA	1965K	2766K	120K	2.81	849

---

[1] http://networkx.github.io.
[2] http://snap.stanford.edu/data/index.html.

Truss Decomposition. To explicitly reveal the performance of our distributed truss decomposition algorithm, which is denoted as DisTrussDecomp, we select the distributed truss decomposition algorithm GPTruss [4] as comparisons. Figure 2 shows the execution time measured in rounds of the two algorithms.

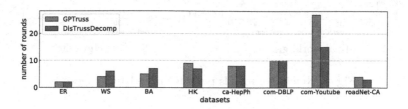

Fig. 2. The number of rounds of distributed truss decomposition.

For the figure, we can find that the execution time is very small (less than 10 rounds) in most graphs. Compared with GPTruss, DisTrussDecomp is more efficient in most of the graphs, especially in dense graphs, since GPTruss needs to convert the graph into a line graph, which generates a much more complex network for a dense graph. In spare graphs, GPTruss and DisTrussDecomp require almost the same number of rounds, as in these graphs, the number of edges is close to the number of nodes.

Truss Maintenance. We denote our distributed truss maintenance algorithm as DisTrussMainten. For each graph, we randomly chose $10 * i$ edges as the dynamic edges for insertion and deletion, where $i = 1, 2, 3, 4, 5, 6$. We analyze the average time for each edge of DisTrussMainten in all datasets for edge insertion and deletion.

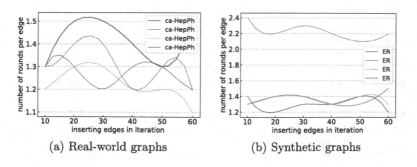

(a) Real-world graphs   (b) Synthetic graphs

Fig. 3. Number of rounds per edge by inserting edges.

Figure 3 and Fig. 4 show the average time per edge of our algorithm in each dataset for edge insertion/deletion. In all graphs, it only takes 1–3 rounds

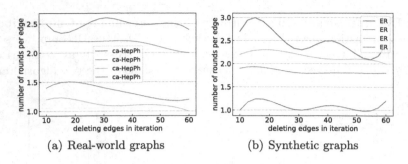

(a) Real-world graphs   (b) Synthetic graphs

Fig. 4. Number of rounds per edge by deleting edges.

to update the trussness when deleting or inserting one edge at one time. For dense graphs, such as **ca-HepPh** and **email-Eu-core**, their average degree is very high. The number of nodes affected is obviously higher than other data sets after inserting edges, so it needs more rounds to converge after the graph changed.

7 Conclusions

We studied the problem of distributed algorithms for truss maintenance in dynamic graphs. By designing a distributed model, we showed that distributed algorithm design is possible for truss decomposition and truss maintenance. Extensive experiments on real-world graphs and synthetic graphs demonstrate the efficiency of the proposed algorithms in reality. It is meaningful to investigate whether efficient distributed solutions can be derived for other fundamental problems in graph analytics.

References

1. Abello, J., Resende, M.G.C., Sudarsky, S.: Massive Quasi-Clique detection. In: Rajsbaum, S. (ed.) LATIN 2002. LNCS, vol. 2286, pp. 598–612. Springer, Heidelberg (2002). https://doi.org/10.1007/3-540-45995-2_51
2. Akbas, E., Zhao, P.: Truss-based community search: a truss-equivalence based indexing approach. PVLDB **10**(11), 1298–1309 (2017)
3. Cai, Z., He, Z., Guan, X., Li, Y.: Collective data-sanitization for preventing sensitive information inference attacks in social networks. IEEE Trans. Dependable Secur. Comput. **15**(4), 577–590 (2018)
4. Chen, P., Chou, C., Chen, M.: Distributed algorithms for k-truss decomposition. In: 2014 IEEE International Conference on Big Data, Big Data 2014, Washington, DC, USA, 27–30 October 2014, pp. 471–480 (2014)
5. Cohen, J.: Trusses: Cohesive subgraphs for social network analysis. Technical report. National Security Agency (2008)
6. Cohen, J.: Graph twiddling in a MapReduce world. Comput. Sci. Eng. **11**(4), 29–41 (2009)

7. Das, A., Svendsen, M., Tirthapura, S.: Incremental maintenance of maximal cliques in a dynamic graph. VLDB J. **28**(3), 351–375 (2019)
8. Hua, Q., et al.: Faster parallel core maintenance algorithms in dynamic graphs. IEEE Trans. Parallel Distrib. Syst. **31**(6), 1287–1300 (2020)
9. Huang, X., Cheng, H., Qin, L., Tian, W., Yu, J.X.: Querying k-truss community in large and dynamic graphs. In: International Conference on Management of Data, SIGMOD, Snowbird, UT, USA, pp. 1311–1322 (2014)
10. Huang, X., Lakshmanan, L.V.S., Yu, J.X., Cheng, H.: Approximate closest community search in networks. PVLDB **9**(4), 276–287 (2015)
11. Jin, H., Wang, N., Yu, D., Hua, Q., Shi, X., Xie, X.: Core maintenance in dynamic graphs: a parallel approach based on matching. IEEE Trans. Parallel Distrib. Syst. **29**(11), 2416–2428 (2018)
12. Luo, Q., Yu, D., Cheng, X., Cai, Z., Yu, J., Lv, W.: Batch processing for truss maintenance in large dynamic graphs. IEEE Trans. Comput. Soc. Syst., 1–12 (2020)
13. Luo, Q., et al.: Distributed core decomposition in probabilistic graphs. In: Tagarelli, A., Tong, H. (eds.) CSoNet 2019. LNCS, vol. 11917, pp. 16–32. Springer, Cham (2019). https://doi.org/10.1007/978-3-030-34980-6_2
14. Malewicz, G., et al.: Pregel: a system for large-scale graph processing. In: Proceedings of the ACM SIGMOD International Conference on Management of Data, SIGMOD, Indianapolis, Indiana, USA, pp. 135–146 (2010)
15. Montresor, A., Pellegrini, F.D., Miorandi, D.: Distributed k-core decomposition. IEEE Trans. Parallel Distrib. Syst. **24**(2), 288–300 (2013)
16. Rossi, M.G., Malliaros, F.D., Vazirgiannis, M.: Spread it good, spread it fast: identification of influential nodes in social networks. In: Proceedings of the 24th International Conference on World Wide Web Companion, WWW, pp. 101–102. ACM (2015)
17. Sariyüce, A.E., Pinar, A.: Fast hierarchy construction for dense subgraphs. PVLDB **10**(3), 97–108 (2016)
18. Seidman, S.B.: Network structure and minimum degree. Soc. Netw. **5**(3), 269–287 (1983)
19. Shao, Y., Chen, L., Cui, B.: Efficient cohesive subgraphs detection in parallel. In: SIGMOD Conference, pp. 613–624. ACM (2014)
20. Sheng, H., Zheng, Y., Ke, W., Yu, D., Xiong, Z.: Mining hard samples globally and efficiently for person re-identification. IEEE Internet Things J. **PP**(99), 1 (2020)
21. Wang, J., Cheng, J.: Truss decomposition in massive networks. PVLDB **5**(9), 812–823 (2012)
22. Wang, N., Yu, D., Jin, H., Qian, C., Xie, X., Hua, Q.: Parallel algorithm for core maintenance in dynamic graphs. In: Lee, K., Liu, L. (eds.) 37th IEEE International Conference on Distributed Computing Systems, ICDCS, pp. 2366–2371. IEEE Computer Society (2017)
23. Yu, D., Zhang, L., Luo, Q., Cheng, X., Yu, J., Cai, Z.: Fast skyline community search in multi-valued networks. Big Data Anal. Mining **3**(3), 171–180 (2020)
24. Zhang, Y., Yu, J.X.: Unboundedness and efficiency of truss maintenance in evolving graphs. In: SIGMOD, pp. 1024–1041 (2019)
25. Zhou, R., Liu, C., Yu, J.X., Liang, W., Zhang, Y.: Efficient truss maintenance in evolving networks. CoRR abs/1402.2807 (2014)

Short-Term Load Forecasting Based on CNN-BiLSTM with Bayesian Optimization and Attention Mechanism

Kai Miao[1], Qiang Hua[2], and Huifeng Shi[1(✉)]

[1] Department of Mathematics and Physics, North China Electric Power University,
Baoding 071003, China
404502485@qq.com, shihf@ncepu.edu.cn
[2] College of Mathematics and Information Science, Hebei University,
Baoding 071002, China
huaq@hbu.edu.cn

Abstract. Short-term power load forecasting is quite vital in maintaining the balance between power production and power consumption of the power grid. Prediction accuracy not only affects the power grid construction, but also influences the economic development of the power grid. This paper proposes a short-term load forecasting based on Convolutional Neural Networks and Bidirectional Long Short-Term Memory (CNN-BiLSTM) with Bayesian Optimization (BO) and Attention Mechanism (AM). The BiLSTM is good at time series forecasting, and the Attention Mechanism can help the model to focus on the important part of the BiLSTM output. In order to make the forecasting performance of the model as good as possible, the Bayesian Optimization is used to tune the hyperparameters of the model. The input of the model is history load, time slot, and meteorological factors. In order to eliminate the seasonal influence, the data set is divided into four subsets with respect to four seasons. The performance of the proposed model is compared with other forecasting models by MAE, RMSE, MAPE, and R^2 score. The experiment results show that the proposed model fits the actual values best and has the best forecasting performance among the contrast models.

Keywords: Short-term load forecasting · Bayesian Optimization Algorithm · Convolutional Neural Networks · Bidirectional long short-term memory network · Attention Mechanism

1 Introduction

As the basis of power system planning and operation, power load forecasting plays a critical role in the actual job. The significance of the prediction lies in keeping a balance between supply and demand of power grid and making effective plans and distribution of energy to minimize energy waste. Therefore, accurate

© Springer Nature Switzerland AG 2021
Y. Zhang et al. (Eds.): PDCAT 2020, LNCS 12606, pp. 116–128, 2021.
https://doi.org/10.1007/978-3-030-69244-5_10

load forecasting is indispensable. However, with the rapid development of the economy and the frequent occurrence of extreme temperatures caused by global warming, the power load consumption continues to intensify, and its fluctuations are also increasing, which makes some traditional load prediction methods less accurate than before. At the same time, the rise of artificial intelligence algorithms provides a more reliable and accurate solution for load forecasting [1]. In the past few decades, many short-term load forecasting schemes have been applied in practice, which can be roughly divided into three types: conventional statistical models, artificial intelligence models, and hybrid forecasting models.

Traditional statistical forecasting methods include regression analysis and time series analysis. Regression analysis achieves the purpose of prediction by determining the correlation between variables. Multiple linear regression has long been used for load forecasting, but ordinary linear prediction methods cannot involve complex nonlinear parts of power load. Therefore, Torkzadeh [2] combined multiple linear regression and nonlinear principal component analysis for load forecasting. Time series is the most commonly used analysis method for load forecasting. Its principle is to establish the functional relationship between variables and time by analyzing the autocorrelation, periodicity and fluctuation trend of load data, and then use this function to carry out load forecasting. Hagan [3] first proposed the application of time series analysis to short-term load forecasting in 1987. Huang and Shih [4] embedded a non-Gaussian process into autoregressive moving average model (ARMA) to improve the performance of the ARMA model in short-term load forecasting.

In order to better solve the nonlinear problem of time series, many scholars have applied some artificial intelligence algorithms to load forecasting, such as expert system, fuzzy logic, support vector machine, neural network and so on. Among these algorithms, neural network has better prediction performance. The neural network has the characteristics of self-adaptive, self-organizing and nonlinear mapping, which makes up the defects of other artificial intelligence algorithms in dealing with intuitive and unstructured information, and greatly accelerates the progress of load forecasting level in this field.

In the beginning, Artificial Neural Network (ANN) was used for short-term load forecasting. However, ANN cannot be modeled and trained based on the relationship between predictor variables and target variables. So Recurrent Neural Network (RNN) [5] was proposed to solve the problem. But RNN is prone to suffer from vanishing gradient and gradient explosion when handling the problem of long-term dependency. Therefore, Long Short-Term Memory (LSTM) was put forward by Hochreiter and Schmidhuber [6]. LSTM is a variant of RNN, it replaces ordinary hidden nodes with memory modules to ensure that the gradient does not disappear or explode after passing through many time steps to conquer some of the difficulties encountered in traditional recurrent neural network training. Kong [7] proved the superiority of LSTM over other algorithms in short-term load forecasting. RNN and LSTM can only predict the output of next time based on the information of previous. Sometimes, the output is not only related to previous state, but also be concerned with future information. As a result, Bidirectional RNN (BiRNN) and Bidirectional LSTM (BiLSTM) were proposed. Hu [8] used a hybrid model based on BiLSTM to predict the urban water demand. Experiments proved that the hybrid model has lower prediction error and shorter training time than other models.

In real situation, it is difficult to accurately fit the actual forecasting model by relying on a single neural network because of some complex influencing factors. In order to get a more accurate prediction, hybrid models were emerged. These models combine many advantages of different single models. Alhussein [9] extracted features of load data with CNN and input the feature vectors to LSTM for sequence learning. Pan [10] integrated the Temporal Attention Mechanism (TAM) into the LSTM model to further improve solar generation forecasting accuracy. Wang [11] employed CNN-BiLSTM with Attention Mechanism (AM) for predictive maintenance. In this model, AM was used to highlight the critical features after CNN-BiLSTM. Kim [12] combined Particle Swarm Optimization (PSO) with CNN-LSTM network for forecasting energy consumption. This model could determine different kinds of hyperparameters automatically by using PSO. He [1] introduced BO into the mixed model of LSTM and variational mode decomposition for hyperparameter optimization.

In addition to optimizing the structure of the model, the load forecasting accuracy can also be improved by analyzing relevant factors that affect power load. The main relevant factors of short-term load forecasting include meteorology, geography, time, date and historical load data. Shi [14] proposed to input time variable, meteorological factors of current and previous into model for short-term load forecasting. He also divided the load data by seasons to eliminate seasonal effects. Zhu et al. [15] selected similar historical load data as exogenous variable of input to avoid errors caused by mutations of other influence factors. The electricity price can also cause power load fluctuations because of the implementation of peak valley electricity price policy. A complete power system is composed of many independent users, and different users choose different times and ways of power usages, which leads to the random and uncertain fluctuation of power load.

In order to accelerate training speed, save training costs and improve performance of neural network model, the hyperparameters need to be tuned. Commonly used tuning methods are grid search and random search. The grid search and random search are very computationally intensive. In the case of limited calculation costs, the performance of the two methods may not be as good as the results obtained by modeling engineer through personal experience. Before choosing the next hyperparameter to evaluate, these two methods will not consider about previous training results, that is, they are completely unaffected by past evaluations. Consequently, the grid search and random search may spend a lot of time evaluating the wrong hyperparameters. In contrast, the BO algorithm will take into account past evaluation results to find better hyperparameters in shorter time. Pelikan [16] first proposed the BO algorithm in 1999. With deeping reserach, this effective algorithm has been applied in many fields.

The dominating contributions of this paper are as follows:

1. Using CNN-BiLSTM to replace LSTM.

Compared with the commonly used time series model LSTM, CNN-BiLSTM can not only grasp the main features of the input data, but also combine the forward and backward states to make better load prediction.

2. Integrating the Attention Mechanism to BiLSTM

For single BiLSTM, it is hard to obtain a reasonable vector representation when the input sequence is too long. Thus, this paper applied Attention Mechanism to selectively focus on the input sequence and associate it with the output sequence of BiLSTM.

3. Optimizing the hyperparameters by Bayesian Optimization

The Bayesian Optimization is more rapid and accurate in hyperparameter tuning than the traditional methods which previous load forecasting model used

This paper is organized as follows: Section 2 introduces the principles of CNN, LSTM, BiLSTM, AM, BO, and the structure of the hybrid model proposed by this paper. Section 3 proposes the four evaluation indexes of forecasting performance. Section 4 describes the data set, comparison benchmarks, and results and discussion of the case study. Section 5 draws the conclusions of this paper.

2 Methodologies

2.1 Convolutional Neural Networks (CNN)

Convolutional Neural Network (CNN) is a feed-forward neural network, which has excellent performance in image feature extraction and classification. The first CNN is the time delay network proposed by Alexander Waibel et al. in 1987. CNN consists of input, hidden and output layer. The hidden layer of CNN has three common structures, including convolutional layer, pooling layer and fully connected layer.

The model structure is shown in Fig. 1. Let $X_t, t = 1, 2, \ldots, n$ be the time series and $X_t \in R^k$ be the k-dimensional vector at time t. In the convolution layer, the filter is used to perform convolution operations in a window of width h to get a new feature c_t. The formula of convolution is as follows.

$$c_t = f(W X_{t:t+h-1} + b) \tag{1}$$

where $W \in R^{hk}$ is the weight term, $X_{t:t+h-1}$ refer to the t-th to $(t + h - 1)$-th time series, b is the bias term and f is an activation function. When a filter convolutes through the whole n time series with window h, $n - h + 1$ new features are generated to a feature map

$$c = [c_1, c_2, \ldots, c_{n-h+1}] \tag{2}$$

Then the feature map is input into the maxpooling layer to pick the max value. The role of the maxpooling layer is maintaining the most important features while reducing parameters and calculations, preventing overfitting, and improving model generalization ability.

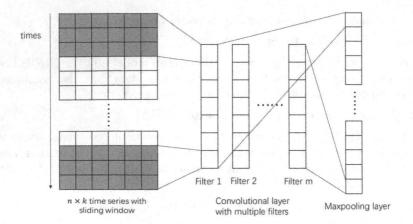

Fig. 1. CNN structure

2.2 Long Short-Term Memory (LSTM)

Long short-term memory (LSTM) is an enhanced model based on Recurrent Neural Network (RNN). The characteristic of LSTM is replacing ordinary hidden node with memory module to ensure that the gradient does not disappear or explode after passing through many time steps.

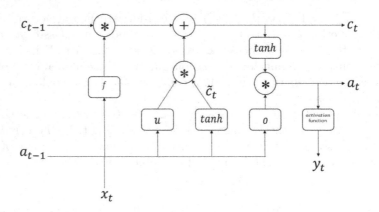

Fig. 2. LSTM structure

Figure 2 shows the architecture of LSTM, which is composed of input gate, forget gate(f), update gate(u) and output gate(o). The input of input gate are last active output value a_{t-1}, last memory cell output value c_{t-1} and time series value at the moment x_t. The output of LSTM is calculated by formula 3–9

$$F_f = \sigma(W_f[a_{t-1}, x_t] + b_f) \tag{3}$$

$$F_u = \sigma(W_u[a_{t-1}, x_t] + b_u) \tag{4}$$

$$\widetilde{c}_t = tanh(W_c[a_{t-1}, x_t] + b_c) \tag{5}$$

$$c_t = F_u \cdot \widetilde{c}_t + F_f \cdot c_{t-1} \tag{6}$$

$$F_o = \sigma(W_o[a_{t-1}, x_t] + b_o) \tag{7}$$

$$a_t = F_o \cdot tanh(c_t) \tag{8}$$

$$y_t = g(a_t) \tag{9}$$

where σ is the sigmoid function, $tanh$ is the tanh functions, W_f, W_u, W_c, W_o are the weight terms, b_f, b_u, b_c, b_o are the bias terms. g is one of the activation functions, such as sigmoid function, tanh function, ReLu function and so on.

2.3 Bidirectional LSTM (BiLSTM)

In some cases, the prediction may need to be determined jointly by the previous inputs and following inputs. However, LSTM model unable to encode information from back to front. Therefore, a Bidirectional LSTM was proposed. BiLSTM is composed of two LSTMs, one forward and one backward. The structure of BiLSTM is shown in the Fig. 3.

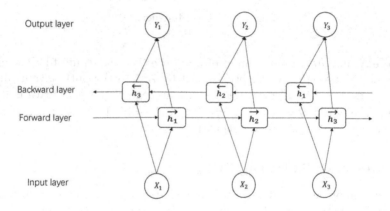

Fig. 3. BiLSTM structure

The output at the corresponding moment of the forward layer and the backward layer are combined to obtain the final output as follows.

$$\overrightarrow{h_t} = \sigma \overrightarrow{W} [\overrightarrow{h_{t-1}}, x_t] \tag{10}$$

$$\overleftarrow{h_t} = \sigma \overleftarrow{W} [\overleftarrow{h_{t-1}}, x_t] \tag{11}$$

$$Y_t = f(W[\overrightarrow{h_t}, \overleftarrow{h_t}]) \tag{12}$$

2.4   Attention Mechanism (AM)

AM is a resource allocation scheme that allocates computing resources to more important tasks and solves information overload problems when computing power is limited. Therefore, it can improve the efficiency and accuracy of model. The structure of AM-LSTM is shown in the Fig. 4.

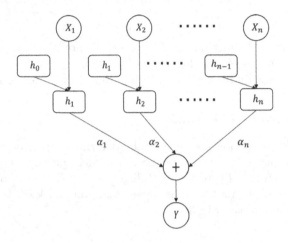

Fig. 4. AM-LSTM structure

Where X_t and h_{t-1} are the input of LSTM, h_t is the output of LSTM hidden layer, α_t is the probability weight assigned by AM to the output value of each hidden layer. $\alpha_t \in R^k$, α_t^p measures the importance of the p-th feature at time t. The updated h_t is $\tilde{h}_t = (\alpha_t^1 x_t^1, \alpha_t^2 x_t^2, \cdots, \alpha_t^k x_t^k)$. Finally, $\tilde{h}_t$ is passed to a fully connected layer to get the final output Y.

2.5   Bayesian Optimization (BO)

The main idea of Bayesian Optimization is to use Bayesian theorem to estimate the posterior distribution of the objective function, and then select the next sampling hyperparameter combination according to the distribution. It makes full use of the previous sampling point to learn the shape of objective function and find the global optimal hyperparameter. Let $X = x_1, x_2, \cdots, x_n$ be a set of hyperparameter combinations, $f(x)$ be a objective function related to the hyperparameter x. What BO do is finding $x \in X$ s.t. $x^* = \underset{x \in X}{\operatorname{argmax}} f(x)$.

3   Evaluation Indexes

In order to evaluate the forecasting performance, four evaluation indexes were used, including mean absolute error (MAE), root mean square error (RMSE), mean absolute percentage error (MAPE) and coefficient of determination (R^2 score). They can be expressed as follows:

$$MAE = \frac{1}{m}\sum_{i=1}^{m}|y_i - \hat{y}_i| \tag{13}$$

$$RMSE = \sqrt{\frac{1}{N}\sum_{i=1}^{m}(y_i - \hat{y}_i)^2} \tag{14}$$

$$MAPE = \frac{1}{m}\sum_{i=1}^{m}\frac{|y_i - \hat{y}_i|}{y_i}100\% \tag{15}$$

$$R^2 = 1 - \frac{\sum_{i=1}^{m}(y_i - \hat{y}_i)^2}{\sum_{i=1}^{m}(y_i - \bar{y}_i)^2} \tag{16}$$

Where m is the number of data, y_i is the actual value, $\bar{y}_i$ is the mean value of actual values, $\hat{y}_i$ is the forecasting value.

4   Experiments

4.1   Data Description

The experimental data of this research is the one-hour load data and some meteorological factors provided by a city in Hebei, China. The data set has 8760 records from Jan 1st to Dec 31th in one year, each record contains load, time slot, temperature, relative humidity, air pressure, wind speed, and precipitation. This paper divided the data set into four subsets (Jan 12th Mar 14th, Mar 12th May 22th, Jun 21th Aug 25th, Sep 25th Nov 22th) with respect to four seasons. The trend chart of the load data set at intervals of days is shown in Fig. 5.

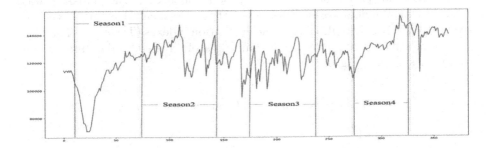

Fig. 5. Daily load of one year

When the correlation coefficient between load and an influencing factor is greater than 0.3, the factor will be selected. The significant factors of season 1 are time slot and temperature. For other seasons, the significant factors are time slot, temperature, and relative humidity. Besides, this paper employed two types of recent load for four seasons as the significant factors. One is the load data at the same moment in the previous and penultimate day. The other is the load data at the previous and penultimate moment in the previous day. Furthermore, in order to distinguish the peak and trough of electricity consumption during the day, we used 0, 1 to stand for trough and peak, respectively. For each season, the data set with above factors were all divided into the training set, validation set, and test set. The validation set were the first six of the last seven days, the test set were the last day, and the rest were the training set.

4.2   Results and Discussion

After data preprocessing, BO was used to tune the hyperparameters of forecasting model to minimize the loss function of the validation set. The hyperparameters of CNN-BiLSTM-AM model in 4 seasons and the minimize validation loss were displayed in Table 1. The filter1 and filter2 are the number of filters in the first and second convolutional layers. The BiLSTM unit and dense unit are the number of units in the BiLSTM layer and dense layer.

Table 1. The hyperparameters of CNN-LSTM-AM model in season 1–4

Seasons	Filter1	Filter2	BiLSTM unit	Dense unit	Validation loss	Test loss
1	128	128	480	480	0.000210	0.000241
2	145	145	512	512	0.000280	0.000294
3	136	136	526	526	0.000145	0.000185
4	148	148	490	490	0.000216	0.000220

The forecasting performance of the proposed model (CNN-BiLSTM-AM) was compared with Random Forest Regressor (RFR), Gradient Boosting Regressor (GBR), BiLSTM, CNN-BiLSTM by the evaluation indexes (Table 2, 3, 4 and 5). The comparison of fitted curves and actual values of different models from season 1 to 4 were shown in Fig. 6.

Table 2. The performance of different models for season 1

Models	MAE	RMSE	MAPE	R^2
RFR	77.359	100.177	1.449	0.976
GBR	77.614	95.756	1.469	0.978
BiLSTM	57.722	79.858	1.098	0.985
CNN-BiLSTM	56.926	75.879	1.073	0.986
CNN-BiLSTM-AM	50.172	67.947	0.946	0.989

Table 3. The performance of different models for season 2

Models	MAE	RMSE	MAPE	R^2
RFR	111.189	140.466	2.002	0.980
GBR	110.716	138.991	1.974	0.980
BiLSTM	102.526	131.461	1.764	0.982
CNN-BiLSTM	65.949	78.142	1.219	0.994
CNN-BiLSTM-AM	64.349	76.656	1.137	0.994

Table 4. The performance of different models for season 3

Models	MAE	RMSE	MAPE	R^2
RFR	56.971	77.797	1.079	0.988
GBR	72.681	87.912	1.373	0.985
BiLSTM	59.509	75.077	1.133	0.989
CNN-BiLSTM	55.206	76.065	1.014	0.989
CNN-BiLSTM-AM	46.393	59.394	0.864	0.993

Table 5. The performance of different models for season 4

Models	MAE	RMSE	MAPE	R^2
RFR	139.166	195.878	2.152	0.964
GBR	94.268	131.737	1.475	0.984
BiLSTM	66.291	88.028	1.054	0.993
CNN-BiLSTM	61.323	77.829	0.992	0.994
CNN-BiLSTM-AM	56.032	68.005	0.884	0.996

It can be seen from the above tables and pictures, the proposed model, CNN-BiLSTM-AM had the lowest values of MAE, RMSE, MAPE, and the highest R^2 score no matter in which season. From the results, a conclusion can be drawn that the proposed model fit the actual value best and have the best forecasting performance. Compared to the traditional machine learning (ADA, RFR and GBR), the deep learning is better at load prediction. Among these deep learning methods, the performance of a single model (BiLSTM) is not as good as the hybrid model (CNN-BiLSTM and CNN-BiLSMT-AM). Model with AM can further accurately predict the electricity load. Therefore, it can be inferred that the hybrid model proposed in this paper is effective.

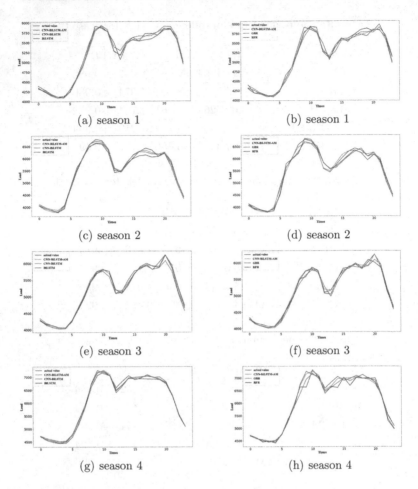

Fig. 6. Comparison of forecasting performance among hybrid model and other models for season 1–4

Since the load data belongs to time series, there is a long-term dependence problem in the load forecasting. This problem caused traditional machine learning methods to fail to learn useful information, and ultimately failed to obtain accurate load forecasting. BiLSTM solved this problem with its forget gate, so the forecasting performance of BiLSTM is better than ADA, RFR and GBR in the data set of this paper. Before inputting data to BiLSTM, CNN can disassemble the data to extract important features. So the CNN-BiLSTM model is better than the single BiLSTM model. Compared with other models, the CNN-BiLSTM-AM model is able to combine the previous load data with future related factors and focus on the most important part. Therefore, The CNN-BiLSTM-AM model had a best performance in load forecasting.

5   Conclusions

In this paper, a hybrid model, CNN-BiLSTM-AM, was proposed to forecast the future 24 h electricity load. The CNN layers were used to capture the data features to maxpooling layer to choose the maximum feature. Then the chosen features were passed to BiLSTM layer with AM to output the final load forecasting of next 24 h. BO was used to tune the hyperparameters of the model to get a better accuracy of forecasting. The five contrast models, ADA, RFR, GBR, BiLSTM, and CNN-BiLSTM were applied to the same data set with the proposed model in the experiments. And the forecasting results showed that the CNN-BiLSTM-AM model can combine the advantages of each layer to get the best load forecasting performance.

References

1. He, F., Zhou, J., Feng, Z., Liu, G., Yang, Y.: A hybrid short-term load forecasting model based on variational mode decomposition and long short-term memory networks considering relevant factors with Bayesian optimization algorithm. Appl. Energy **237**, 103–116 (2019)
2. Torkzadeh, R., Mirzaei, A., Mirjalili, M.M., Anaraki, A.S., Sehhati, M.R., Behdad, F.: Medium term load forecasting in distribution systems based on multi linear regression and principal component analysis: a novel approach. In: 19th Conference on Electrical Power Distribution Networks (EPDC), pp. 66–70. IEEE, Tehran (2014)
3. Hagan, M.T., Behr, S.M.: The time series approach to short term load forecasting. IEEE Trans. Power Syst. **2**(3), 785–791 (1987)
4. Huang, S.-J., Shih, K.-R.: Short-term load forecasting via ARMA model identification including non-Gaussian process considerations. IEEE Trans. Power Syst. **18**(2), 673–679 (2003)
5. Vermaak, J., Botha, E.C.: Recurrent neural networks for short-term load forecasting. IEEE Trans. Power Syst. **13**(1), 126–132 (1998)
6. Hochreiter, S., Schmidhuber, J.: Long short-term memory. Neural Comput. **9**(8), 1735–1780 (1997)
7. Kong, W., Dong, Z.Y., Jia, Y., Hill, D.J., Xu, Y., Zhang, Y.: Short-term residential load forecasting based on LSTM recurrent neural network. IEEE Trans. Smart Grid **10**(1), 841–851 (2019)
8. Hu, P., Tong, J., Wang, J., et al.: A hybrid model based on CNN and Bi-LSTM for urban water demand prediction. In: Congress on Evolutionary Computation (CEC), Wellington, New Zealand, pp. 1088–1094. IEEE (2019)
9. Alhussein, M., Aurangzeb, K., Haider, S.I.: Hybrid CNN-LSTM model for short-term individual household load forecasting. IEEE Access **8**, 180544–180557 (2020)
10. Pan, C., Tan, J., Feng, D., Li, Y.: Very short-term solar generation forecasting based on LSTM with temporal attention mechanism. In: 5th International Conference on Computer and Communications (ICCC), Chengdu, China, pp. 267–271. IEEE (2019)
11. Wang, M., Cheng, J., Zhai, H.: Life prediction for machinery components based on CNN-BiLSTM network and attention model. In: 5th Information Technology and Mechatronics Engineering Conference (ITOEC), Chongqing, China, pp. 851–855. IEEE (2020)

12. Kim, T., Cho, S.: Particle swarm optimization-based CNN-LSTM networks for forecasting energy consumption. In: Congress on Evolutionary Computation (CEC), Wellington, New Zealand, pp. 1510–1516. IEEE (2019)

13. Lu, J., Zhang, Q., Yang, Z., et al.: A hybrid model based on convolutional neural network and long short-term memory for short-term load forecasting. In: Power and Energy Society General Meeting (PESGM), Atlanta, GA, USA, pp. 1–5. IEEE (2019)

14. Niu, D.X., Shi, H.F., Wu, D.D.: Short-term load forecasting using Bayesian neural networks learned by Hybrid Monte Carlo algorithm. Appl. Soft Comput. **12**(6), 1822–1827 (2012)

15. He, Y.J., Zhu, Y.C., Gu, J.C., et al.: Similar day selecting based neural network model and its application in short-term load forecasting. In: International Conference on Machine Learning and Cybernetics, Guangzhou, China, pp. 4760–4763, IEEE (2005)

16. Pelikan, M., Goldberg, D.E., Cantu-Paz, E.: BOA: The Bayesian optimization algorithm. In: Proceedings of the Genetic and Evolutionary Computation Conference (GECCO 1999), pp. 525–532 (1999)

On the Non-ergodic Convergence Rate of the Directed Nonsmooth Composite Optimization

Yichuan Dong[1], Zhuoxu Cui[2(✉)], Yong Zhang[3], and Shengzhong Feng[1]

[1] National Supercomputing Center in Shenzhen, Guangzhou, China
{dongyc,fengsz}@nsccsz.cn
[2] Wuhan University, Wuhan, China
zhuoxucui@whu.edu.cn
[3] Shenzhen Institute of Advanced Technology, Chinese Academy of Sciences, Shenzhen, People's Republic of China
zhangyong@siat.ac.cn

Abstract. This paper considers the distributed "nonsmooth+ nonsmooth" composite optimization problems for which n agents collaboratively minimize the sum of their local objective functions over the directed networks. In particular, we focus on the scenarios where the sought solutions are desired to possess some structural properties, e g., sparsity. However, to ensure the convergence, most existing methods produce an ergodic solution via the averaging schemes as the output, which causes the desired structural properties of the output to be destroyed. To address this issue, we develop a new decentralized stochastic proximal gradient method, termed DSPG, in which the nonergodic (last) iteration acts as the output. We also show that the DSPG method achieves the nonergodic convergence rate $O(\log(T)/\sqrt{T})$ for generally convex objective functions and $O(\log(T)/T)$ for strongly convex objective functions. When the structure-enhancing regularization is absent and the simple and suffix averaging schemes are used, the convergence rates of DSPG reach $O(1/\sqrt{T})$ for generally convex objective functions and $O(1/T)$ for strongly convex objective functions, showing improvement relative to the rates $O(\log(T)/\sqrt{T})$ and $O(\log(T)/T)$ provided by the existing methods. Simulation examples further illustrate the effectiveness of the proposed method.

Keywords: Nonergodic convergence distributed nonsmooth optimization · Directed graphs · Multiagent networks · Distributed nonsmooth optimization

Supported by National Supercomputing Center in Shenzhen.

Electronic supplementary material The online version of this chapter (https:// doi.org/10.1007/978-3-030-69244-5_11) contains supplementary material, which is available to authorized users.

© Springer Nature Switzerland AG 2021
Y. Zhang et al. (Eds.): PDCAT 2020, LNCS 12606, pp. 129–140, 2021.
https://doi.org/10.1007/978-3-030-69244-5_11

1   Introduction

In recent years, there has been a growing interest in distributed optimization (DO) due to its wide range of applications such as sensor networks [1], machine learning [18,19], signal processing [20], microgrids [21] and brain science [22]. For such problems, DO aims to design a decentralized algorithm in which the computation task is allocated to multiple agents, and these agents can only exchange information with their immediate neighbors. It is natural to handle this kind of problems from the viewpoint of graph theory, where the agents are the points and the connection between them are the edges. The core problem of DO is to realize the decentralized optimization where the network agents cooperatively solve the optimization problem by optimizing their own (local) objective functions; i.e., the network agents act in the absence of the whole picture of the true (global) objective function to be optimized [4,23,24].

In this work, we aim to design a decentralized algorithm for solving the following composite optimization problem:

$$\min_{x \in X} F(x) := \sum_{i=1}^{n} [f_i(x) + r_i(x)] \tag{1}$$

where X is a convex compact set in the Euclidean space $\mathbb{R}^d$, $f_i : X \to \mathbb{R}$ are convex but possibly nonsmooth functions that often are taken as the average of the loss functions in machine learning, and $r_i : X \to \mathbb{R}$ are convex and nonsmooth regularizers, e.g., sparsity promoting ℓ_1-norm [16].

We assume that there exists a collection of n agents connected by a network defined by a directed graph $\mathcal{G} = \{\mathcal{V}, \mathcal{E}(t)\}$ with $|\mathcal{V}| = n$ vertices where $\mathcal{E}(t)$ denotes the union of all of the edges in graph $\mathcal{G}$ in round $t > 0$. Each agent $i \in \mathcal{V}$ can only obtain the information about its local functions f_i and r_i privately and can communicate with its immediate neighbors.

Problem (1) covers a wide range of optimization problems frequently encountered in machine learning. For example, given a sequences of examples $\{(a_i, b_i) \in \mathbb{R}^d \times \mathbb{R}\}_{i=1}^{k}$ and allocated into n parts uniformly with index set $\{\Omega_i\}_{i=1}^{n}$, if the ith agent holds examples $\{(a_j, b_j)\}_{j \in \Omega_i}$ privately and chooses $f_i(x) = \frac{1}{|\Omega_i|} \sum_{j \in \Omega_i} \max\{0, 1 - b_j \langle x, a_j \rangle\}$ and $r_i(x) = \lambda \|x\|^2$ with $\lambda > 0$, then Problem (1) reduces to the distributed support vector machine (SVM) [17]; otherwise, if we let $f_i(x) = \frac{1}{|\Omega_i|} \sum_{j \in \Omega_i} (\langle x, a_j \rangle - b_j)^2$ and $r_i(x) = \lambda \|x\|_1$, then we obtain the distributed Lasso [16].

Formally, in SGD-PS, the ith agent sequentially queries the stochastic first-order oracle ($\mathcal{SFO}$) and obtains an unbiased stochastic gradient $G_i(z_{t,i}, \xi_{t,i})$ of f_i at $z_{t,i}$, where $z_{t,i}$ is a weighted combination of $\{x_{t,i}\}_{i=1}^{n}$ and $\xi_{t,i}$ is a random variable, and then updates $x_{t+1,i}$ iteratively as follows:

$$x_{t+1,i} = \sum_{j=1}^{n} w_{i,j}(t) x_{t,j} - \eta_t \left(G_i(z_{t,i}, \xi_{t,i}) + \nabla r_i(z_{t,i}) \right) \tag{2}$$

where $w_{i,j}(t)$ is the weight of the underlying graph from agent j transported to i at round t, η_t is the stepsize and $\nabla r_i(z_{t,i})$ denotes one of the subgradients of r_i at $z_{t,i}$.

In this paper, we first propose a DSPG algorithm in which the computational task is allocated into multiple agents to solve the "nonsmooth+nonsmooth" optimization problems (1). Further, in the case where the structural-enhancing regularization is considered, we investigate the nonergodic convergence rate for the proposed DSPG. When the structural properties are not that important, we show that averaging schemes can accelerate the rate of DSPG. Specifically, the main contributions of the paper are summarized as follows:

- Considering the case where f_i and r_i are all convex but possibly nonsmooth, we propose a distributed SPG-based method, termed DSPG, for solving problem (1). Inheriting the advantages of the SPG, the DSPG can exploit the problem's structure well and simultaneously implement parallel computing. Therefore, we address "nonsmooth+nonsmooth" composite optimization problems, e.g., ℓ_1-regularized SVM, which do not appear to be handled well by the existing distributed methods.
- In the case where the structural-enhancing regularization is considered, to share the specific structure of the sought solutions, we investigate the nonergodic convergence analysis for DSPG and prove that it achieves the nonergodic rate $O(\log(T)/\sqrt{T})$ for generally convex objective functions and the rate of $O(\log(T)/T)$ for strongly convex objective functions.
- When the structural properties are not that important, combined with the simple and suffix averaging scheme, the convergence rates of DSPG achieve $O(1/\sqrt{T})$ for generally convex objective functions and $O(1/T)$ for strongly convex objective functions, respectively, improving on the rates of $O(\log(T)/\sqrt{T})$ and $O(\log(T)/T)$ reported for the existing algorithms. To the best of our knowledge, without considering the scalars, these two rates are the optimal rates for the stochastic first-order methods under the generally and strongly convex assumptions, respectively.

We also numerically evaluate our algorithm through a series of experiments on a number of benchmark datasets; these results are consistent with our theoretical findings.

2   DSPG Development

In this section, we consider the case where the agents jointly solve the problem (1) over the directed graphs. We first describe the DSPG algorithm, namely, we provide an interpretation for it step by step and then give the formal convergence results.

2.1   DSPG Algorithm

We now begin to consider the DSPG over directed graphs. We first review the common distributed optimization methods, realized by a communication protocol that allows the nodes to compute averages and other aggregates in the network with directed communication links over directed graphs. Briefly, the common distributed optimization methods for solving Problem (1) can be mainly realized by repeating the following steps 2)–3):

1) Let each agent i obtain an approximate copy $x_i \in X$ of the real vector $x \in X$;
2) For each agent i, communicating with its neighborhood, agent i obtains the weighted average, denoted as y_i;
3) Starting from y_i, each agent i makes a step forward in the direction that decreases its local object function $f_i + r_i$.

Algorithm 1. Distributed Stochastic Proximal Gradient.

1: **Input:** Given the initial point $x_{0,i} = 0$, $\forall i \in [n]$, iteration limit T and stepsize $\{\eta_t\}_{t=1}^{T}$;
2: Set for every $i \in [n]$,
3: **for** $t = 1, 2, \ldots, T$ **do**
4:     $y_{t,i} = \sum_{j \in \mathcal{N}_i^{in}(t-1)} w_{i,j}(t) x_{t-1,j}$
5:     $x_{t,i} = \arg\min_{u \in X}\{\langle G_i(x_{t-1,i}, \xi_{t,i}), u\rangle + \frac{1}{2\eta_t}\|u - y_{t,i}\|^2 + r_i(u)\}$
6: **end for**
7: Sample s from $[n]$ uniformly;
8: **Output:** $x_{T,s}$.

For step 3), such as in GD-PS, a subgradient descent step is usually taken to decrease $f_i + r_i$. Therefore, we can lump r_i into f_i and treat them as an unregularized problem, which indicates that GD-PS is not effective for the problem with structural-enhancing regularization. Compared with the subgradient decent, the proximal gradient method may possess some merits, such as a) it can exploit the problem's structure well [9]; and b) there are more iterations located at the points of nondifferentiability that are often the true minima of the objective function [10]. Meanwhile, we consider the case where the exact subgradients of f_i are not available. Thus, we can only access the stochastic proximal gradient (SPG) method. Formally, based on the SPG, we give the following distributed optimization algorithm: In Algorithm 1, the $4th$ step corresponds to 2) mentioned above, and $w_{i,j}(t)$ is the weight from agent j to i at time t. Here, the weight matrix $W(t)$ is defined as

$$w_{i,j}(t) = [W(t)]_{i,j} := \begin{cases} 1 - \alpha_i(t)|\mathcal{N}_i^{out}(t)|, & \text{if } i \in \mathcal{N}_i^{in}(t) \\ \alpha_j(t), & \text{if } i \neq j \in \mathcal{N}_i^{in}(t) \\ 0, & \text{otherwise} \end{cases} \tag{3}$$

where $\alpha_i(0)$ is initialized with $1/n$, $\forall i \in [n]$, and is updated with the following rule

$$\alpha_i(t+1) := \frac{1}{2}\alpha_i(t) + \frac{1}{|\mathcal{N}_i^{out}(t)|} \sum_{j \in \mathcal{N}_i^{in}(t)} \frac{1}{2}\alpha_j(t)$$

which is deduced by the balancing weights rule

$$\alpha_i(t)|\mathcal{N}_i^{out}(t)| = \sum_{j \in \mathcal{N}_i^{in}(t)} \alpha_j(t). \tag{4}$$

From Proposition 1 of [8], we know that $W(t)$ is a column stochastic matrix; i.e., $\sum_{j=1}^{n}[W(t)]_{i,j} = 1$, $\forall i \in [n]$. The 5th step is to implement the SPG method for each agent. Note that here, in Algorithm 1, the output is chosen as the nonergodic (last) iteration of a certain agent that is conducive to the preservation of the iterations' specific structure, e.g., sparsity.

2.2 Theoretical Aspects of DSPG

In this section, we will give the formal convergence results of Algorithm 1 under strongly and generally convex assumptions, respectively. Note that in the subsequent contexts, we use $\overline{x}$ to denote the simple average of $(x_1, \ldots, x_n)$, i.e., $\overline{x} := \frac{1}{n} \sum_{i=1}^{n} x_i$. Now, let us first derive the convergence rate of Algorithm 1 under strongly convex assumption.

Theorem 21. *Suppose that f_i are σ_i-strongly convex, $\forall i \in [n]$, graph $\mathcal{G}(t)$ is strongly connected and Assumptions (A1)–(A4) hold. Considering Algorithm 1 with dynamic stepsize $\eta_t = \frac{n}{2t \sum_{i=1}^{n} \sigma_i}$, it holds that*

$$
\begin{aligned}
\mathbb{E}[F(x_{T,s})] - F(x^*) \\
\leq \frac{2D_2(2\log(2) + 8)\sigma}{T} + \frac{16(1 + \log((T+1)/2))D_2\sigma}{T} \\
+ \frac{D_1\sigma(1 + \log(T))}{T} + \frac{4K\overline{C}}{T(1-\theta)\sigma}
\end{aligned}
$$

for any $s \in [n]$ and $T > 4T_0 + 4$, where x^* is the optimal solution of F on X, T_0 is the minimum positive integer such that $\theta^{\lfloor T_0/2 \rfloor} \leq \frac{1}{\lfloor T_0/2 \rfloor + 1}$, $K > 0$, $\theta \in (0, 1)$ introduced in Lemma 31, $D_2 = \max(\|\overline{x}_{T_0} - x^*\|, 12D_1)$ and

$$
D_1 := \frac{4n(1-\theta)\widetilde{C} + (1-\theta)\overline{C} + 24Kn\overline{C}}{(1-\theta)\sigma^2} \tag{5}
$$

with $\widetilde{C} = \sum_{i=1}^{n}(B_i + B_{f_i} + B_{r_i})^2$, $\overline{C} = [\sum_{i=1}^{n}(B_i + B_{f_i} + B_{r_i})]^2$ and $\sigma := \sum_{i=1}^{n} \sigma_i$.

Next, we consider the convergence rate of Algorithm 1 with ergodic output. By using the suffix averaging scheme (6) introduced in [5,6], first we obtain the following result:

Theorem 22. *Suppose that f_i are σ_i-strongly convex, $\forall i \in [n]$, graph $\mathcal{G}(t)$ is strongly connected and Assumptions (A1)–(A4) hold. Considering Algorithm 1 with dynamic stepsize $\eta_t = \frac{n}{2t \sum_{i=1}^{n} \sigma_i}$ and suffix averaging scheme, i.e.,*

$$
\widehat{x}_T = \frac{1}{\lfloor T/2 \rfloor + 1} \sum_{t=T-\lfloor T/2 \rfloor}^{T} \overline{x}_t, \tag{6}
$$

it holds that

$$
\mathbb{E}[F(\widehat{x}_T)] - F(x^*) \leq \frac{2D_2(8 + 2\log(2)) \sum_{i=1}^{n} \sigma_i}{T}
$$

for any $T > 4T_0 + 4$, where x^* is the optimal solution of F on X, T_0 is the minimum positive integer such that $\theta^{\lfloor T_0/2 \rfloor} \leq \frac{1}{\lfloor T_0/2 \rfloor + 1}$, $K > 0$, $\theta \in (0,1)$ introduced in Lemma 31, $D_2 = \max(\|\bar{x}_{T_0} - x^*\|, 12D_1)$ and D_1 is given in (5).

In the above theorem, the convergence rate has been enhanced from $O(\log(T)/T)$, obtained by the existing methods, to $O(1/T)$ for strongly convex objective functions, matching the near optimal convergence rate of the stochastic first-order methods under strongly convex assumption. More importantly, this result reveals that although the fusion center is avoided in decentralized methods, the optimal convergence rate can also be theoretically achieved.

3   Proof Sketch

In this section, we briefly sketch the main ideas of the proof of the results described in Sect. 2. We argue that Algorithm 1 can achieve nonergodic convergence rate $O(\log(T)/\sqrt{T})$ for generally convex functions and $O(\log(T)/T)$ for strongly convex functions; further, based on the averaging schemes, the convergence rate can be enhanced to $O(1/\sqrt{T})$ and $O(1/T)$, respectively.

First, let us define the following matrix, and show a useful result with respect to it.

$$W(t:s) := W(t)W(t-1)\cdots W(s)$$

where $W(t)$ is the weighted matrix defined in (3). For the above matrix, the following lemma holds.

Lemma 31 *[Lemma 3 in [8]]. Let the graph $\{\mathcal{G}(t)\}$ be strongly connected. It holds that*

$$\left| [W(t:s)]_{i,j} - \frac{1}{n} \right| \leq K\theta^{t-s+1} \tag{7}$$

for any s with $t \geq s$, where K is a positive constant and $\theta \in (0,1)$.

Then, let us give some useful lemmas as follows:

Lemma 32. *Suppose that f_i are σ_i-strongly convex, $\forall i \in [n]$, graph $\mathcal{G}(t)$ is strongly connected and Assumptions (A1)–(A4) hold. Considering Algorithm 1 with dynamic stepsize $\eta_t = \frac{n}{2t\sum_{i=1}^{n}\sigma_i}$, we have*

$$\mathbb{E}[\|\bar{x}_T - x^*\|^2] \leq \frac{\max(\|\bar{x}_{T_0} - x^*\|, 12D_1)}{T - T_0}. \tag{8}$$

for any $T > T_0$, where x^* is the optimal solution of F on X, T_0 is the minimum integer such that $\theta^{\lfloor T_0/2 \rfloor} \leq \frac{1}{\lfloor T_0/2 \rfloor + 1}$, $K > 0$, $\theta \in (0,1)$ introduced in Lemma 31, and D_1 given in (5).

Lemma 32 shows that the averaging of iterations $\{\bar{x}_t\}$ converges to the optimal solution of F with sublinear rate under the expectation.

Lemma 33. *Suppose that f_i are σ_i-strongly convex, $\forall i \in [n]$, graph $\mathcal{G}(t)$ is strongly connected and Assumptions (A1)–(A4) hold. Considering Algorithm 1 with dynamic stepsize $\eta_t = \frac{n}{2t \sum_{i=1}^n \sigma_i}$ and defining $\Gamma_k := \frac{1}{k+1} \sum_{t=T-k}^{T} F(\overline{x}_t)$, it holds that*

$$\mathbb{E}[\Gamma_k - F(x^*)] \leq \frac{D_2 \sum_{i=1}^n \sigma_i}{(k+1)} \left(8 + \log \left(\frac{T-1}{T-k-2} \right) \right)$$

for any $k \in \{1, \dots, \lfloor T/2 \rfloor\}$ and $T > 4T_0 + 4$, where x^* is the optimal solution of F on X, T_0 is the minimum positive integer such that $\theta^{\lfloor T_0/2 \rfloor} \leq \frac{1}{\lfloor T_0/2 \rfloor + 1}$, $\theta \in (0, 1)$ introduced in Lemma 31, $D_2 = \max(\|\overline{x}_{T_0} - x^*\|, 12D_1)$ and D_1 given in (5).

The proofs for Lemmas 32 and 33 are shown in supplementary materials.

4   Experiments

In this section, we test the proposed DSPG, and compared it with SGD-PS, to solve the SVM problems over the time-varying directed graph, which contains 30 nodes (corresponding to 30 agents) and the topology of directed graph in Fig. 1. Meanwhile, to find a sparse solution, we consider the elastic net regularization, i.e., $r_i(x) = \lambda_1 \|x\|_1 + \frac{\lambda_2}{2} \|x\|^2$, which is deemed to be more robust than ℓ_1-norm. Given a sequences of examples downloaded from the LIBSVM[1] and UCI data repository[2] (please see Table 1 for details), we divide them into two sets, with the first set used for training and the other for testing. For the training set, we uniformly allocate the examples into 30 parts and adopt the elastic net regularized hinge loss on the ith part as the ith agent's objective function.

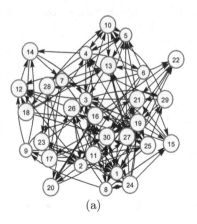

(a)

Fig. 1. Time-varying directed graphs.

[1] https://www.csie.ntu.edu.tw/~cjlin/libsvm/.
[2] http://archive.ics.uci.edu/ml/datasets.html.

Table 1. Data sets used in the experiments.

Dataset	#Samples	#Features
rcvl-2	20242	47236
real-sim	72309	20958
mnist	10000	657
connect-4	67557	126
mushrooms	8124	112

Unless noted otherwise, we will adopt 3-fold validation to tune the regularization parameters λ_1 and λ_2. With the tuned parameters, the algorithms are run on the randomly permuted training examples and then are tested on the test sets. To reduce statistical variability, the results are averaged over 10 permuted runs. Average classification error (ACE) (%, mean $\pm$ std) is used to evaluate the performance. All algorithms are initialized at $x_{0,i} = 0$, $\forall i \in [n]$ and no projections are performed; i.e., X is the whole space $\mathbb{R}^d$, as also adopted in [5,6]. For different experimental settings, three output schemes (please see Table 2 for details) are considered for SGD-PS and DSPG. The first and fourth rows of the Table 2 are the nonergodic outputting schemes. SGD-PS+ and SGD-PS++ are discussed in [7], respectively, and DSPG+ and DSPG++ are discussed in the previous sections. In addition to the training loss as discussed in [7], we examine the test error of all of the algorithms with different training set size N_{Tr}, namely, $N_{Tr} = \{0.1N, 0.2N, 0.3N, 0.4N, 0.5N, 0.6N\}$, where N denotes the total number of the examples of a dataset. At last, we choose the maximum iteration $T = 2000$ and "batch size" $= 1$ to compute the stochastic gradients of the loss functions.

Table 2. Outputting schemes for the experiments.

Schemes	Outputs
DSPG	$\widehat{x}_t = x_{t,i}$
DSPG+	$\widehat{x}_t = \frac{1}{t}\overline{x}_t + \left(1 - \frac{1}{t}\right)\widehat{x}_{t-1}$
DSPG++	$\widehat{x}_t = \frac{1}{\lfloor t/2 \rfloor + 1}\sum_{s=t-\lfloor t/2 \rfloor}^{t} \overline{x}_s$
SGD-PS	$\widehat{x}_t = x_{t,i}$
SGD-PS+	$\widehat{x}_t = (\eta_t z_{t,i} + \phi(t)\widehat{x}_{t-1})/\phi(t+1)$, $\phi(t) = \sum_{s=1}^{t} \eta_t$
SGD-PS++	$\widehat{x}_t = (t z_{t,i} + \phi(t)\widehat{x}_{t-1})/\phi(t+1)$, $\phi(t) = t(t-1)/2$

4.1 DSPG vs. SGD-PS Under the Strongly Convex Condition

First, we compare the behaviors of the proposed DSPG and SGD-PS with different outputting schemes under the strongly convex condition. As mentioned above, we consider the elastic net regularized hinge loss, which means that the objective function for each agent is λ_2-strongly convex. Therefore, we choose $\frac{1}{2\lambda_2 t}$ as the stepsize, which is determined by cross-validation.

The first row of Fig. 2 shows how DSPG and SGD-PS perform using the non-ergodic and ergodic (DSPG++ and SGD-PS++ showed in Table 2) outputting schemes. We can see that the nonergodic DSPG can achieve the same or even slightly better performance with respect to ACE than the nonergodic SGD-PS while using far fewer attributes of the data (see Table 3). From the viewpoint of avoiding data redundancy and saving storage space in big data, this approach is reasonable. With respect to the stability, we find that our proposed nonergodic DSPG performs better than nonergodic SGD-PS, especially on data sets connect-4 and MNIST, confirming our previous claim; i.e., the iterations produced by the SPG-based methods are relatively more rich at the nondifferentiable minimum point. Meanwhile, DSPG++ achieves consistently better performance over all the five data sets used in the experiments, which strongly confirms our theoretical analysis.

Table 3. Average support of the outputs of DSPG and SGD-PS with the training set size $N_{Tr} = 0.6N$.

Methods	rcv1-2	real-sim	connect-4	MNIST	mushrooms
DSPG	6062	6203	95	488	88
DSPG++	32561	20414	125	604	112
SGD-PS	31239	20601	126	613	112
SGD-PS++	31313	20598	126	617	112

The second row of Fig. 2 presents data to verify whether the suffix averaging scheme can accelerate the convergence rate in practice. For fair comparison, for the parameter values, we choose the average values of the parameters of DSPG and DSPG++ determined by cross-validation, so that they can hold the same objective function. We can see that the objective function value of DSPG++ is far lower than that of DSPG after only one iteration over all of the five data sets, in agreement with the theoretical analysis.

4.2 DSPG vs. SGD-PS Under the Generally Convex Condition

In this subsection, we compare the behaviors of the ergodic outputs of the SGD-PS and SPG-PS algorithms using the averaging schemes (please see Table 2 for details). Based on the previous results, we know that using the averaging schemes, we can enhance the convergence rate from $O(\log(T)/\sqrt{T})$ (SGD-PS+) to $O(1/\sqrt{T})$ (SPG-PS+) for arbitrary convex functions and from $O(\log(T)/T)$ (SGD-PS++) to $O(1/T)$ (SPG-PS++) for strongly convex functions.

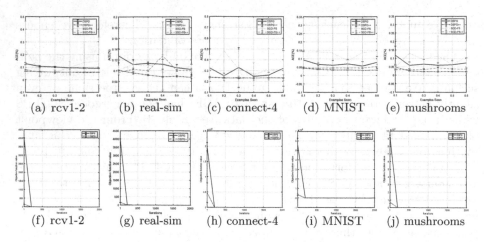

Fig. 2. Performance characteristics of DSPG and SGD-PS under strongly convex condition.

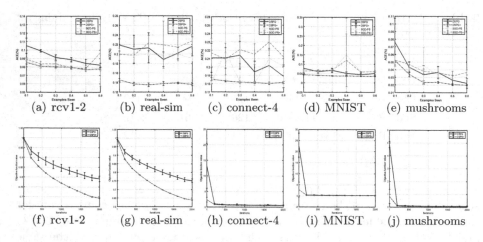

Fig. 3. Performance characteristics of DSPG and SGD-PS under generally convex condition.

Note here that we choose $\lambda = 0$ throughout this subsection and choose $\gamma = 0$ for the arbitrary convex case and $\gamma = 0.1$ for the strongly convex case. We can see from Fig. 3 that the function values of SPG-PS+ and SPG-PS++ are all decreasing faster than those for SGD-PS+ and SGD-PS++. This result is also in accordance with the theory.

5 Conclusion

In this paper, we focused on solving the nonsmooth composite optimization problems for directed graphs. To preserve some structural properties of the sought solution, we proposed a SPG-based decentralized method, termed DSPG. We also presented its nonergodic convergence rates under both generally and strongly convex assumptions. When the structural properties are not that important, with the averaging schemes, we proved that our proposed DSPG can reach the near optimal rates under both generally and strongly convex assumptions. Finally, we also provided a numerical example to illustrate the superior performance of DSPG.

There are several extensions for this workfirst of all, when the objective functions are time-varying, it would be interesting to extend our methods to the online version; secondly, it would be also interesting to consider the case where all agents are allowed to work asynchronously, for which more iterations can be executed by the fast agents, reducing the idle time; thirdly, the results of this work are based on the assumption that the objective functions are convex. However, the successful development of nonconvex models such as deep neural networks [3,25] motivates the development of efficient methods for nonconvex optimization. Thus, it is meaningful to extend the results to nonconvex cases.

References

1. Lesser, V., Ortiz, C.L., Tambe, M.: Distributed Sensor Networks: A Multiagent Prespective. Kluwer, Norwell (2003)
2. Xie, S., Guo, L.: Analysis of distributed adaptive filters based on diffusion strategies over sensor networks. IEEE Trans. Autom. Control 6(3), 3643–3658 (2018)
3. LeCun, Y., Bengio, Y., Hinton, G.: Deep learning. Nature 5(21), 436–441 (2015)
4. Author, F.: Article title. Journal 2(5), 99–110 (2003)
5. Rakhlin, A., Shamir, O., Sridharan, K.: Making gradient descent optimal for strongly convex stochastic optimization. In: Proceedings of the International Conference on Machine Learning, vol. 2, no. 5, pp. 1571–1578 (2012)
6. Shamir, O., Zhang, T.: Stochastic gradient for non-smooth optimization: Convergence results and optimal averaging schemes. In: Proceedings of the International Conference on Machine Learning, vol. 2, no. 1, pp. 71–79 (2013)
7. Nedic, A., Olshevsky, A.: Stochastic gradient-push for strongly convex function on time-varying directed graphs. IEEE Trans. Autom. Control 6(1), 3936–3947 (2016)
8. Makhdoumi, A., Ozdaglar, A.E.: Graph balancing for distributed subgradient methods over directed graphs. In: Proceedings of the International Conference on Machine Learning, vol. 2, pp. 71–79 (2013)
9. Xiao, L.: Dual averaging method for regularized stochastic learning and online optimization. J. Mach. Learn. Res. 1, 99–110 (2003)
10. Duchi, J.C., Singer, Y.: Efficient online and batch learning using forward backward splitting. J. Mach. Learn. Res. 1, 2899–2934 (2009)
11. Parikh, N., Boyd, S.: Proximal algorithms. Foundations Trends Optim. 1(3), 127–239 (2013)

12. Ghadimi, S., Lan, G., Zhang, H.: Mini-batch stochastic approximation methods for nonconvex stochastics composite optimization. Math. Program. **1**(55), 267–305 (2016)
13. Kempe, D., Dobra, A., Gehrke, J.: Gossip-based computation of aggregate information. In: Proceedings of IEEE Symposium on Foundations of Computer Science, vol. 4, no. 4, pp. 482–491 (2003)
14. Kushner, H.J., Yin, G.: Stochastic Approximation and Recursive Algorithms and Applications, vol. 3, pp. 99–110. Springer, Cham (2003)
15. Bottou, L.: Stochastic gradient descent tricks. In: Montavon, G., Orr, G.B., Müller, K.-R. (eds.) Neural Networks: Tricks of the Trade. LNCS, vol. 7700, pp. 421–436. Springer, Heidelberg (2012). https://doi.org/10.1007/978-3-642-35289-8_25
16. Tibshirani, R.: Regression shrinkage and selection via the lasso. J. Roy. Stat. Soc. B **5**(8), 267–288 (1996)
17. Hearst, M.A., Dumais, S.T., Osman, E., Platt, J., Scholkopf, B.: Support vector machines. IEEE Intell. Syst. Their Appl. **1**(3), 18–28 (1998)
18. Hong, M., Chang, T.H.: Stochastic proximal gradient consensus over random networks. IEEE Trans. Signal Process. **6**(5), 2933–2948 (2017)
19. Scaman, K., Bach, F., Bubeck, S., Yin, T.L., Massoulie, L.: Optimal algorithms for non-smooth distributed optimization in networks. In: Proceedings of the Advances in Neural Information Processing Systems, vol. 2, no. 5, pp. 2745–2754 (2018)
20. Xue, D., Hirche, S.: Distributed topology manipulation to control epidemic spreading over networks. IEEE Trans. Signal Process. **6**(7), 1163–1174 (2019)
21. Lu, X., Lai, J., Yu, X., Wang, Y., Guerrero, J.M.: Distributed coordination of islanded microgrid clusters using a two-layer intermittent communication network. IEEE Trans. Ind. Inform. J. **1**(4), 3956–3969 (2018)
22. Smith, G.B., Hein, B., Whitney, D.E., Fitzpatrick, D.: Distributed network interactions and their emergence in developing neocortex. Nat. Neurosci. **2**(1), 1600–1608 (2018)
23. Yu, D., Zou, Y., Yu, J., Dressler, F., Lau, F.C.: Implementation abstract MAC Layer in Dynamic Dressler. IEEE Trans. Mobile Comput. **1599** (2020). https://doi.org/10.1109/TMC.2020.297
24. Yu, D., Zou, Y., Yu, J., Dressler, F., Lau, F.C.: Stable local broadcast in Multihop wireless networks under SINR. IEEE/ACM Trans. Netw. J. **2**(6), 1278–1291 (2018)
25. Cai, Z., Zheng, X.: A private and efficient mechanism for data uploading in smart cyber physical systems. IEEE Trans. Netw. Sci. Eng. (TNSE) J. **7**(2), 766–775 (2020)

6D Pose Estimation Based on the Adaptive Weight of RGB-D Feature

Gengshen Zhang[1,2], Li Ning[1,2(✉)], and Liangbing Feng[3]

[1] Shenzhen Institutes of Advanced Technology, Chinese Academy of Sciences, Beijing, China
li.ning@siat.ac.cn
[2] University of Chinese Academy of Sciences, Beijing, China
[3] Shenzhen CosmosVision Technology Co., LTD., Shenzhen, China

Abstract. In the task of 6D pose estimation by RGB-D image, the crucial problem is how to make the most of two types of features respectively from RGB and depth input. As far as we know, prior approaches treat those two sources equally, which may overlook that the different combinations of those two properties could have varying degrees of impact. Therefore, we propose a Feature Selecting Mechanism (FSM) in this paper to find the most suitable ratio of feature dimension from RGB image and point cloud (converted from depth image) to predict the 6D pose more effectively. We first conduct artificial selection in our Feature Selecting Mechanism (FSM) to prove the potential for the weight of the RGB-D feature. Afterward, the neural network is deployed in our FSM to adaptively pick out features from RGB-D input. Through our experiments on the LINEMOD dataset, YCB-Video dataset, and our multi-pose synthetic image dataset, we show that there is an up to 2% improvement in the accuracy by utilizing our FSM, compared to the state-of-the-art method.

Keywords: Deep learning · 6D pose estimation · RGB-D image

1 Introduction

In daily life, we have closely carried on human-object interaction. In those interactions, the task for 6D object pose estimation takes a crucial role. For instance, when we reach out for a pen on the desk, it is significant for the estimation for where is the pen from our position and what is the posture at its location. So is the machine. The machine utilizes what it sees to calculate the 6D pose of objects and make decisions, such as grasping [1,2] and augmented reality [3,4] *etc.*

This work is supported by NSFC 12071460, Shenzhen research grant (KQJSCX2018-0330170311901, JCYJ20180305180840138 and GGFW2017073114031767).

© Springer Nature Switzerland AG 2021
Y. Zhang et al. (Eds.): PDCAT 2020, LNCS 12606, pp. 141–153, 2021.
https://doi.org/10.1007/978-3-030-69244-5_12

To estimate 6D pose, we can only feed RGB image in the machine [5,6], or attempt to make use of RGB-D image [7–9]. Nevertheless, with the prevalence of 3D computer vision, it is a wise choice to leverage the combination between 2D image and 3D data (such as point cloud) in some tasks. As shown in our title, RGB-D input will be applied in this paper.

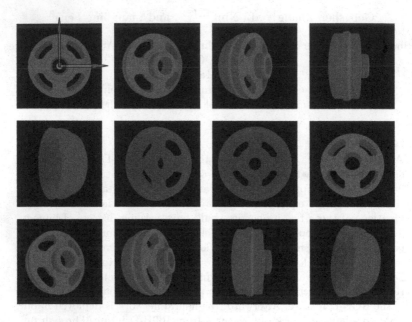

Fig. 1. The above images display the state of an industrial component in various postures on our multi-pose synthetic image dataset. We suppose that the first image in the upper-left is the original pose and its coordinate system is appointed as the rotation and translation axis. The red arrow that points to the top is X axis, the blue one is Y axis, and the yellow one vertically penetrating to the inside of this page is Z axis. And the rest of the poses are transformed from the original pose. (Color figure online)

However, to the best of our knowledge, existing methods [9,10] fuse the RGB image and point cloud in the proportion of 1 : 1. As we know, the RGB image and point cloud are diverse data formats in computer vision, which embody complementary feature information. To be specific, the RGB image provides appearance feature information, and the point cloud possesses geometry feature information. Even though those two kinds of information jointly contribute to the mission for the 6D pose estimation, the impact from each side may not be the same. Citing a vivid example, there are several rocks on the peak, which are in variant colors and shapes. We would like to pick out the rock that is the fastest to roll down the hillside. Although the properties of both color and shape are under consideration, apparently, the shape feature is more useful in this task. Therefore, for the sake of finding the optimal solution, we ought to think over increasing the share of more beneficial feature information.

Based on the above statement, in this paper, we come up with a deep learning network to estimate the 6D pose using an adaptive weight of RGB-D input. The crux of our method is to combine the RGB value with point cloud (converted from depth image) at a certain proportion in each pixel. Compared with prior researches, we take into account the ratio of heterogeneous features, before joint them together. By this means, more useful features are employed sufficiently, which facilitates better achievement.

We evaluate our network on the LINEMOD dataset [11] and the YCB-Video dataset [12], which are two of the popular benchmarks for the 6D pose estimation task. The results illustrate that the pose accuracy of our proposed method (96.3%) surpasses the state-of-the-art DenseFusion [9] by 2% on the LINEMOD dataset, and for the YCB-Video dataset, there is up to 1.0% increase as well. Furthermore, we generate a multi-pose synthetic image dataset of one industrial component with 6859 postures by OpenGL, which is shown in the Fig. 1. This dataset is applied for assessing the generalization ability of our network, and the accuracy for this is 98.9%, which is also satisfactory.

To sum up, the main contributions of this paper are the following two points: First, we propose adaptively adjusting the input weight of 2D RGB image and 3D point cloud in the 6D object pose estimation end-to-end deep learning network to make full use of the advantageous features. Second, we introduce our multi-pose synthetic image dataset of one industrial component for 6D pose estimation.

2   Related Work

The classical methods seek help for extracting features from RGB images [13,14] or with the corresponding depth maps [11,15–17] to find the nearest template. For example, Wu *et al.* [13] extracts SIFT keypoints from RGB image and matches them with the VIP feature from 3D model template. While Hinterstoisser *et al.* [15–17] combines the information from RGB and depth images to define a template. Despite high precision and real-time capability, there are two main drawbacks in those methods. In the first place, those methods are almost impossible to deal with occluded objects. Besides, as the complexity of this assignment increases, the number of templates is bound to grow swiftly.

For the above reason, some researchers have a try at machine learning techniques [7,10,18,19,22,23]. Brachmann *et al.* [10] uses a random forest to directly predict object 3D coordinate, then to estimate and refine the 6D pose. Wohlhart *et al.* [18] learns the map from the input image to a descriptor by Convolutional Neural Network (CNN), and finds the nearest neighbor to obtain the pose. Kehl *et al.* [7] regresses feature from local patch cropped from original RGB-D images by neural network. After that, it searches k nearest neighbors in the codebook of synthetic images to vote for the 6D pose. PointFusion [19] combines the RGB-D feature to predict a 3D bounding box. While those approaches, don't straightforwardly acquire the 6D pose, they guide and inspire the following scholars.

Subsequent methods propose directly infering 6D pose by loading RGB images [12] or RGB-D images [8,9] into an end-to-end deep Convolutional Neural

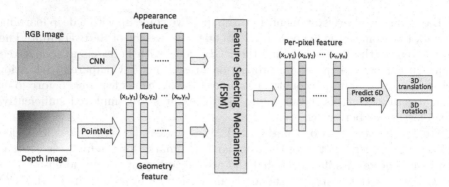

Fig. 2. N pixels, like (x_1, y_1) (x_2, y_2) $\cdots$ (x_n, y_n), in RGB image pass through CNN, and we get appearance feature. Simultaneously, those N pixels in depth image convert to N points in a point cloud. Then feed the N points in PointNet to extract geometry feature. Finally, let our feature selecting mechanism (FSM) picks out a preferable per-pixel feature to predict the 6D pose.

Network (CNN). PoseCNN [12] creates a powerful network to predict 6D pose by a single RGB image. Li *et al.* [8] takes the depth data into consideration. And it should be noted that DenseFusion [9] separately processes the RGB input and depth input, and fuses them in each pixel by the proportion of 1 : 1 to shape per-pixel feature for the 6D pose estimation assignment and achieves a fairly good result.

Compared to the previous methods, what we are most concerned about in this paper is the following two questions: (1) whether there is influence from different combinations of RGB and depth source, and (2) how to make the best of the heterogeneous information from those two sources for each object. We will answer them soon later. And it shows that we have an advantage over DenseFusion's proportion of 1 : 1 in this paper, which implies that finding the right weight between 2D and 3D input is promising.

3 Model

Our model is to estimate 6D pose by a proper ratio of RGB-D input. In the field of 3D computer vision, the 6D pose consists of 3D translation and 3D rotation. 3D translation is the distance we move along X, Y and Z axis from the origin of camera coordinate system to the origin of object coordinate system, which is denoted as a 3D vector $t \in \mathbb{R}^3$. And 3D rotation could be seen as the angle we rotate on X, Y and Z axis from the camera space to the object space, which is represented by a 3×3 matrix $R \in SO(3)$. Thus, each 6D pose could be marked as $[R|t]$.

Concerning the RGB-D input of various objects, influence from one side may be more powerful than the others. Therefore, an equivalent share of those two sources may not be suitable for all circumstances. To realize the above view, we

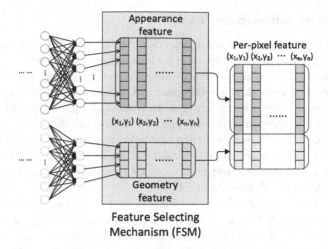

Fig. 3. The artificial selection is utilized in our Feature Selecting Mechanism (FSM). And we exemplify how to obtain the 2 : 1 ratio of appearance feature to geometry feature in this figure.

propose a Feature Selecting Mechanism (called FSM) before combining appearance feature with geometry feature, which is illustrated by Fig. 2. Specifically, the FSM picks out the feature for the desired dimension from two kinds of sources in each pixel to compose the per-pixel feature. When we are in the process of designing the FSM, several schemes are coming into our thought, such as (1) artificial selection, and (2) traditional autoencoder (AE) [20] or convolutional autoencoder (CAE) [21].

3.1   Artificial Selection

First of all, for the sake of verifying our idea to be potential, we manually select the feature for the needed dimension in the Feature Selecting Mechanism (FSM). For details, we take a shot at modifying the feature dimension by separately adjusting the number of output neurons in their previous convolution layer and making the proportion be 1 : 2 or 2 : 1. Figure 3 depicts how we fuse the one-share geometry feature with double-shot appearance feature, and it will be the similar treatment for other cases.

This mode is restricted to a fixed weight for various objects, which reduces the flexibility of the network. Since in different situations, it is uncertain which is more significant between geometry and appearance feature. And even if the geometry information is more beneficial in some cases, the ratio is mysterious. Nonetheless, this trial plays an illuminating role in our research and motivates us into further exploration.

3.2   Variant of Autoencoder (AE)

When it comes to feature dimension selection, the Principal Component Analysis (PCA) is a popular representative, which calculates the direction of the main features in the feature set as the new dimension, and projects the entire original feature set into the new-dimensional space. PCA could describe information as complete as possible with fewer features and reach the global optimal solution. But then, it is inappropriate to be deployed in the middle of our end-to-end neural network model. Hence, we turn our attention to the autoencoder (AE) [20,21], which is a sort of neural network with a similar function to PCA. Autoencoder (AE) acquires the local optimal result by the back-propagation algorithm, which makes AE more flexible than PCA.

The Fig. 4 portrays the schematic diagram of using a variant of autoencoder (AE) in our Feature Selecting Mechanism (FSM). We insert a network as an encoder at the back of the two same-dimensional features to choose more useful features for the first time, which yield various weights on each pixel, called local weighted feature. After that we feed the local weighted feature into another

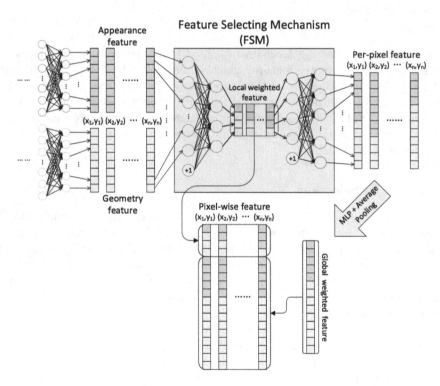

Fig. 4. The variant of autoencoder (AE) is used in our Feature Selecting Mechanism (FSM). And in this scheme, we illustrate that the final feature ratio in each pixel could be 5 : 11, 3 : 5 or 7 : 9, not naive 1 : 2, 2 : 1 or 1 : 1. In reality, due to the back-propagation algorithm, the proportion would be even more complicated, which depends on the knowledge that our FSM has learned.

network as a decoder to get our per-pixel feature, which is a little more complex combination of those two kinds of features. Then via multi-layer perceptron (MLP) and average pooling layer, we obtain the global proportion on the whole pixels, which is global weighted feature. Ultimately the local weighted feature and the global weighted feature are concatenated into the pixel-wise feature. This means that there is the fittest and extremely sophisticated combination approach in each pixel. Moreover, with the help of back-propagation algorithm, more valuable features remain and offer us more profitable prerequisites to predict 6D pose later.

3.3 6D Pose Estimation

From our cherry-picked high-dimensional pixel-wise feature, we could regress to the predicted 6D pose $[\tilde{R}|\tilde{t}]$. As for the loss function, we refer to Wang *et al.* [9], and minimize the average distance between the points in the 3D model with the above predicted 6D pose and those corresponding points with the ground truth 6D pose. And the details about the comparison between the predicted 6D pose and the ground truth 6D pose will be explicated in Sect. 4.2.

4 Experiments

In this section, we will demonstrate how to untangle the above two questions we remain early: (1) To confirm the 6D pose estimation task is indeed related to the weight of RGB-D input, we first conduct artificial selection in our Feature Selecting Mechanism (FSM). In this section, we set the fixed ratio of appearance feature to geometry feature be 1 : 2 and 2 : 1. (2) Then, we apply the variant of autoencoder (AE) in our Feature Selecting Mechanism (FSM), which adopts the neural network to choose feature by itself. We evaluate our method on the LINEMOD dataset [11], YCB-Video dataset [12] and our multi-pose synthetic image dataset to make a comparison with the state-of-the-art method [9] in 6D pose estimation by RGB-D images. Due to the page limitations of this paper, the detailed results of YCB-Video dataset are not presented in this paper, and it will be discussed in the full version.

4.1 Datasets

The LINEMOD dataset contains 13 low-textured objects and 15,800 RGB-D images in all, including ape, bench vice, camera, can, cat, driller, duck, hole puncher, iron, lamp, phone and two symmetrical objects, egg box and glue. Each image segments out one object and is labeled with the ground truth 6D pose of this object. Following the prior work [9], 2,373 images, 15% of the whole dataset, are split from the dataset for training, where there is almost the same number of images for the 13 objects.

Furthermore, our multi-pose synthetic image dataset is composed of one industrial component with 6,859 different postures, which are generated by sequentially rotates on one of the X, Y and Z axis by 10°. Specifically, the industrial component rotates on one axis until it completes 180° of rotation, and then the transformation on this axis returns to the initial state while it rotates on the next axis by 10°. Due to the minor difference between poses, the 6D pose estimation task on this dataset could be challenging.

4.2 Metrics

For the LINEMOD dataset and our multi-pose synthetic image dataset, we apply the ground truth pose $[R|t]$ to each point in 3D model to get the target model $(RX + t)$, and utilize predict pose $[\tilde{R}|\tilde{t}]$ to define the predict model $(\tilde{R}X + \tilde{t})$. For asymmetric object, we compute the M points' average distance (ADD), proposed by [11], between the same point in target model and predict model:

$$ADD = \frac{1}{M} \sum_{0 \leq i < M} \left\| (Rx_i + t) - \left(\tilde{R}x_i + \tilde{t} \right) \right\|. \tag{1}$$

As regards symmetric objects, similar to egg box and glue, the ADD-S [11] is requested. The ADD-S seek out the closest to $i-th$ point in target model from M points of predict model:

$$ADD - S = \frac{1}{M} \sum_{0 \leq i < M} \min_{0 \leq j < M} \left\| (Rx_i + t) - \left(\tilde{R}x_j + \tilde{t} \right) \right\|. \tag{2}$$

The threshold for ADD and ADD-S is the 10% of the object's diameter (the unit of measure is meter), the same as [9]. Below the threshold, the predicted value is acceptable for us.

4.3 Implementation Details

The CNN module is PSPNet with Resnet-18 to extract appearance feature. At the same time, the PointNet deals with geometry feature. Those two kinds of feature are blended in our network Feature Selecting Mechanism (FSM) to get per-pixel feature and then pass through several multi-layer perceptron (MLP) as the pose estimation decoder to predict 6D pose.

4.4 Result on LINEMOD Dataset

In Table 1, we compare several methods on LINEMOD dataset in this section.

(1) DenseFusion [9] makes use of appearance feature and geometry feature in an equal share.
(2) Ours (artificial selection) manually adjusts the ratio of appearance feature to geometry feature (abbr. RGB : Point) be 1 : 2 and 2 : 1 in our Feature Selecting Mechanism (FSM).
(3) Ours (variant of AE) deploys neural network in our FSM to automatically pick up more beneficial information from the appearance feature and geometry feature to form a per-pixel feature.

Table 1. The accuracy rate of 6D pose on LINEMOD dataset. The egg box and glue in bold are symmetric objects. The accuracy rate in bold is the highest in its row.

	DenseFusion [9]	Ours (artificial selection, RGB : Point = 1 : 2)	Ours (artificial selection, RGB : Point = 2 : 1)	Ours (variant of AE)
Ape	92.3	92.8	92.9	**93.4**
Bench vice	93.2	96.7	91.9	**96.9**
Camera	94.4	95.4	95.2	**95.7**
Can	93.1	**96.2**	91.6	96.0
Cat	96.5	**98.0**	97.4	97.3
Driller	87.0	92.5	93.1	**93.3**
Duck	92.3	94.2	94.3	**95.7**
Egg box	99.8	**99.9**	99.8	99.8
Glue	**100.0**	99.5	99.7	99.5
Hole puncher	92.1	**94.2**	92.5	93.5
Iron	97.0	**99.1**	96.1	97.8
Lamp	95.3	96.6	96.1	**97.3**
Phone	92.8	95.6	**96.0**	95.7
ALL MEAN	94.3	96.2	95.1	**96.3**

From the Overall Perspective. In Table 1, we can see that our three scenarios perform better than the state-of-the-art method [9] on the majority of objects, and the accuracy rate of each object in our approaches is all over 91%. Especially, the driller, which is on $6th$ row of objects column in the Table 1, is 87% in the state-of-the-art method, and we raise this index by 5.5% point at least. And the results of those two symmetric objects are very close to each other in the above four experiments. Moreover, the result of our neural network approach, on $5th$ col of Table 1, has a slight upgrade compared to our artificial selection methods (on $3rd$ and $4th$ cols of Table 1), and we will give a detailed analysis for this in the following part. All in all, the average accuracy rate of our best solution, adopting the variant of autoencoder (AE) in our Feature Selecting Mechanism (FSM), for 6D pose estimation by RGB-D images could reach 96.3%, which exceed the powerful state-of-the-art method by 2%.

From the Three Separate Groups. After our artificial selection experiments, we are aware that there is a real need for a suitable weight of features from RGB and point cloud input for each object. And we intend to divide those 13 objects into 3 groups by considering what kind of feature is more valuable in each stuff, as Table 2 exhibits. When someone's accuracy rate of $3rd$ col is higher than $4th$ col by 0.3%, we suppose that the geometry feature of this object is more useful than appearance feature, such as bench vice, can, cat, hole puncher, iron and lamp. In the contrary, *i.e.* the appearance feature is more profitable, so there

Table 2. The accuracy rate of 6D pose on the 3 groups in LINEMOD dataset. The geometry feature of the top 6 objects is more valuable. The appearance feature of the middle 2 things is more profitable. And the last 5 stuffs have hardly any influence by the weight of RGB-D input. The egg box and glue in bold are symmetric objects. The highest average accuracy rate is in bold.

	DenseFusion [9]	Ours (artificial selection, RGB : Point = 1 : 2)	Ours (artificial selection, RGB : Point = 2 : 1)	Ours (variant of AE)
Bench vice	93.2	96.7	91.9	96.9
Can	93.1	96.2	91.6	96.0
Cat	96.5	98.0	97.4	97.3
Hole puncher	92.1	94.2	92.5	93.5
Iron	97.0	99.1	96.1	97.8
Lamp	95.3	96.6	96.1	97.3
MEAN	94.5	**96.8**	94.3	96.5
Driller	87.0	92.5	93.1	93.3
Phone	92.8	95.6	96.0	95.7
MEAN	89.9	94.1	**94.6**	94.5
Ape	92.3	92.8	92.9	93.4
Camera	94.4	95.4	95.2	95.7
Duck	92.3	94.2	94.3	95.7
Egg box	99.8	99.9	99.8	99.8
Glue	100.0	99.5	99.7	99.5
MEAN	95.8	96.4	96.4	**96.8**
ALL MEAN	94.3	96.2	95.1	**96.3**

are 2 things in this group, like driller and phone. The rest of 5 items, namely ape, camera, duck, eggbox, and glue, evince slightly indifference to the weight of RGB-D input, since the difference in accuracy between their appearance and geometry is no more than 0.2%.

As we mentioned above, we split the LINEMOD dataset into 3 groups according to the artificial selection experiments. The top 6 objects in Table 2 need pay more attention to the geometry feature. We could find out that the preferable weight for those 6 objects may be around 1 : 2 for RGB to point cloud, for the accuracy rate of our best solution on the overall result (on 5th col of Table 2) is a little bit (0.3%) less than the fixed 1 : 2 ratio of appearance feature to geometry feature. The driller and phone in the middle of Table 2 rely more on appearance feature, and the proportion of about 2 : 1 may suit them very well. Similar to the above situation, our best solution on the overall result is extremely close to the result of 2 : 1 ratio of RGB-D features. For the remaining 5 objects, at the bottom of Table 2, we could notice that our best solution wins and the average accuracy rate of two artificial selection methods is the same, so the impact from different combinations between appearance and geometry information may be negligible on those stuffs. Despite the average score of our best solution (on $5th$ col of Table 2) not always being the top one on the above three circumstances, it is still outstanding enough, and due to its strong adaptive faculty, it works best in most situations on the whole, achieving 96.3% accuracy on the LINEMOD dataset.

4.5 Result on Our Dataset

This one is a comparison between the state-of-the-art DenseFusion [9] and our proposed approach that deploys the neural network to pick up suitable features to predict 6D pose. In this experiment, the accuracy rate of our approach is up to 98.9%, which is still beyond the state-of-the-art method by 1.9%. We could find out that our network by utilizing Feature Selecting Mechanism (FSM) has a great generalization capacity and powerful adaptive competence to meet the needs of varied objects.

5 Conclusion

In this paper, we put forward a proposal that adjusts the ratio of feature from 2D RGB image and 3D point cloud on the basis of the excellent predecessors, and we come up with the Feature Selecting Mechanism (FSM) before fuse those two heterogeneous features together. The FSM has two forms: artificial selection and neural network. Through the artificial selection method, we find the potential of this issue. With the assistance of our neural network approach, our method could adaptively find the suitable weight of RGB-D features and outperform the state-of-the-art method by up to 2% in 6D pose estimation on LINEMOD dataset, YCB-Video dataset and our multi-pose synthetic image dataset.

References

1. Deng, X., Xiang, Y., Mousavian, A., Eppner, C., Bretl, T., Fox, D.: Self-supervised 6d object pose estimation for robot manipulation. CoRR abs/1909.10159 (2019)
2. Tremblay, J., To, T., Sundaralingam, B., Xiang, Y., Fox, D., Birchfield, S.: Deep object pose estimation for semantic robotic grasping of household objects. CoRR abs/1809.10790 (2018)
3. Marchand, É., Uchiyama, H., Spindler, F.: Pose estimation for augmented reality: a hands-on survey. IEEE Trans. Vis. Comput. Graph. **22**, 2633–2651 (2016)
4. Shahrokni, A., Vacchetti, L., Lepetit, V., Fua, P.: Polyhedral object detection and pose estimation for augmented reality applications. In: Computer Animation 2002, CA 2002, pp. 65–72 (2002)
5. Brachmann, E., Michel, F., Krull, A., Yang, M.Y., Gumhold, S., Rother, C.: Uncertainty-driven 6D pose estimation of objects and scenes from a single RGB image. In: 2016 IEEE Conference on Computer Vision and Pattern Recognition, CVPR 2016, 3364–3372 (2016)

6. Kehl, W., Manhardt, F., Tombari, F., Ilic, S., Navab, N.: SSD-6D: making RGB-based 3D detection and 6D pose estimation great again. In: IEEE International Conference on Computer Vision, ICCV 2017, pp. 1530–1538 (2017)
7. Kehl, W., Milletari, F., Tombari, F., Ilic, S., Navab, N.: Deep learning of local RGB-D patches for 3D object detection and 6D pose estimation. In: Leibe, B., Matas, J., Sebe, N., Welling, M. (eds.) ECCV 2016. LNCS, vol. 9907, pp. 205–220. Springer, Cham (2016). https://doi.org/10.1007/978-3-319-46487-9_13
8. Li, C., Bai, J., Hager, G.D.: A unified framework for multi-view multi-class object pose estimation. In: Ferrari, V., Hebert, M., Sminchisescu, C., Weiss, Y. (eds.) ECCV 2018. LNCS, vol. 11220, pp. 263–281. Springer, Cham (2018). https://doi.org/10.1007/978-3-030-01270-0_16
9. Wang, C., et al.: DenseFusion: 6D object pose estimation by iterative dense fusion. In: IEEE Conference on Computer Vision and Pattern Recognition, CVPR 2019, pp. 3343–3352 (2019)
10. Brachmann, E., Krull, A., Michel, F., Gumhold, S., Shotton, J., Rother, C.: Learning 6D object pose estimation using 3D object coordinates. In: Fleet, D., Pajdla, T., Schiele, B., Tuytelaars, T. (eds.) ECCV 2014. LNCS, vol. 8690, pp. 536–551. Springer, Cham (2014). https://doi.org/10.1007/978-3-319-10605-2_35
11. Hinterstoisser, S., et al.: Model based training, detection and pose estimation of texture-less 3D objects in heavily cluttered scenes. In: Lee, K.M., Matsushita, Y., Rehg, J.M., Hu, Z. (eds.) ACCV 2012. LNCS, vol. 7724, pp. 548–562. Springer, Heidelberg (2013). https://doi.org/10.1007/978-3-642-37331-2_42
12. Xiang, Y., Schmidt, T., Narayanan, V., Fox, D.: PoseCNN: a convolutional neural network for 6D object pose estimation in cluttered scenes. In: Robotics: Science and Systems XIV, Carnegie Mellon University (2018)
13. Wu, C., Fraundorfer, F., Frahm, J., Pollefeys, M.: 3D model search and pose estimation from single images using VIP features. In: IEEE Conference on Computer Vision and Pattern Recognition, CVPR Workshops 2008, pp. 1–8 (2008)
14. Zhu, M., et al.: Single image 3D object detection and pose estimation for grasping. In: 2014 IEEE International Conference on Robotics and Automation, ICRA 2014, pp. 3936–3943 (2014)
15. Hinterstoisser, S., et al.: Multimodal templates for real-time detection of texture-less objects in heavily cluttered scenes. In: IEEE International Conference on Computer Vision, ICCV 2011, pp. 858–865 (2011)
16. Lowe, D.G.: Distinctive image features from scale-invariant keypoints. Int. J. Comput. Vis. **60**, 91–110 (2004)
17. Wu, C., Clipp, B., Li, X., Frahm, J., Pollefeys, M.: 3D model matching with viewpoint-invariant patches (VIP). In: 2008 IEEE Computer Society Conference on Computer Vision and Pattern Recognition, CVPR 2008 (2008)
18. Wohlhart, P., Lepetit, V.: Learning descriptors for object recognition and 3D pose estimation. In: IEEE Conference on Computer Vision and Pattern Recognition, CVPR 2015, pp. 3109–3118 (2015)
19. Xu, D., Anguelov, D., Jain, A.: PointFusion: deep sensor fusion for 3D bounding box estimation. In: 2018 IEEE Conference on Computer Vision and Pattern Recognition, CVPR 2018, pp. 244–253 (2018)
20. Hinton, G.E., Salakhutdinov, R.R.: Reducing the dimensionality of data with neural networks. Science **313**, 504–507 (2006)
21. Masci, J., Meier, U., Cireşan, D., Schmidhuber, J.: Stacked convolutional auto-encoders for hierarchical feature extraction. In: Honkela, T., Duch, W., Girolami, M., Kaski, S. (eds.) ICANN 2011. LNCS, vol. 6791, pp. 52–59. Springer, Heidelberg (2011). https://doi.org/10.1007/978-3-642-21735-7_7

22. Liang, Y., Cai, Z., Jiguo, Yu., Han, Q., Li, Y.: Deep learning based inference of private information using embedded sensors in smart devices. IEEE Netw. Mag. **32**(4), 8–14 (2018)
23. Cai, Z., Xu, Z., Yu, J.: A differential-private framework for urban traffic flows estimation via taxi companies. IEEE Trans. Ind. Inform. (TII) **15**(12), 6492–6499 (2019)

Blockchain-Based Secure Outsourcing of Fully Homomorphic Encryption Using Hidden Ideal Lattice

Mingyang Song, Yingpeng Sang$^{(\boxtimes)}$ iD, Yuying Zeng, and Shunchao Luo

School of Data and Computer Science, Sun Yat-Sen University, Guangzhou, China
{songmy5,zengyy26,luoshch5}@mail2.sysu.edu.cn,
sangyp@mail.sysu.edu.cn

Abstract. The efficiency of homomorphic encryption has always affected its practicality. With the dawn of internet of things, the demand for computation and encryption on lightweight devices is increasing. Complex cryptographic computing is an important burden for lightweight devices, but outsourcing provides great convenience for them. In this paper, based on blockchain, we propose a secure outsourcing scheme for Fully Homomorphic Encryption using Hidden Ideal Lattice (FHEHIL), in which the time-consuming operations (including modular exponentiation and polynomial multiplication) are outsourced. For polynomial multiplication, we propose a secure outsourcing algorithm that reduces the local computation cost to $O(n)$. Previous work based on Fast Fourier Transform can only achieve $O(nlog(n))$ for the local cost. Through security analysis, our scheme achieves the goals of privacy protection against passive attackers and cheating detection against active attackers. Experiments also demonstrate our scheme is more efficient in comparison with the non-outsourcing FHEHIL.

Keywords: Fully homomorphic encryption · Computation outsourcing · Polynomial multiplication · Modular exponentiation · Blockchain

1 Introduction

As the draw of the big data era, there is an increasing demand for large-scale time-consuming computations. Fortunately, with the emergence of cloud computing, computation outsourcing brings convenience to users with limited computing capacity. Resource-constrained users can outsource complex computing tasks for a fee. They do not need to buy expensive high-performance hardware. It not only improves the resource utilization of high-performance equipment but also brings economic benefits to users with limited computing capacity. Nevertheless, this attractive computing scheme also causes security issues (such as privacy leaks). Whether for financial gain or curiosity, the cloud may leak user data, do malicious damage, and forge the computations. On one hand, the outsourced data of users may contain user privacy. The leakage of these data may bring serious losses to users. On the other hand, computing errors caused by cloud hardware failures and software errors, etc. should also be considered. Therefore,

© Springer Nature Switzerland AG 2021
Y. Zhang et al. (Eds.): PDCAT 2020, LNCS 12606, pp. 154–165, 2021.
https://doi.org/10.1007/978-3-030-69244-5_13

the computing outsourcing scheme which can not only protect the privacy of users but also ensure the correct results has become a hot topic of research.

In 2009, Gentry proposed a homomorphic encryption algorithm based on ideal lattice [7] for the first time, providing us with a direction to solve the privacy issues in computation outsourcing. The direction is a secure computation outsourcing mode: Encryption-Outsourcing-Decryption (EOD). However, if we use the common EOD model, the local device should also undertake the computations of secret key generation, encryption, verification, decryption, and so on. These computations are also a burden for the devices with limited computing resources (such as mobile phones, IoT nodes). To make EOD effective, we should reduce the local computing burden.

Due to the limited computing resources of users and the low efficiency of the fully homomorphic encryption, general computation outsourcing using a fully homomorphic encryption algorithm is not practical. Considering the advantages of blockchain technology, in this paper, we outsource some complex computations into a blockchain framework, in order to improve the efficiency of Fully Homomorphic Encryption using Hidden Ideal Lattice (FHEHIL). The contributions of this paper can be summarized as follows.

- We propose a framework of blockchain-based computation outsourcing, in which we can implement secure outsourcing for FHEHIL.
- We propose a secure outsourcing algorithm for polynomial multiplication, which reduces the local computation cost to $O(n)$. Previous work based on Fast Fourier Transform can only achieve $O(nlog(n))$ for the local cost. In addition, the algorithm can not only detect cheating but also identify cheating nodes combing with blockchain, when all the computational nodes collude.
- The secure outsourcing algorithm for polynomial multiplication is employed in FHE-HIL as a basic operation, and the implementation of FHEHIL on the blockchain-based framework can have higher efficiency compared with previous work.

The paper is organized as follows. We introduce the related work in Sect. 2. Section 3 introduces the mathematical symbols and background knowledge about the FHEHIL [3] and the extended Euclidean secure outsourcing algorithm [6]. Section 4 constructs the framework of blockchain-based computation outsourcing. In Sect. 5 we propose the secure outsourcing algorithms for basic operations in FHEHIL including polynomial multiplication and modular exponentiation. Section 6 evaluates the performance of our scheme. Finally, we conclude our work in Sect. 7.

2 Related Work

At present, the research on secure outsourcing mainly has two directions. In one direction, a general outsourcing mechanism is studied. In this mechanism, a fully homomorphic encryption algorithm is designed and the EOD model is utilized to outsource any computation. After the works of Gennaro et al. [7], great progress has been made in this field [10–12]. In the other direction, special outsourcing algorithms are designed for the special scientific computations, e.g. modular exponentiation, solution of large-scale linear equations [8] and Extend Euclidean et al. Because of the wide application in cryptography, the study on modular exponentiation is a hot topic of research. Hohenberger et al. [13] proposed a secure outsourcing scheme for modular exponentiation, in 2005. Chen et al. [14] further improved its efficiency and verifiability. Dijk et al. [15] proposed a scheme that only ensures exponent privacy. Recently, Fu et al. [9] proposed a secure outsourcing scheme of modular exponentiation with hidden exponent and base. It has a stronger detectability.

Polynomial multiplication is also a commonly-used operation in cryptographic schemes, including FHEHIL. Utilizing the Fast Fourier transform (FFT), the local computation of polynomial multiplication can achieve the complexity of $O(nlogn)$. Till now, there is little research on the secure outsourcing of polynomial multiplication. To fill this gap, we propose a secure outsourcing algorithm of polynomial multiplication, whose local complexity is reduced to $O(n)$.

Due to the characteristics of the blockchain and Bitcoin [1], there are lots of research and applications on them in recent years. The blockchain has been widely used in the research of Secure multi-party computation [2]. Zhang [16] proposed a fair payment scheme for cloud computing based on blockchain. Taking advantage of the non-tampering of blockchain, there are also some schemes [17, 18] of outsourced data integrity verification. Recently, Zheng et al. [19] proposed a secure outsourcing scheme for Attribute-based encryption on the blockchain. However, there is still a lack of the general framework of blockchain-based secure outsourcing of encryption. In this paper, we will propose this general framework, which assigns computing tasks based on the credit values and has a system of reward and punishment. In this way, it will significantly reduce the probability of computing failure. Meanwhile, except for the financial incentives, the promotion of node credit also becomes a motivation for participants.

3   Notations and Background

We use upper case bold and italic letters for matrices, use $det(M)$ for the determinant of matrix M. Lower case bold and italic letters respect vectors (such as $v = [v_0, v_1, \ldots, v_n]$, where v_i is the $(i + 1)^{th}$ element). Denote polynomial by lower case italics (e.g. $f(x)$). For a rational number r, $round(r)$ represents the nearest integer to r. The rational vector v can also be rounded to $round(v) = [round(v_0), round(v_1), \ldots, round(v_n)]$. We use $Pol(v)$ for the polynomial whose coefficients are elements of vector v ($Plo(v) = \sum_{i=0}^{n-1} v_i x^i$). For convenience, we use $v(x)$ to denote $Pol(v)$. $\|v\|$ represents the norm of v. We use $Vec(f(x))$ for the vector form of the polynomial $f(x)$. We use $v_1 \times v_2$ for polynomial multiplication on the ring ($v_1 \times v_2 = Vec(v_1(x) \times v_2(x) mod f(x))$), use $v_1 \cdot v_2$ for dot product of v_1 and v_2. We use $\boldsymbol{v_1} \circledast \boldsymbol{v_2}$ for correlation ($\boldsymbol{v_1} \circledast \boldsymbol{v_2} = [v_{1_0} * v_{2_0}, ..., v_{1_n} * v_{2_n}]$). We use $R(v, f)$ for the rotation matrix of v whose i^{th} row is the coefficients of $v(x) \times x^{i-1} mod f(x)$. We use $gcd(a(x), b(x))$ for the Euclidean algorithm on $a(x)$ and $b(x)$, use $xgcd(a(x), b(x))$ for the extended Euclidean algorithm. $deg(f(x))$ is the degree of $f(x)$. We use F_v or $F(v_0, v_1, \ldots, v_n)$ for the Discrete Fourier transform (DFT) of v, use F_v^{-1} or $F^{-1}(v_0, v_1, \ldots, v_n)$ for the inverse DFT of v.

3.1 Fully Homomorphic Encryption Using Hidden Ideal Lattice

The FHEHIL is based on the following observation: It is not necessary to publish lattices to construct a fully homomorphic encryption scheme. Virtually, we can compute vectors that are close to the lattice without knowing the lattice. A lattice is unique if several bounded distance vectors are provided. For this reason, we can directly use these vectors, rather than a bad basis of the lattice. The FHEHIL scheme [3] is described in Algorithm 1. The parameters involved in it are shown in Table 1.

Table 1. Parameters in the FHEHIL algorithm

Parameter	Implication
ζ	The norm of random noise vector
γ	The bit length of norm of generating polynomial
η	The bit length of norm of the random multiplier vector
t	The number of vectors contained in the public key
n	The dimension of the hidden lattice (power of 2)

The most time-consuming computations are polynomial multiplication and the computation of w in the process of key generation. The computational complexity of $w(x)$ is affected by d. If d is a prime, we can execute $xgcd(v(x), f(x))$ once to get $w(x)$ [4]. In other cases, we could use Gentry's method [5] in local computing. In this paper, we use the method [6] to securely outsource the extended Euclidean algorithm. But the local computation is still complex. We will further outsource the time-consuming operations in [6] to increase efficiency.

Algorithm 1. Key generation

Output: $sk = \{d, w\}, pk = [p_1, ..., p_t]$

1: Generate the irreducible polynomial $f(x) = x^n + 1$.
2: Generate a random vector v satisfying $\{v \in Z^n, 2^{\gamma-1} < \|v\| < 2^\gamma, \sum_{i=0}^{n-1} v_i \bmod 2 = 1\}$.
3: $V = R(v, f)$ and $d = |det(V)|$.
4: Generate $t - 1$ random vectors $[g_1, g_2, ..., g_{t-1}]$ satisfying $\{g \in Z^n, 2^{\eta-1} < \|g\| < 2^\eta\}$.
5: Generate a random vector g_t satisfying $\{g_t \in Z^n, \|g_t\| < 2^\eta, \sum_{i=0}^{n-1} g_{t,i} \bmod 2 = 1\}$.
6: Generate $t - 1$ random vectors $[r_1, r_2, ..., r_{t-1}]$ satisfying $\{r \in \{-1, 0, 1\}^n, \|r\| < \zeta\}$.
7: Generate a random vector r_t satisfying $\{r_t \in \{-1, 0, 1\}^n, \|r_t\| < \zeta, \sum_{i=0}^{n-1} r_{t,i} \bmod 2 = 1\}$.
8: Compute t vectors $[p_1, ..., p_t]$ by $p_i = g_i \times v + r_i$.
9: Compute w satisfying $w \times v = d \bmod f$.

In the process of encryption, firstly, generate $t - 1$ integer vectors $[s_1, s_2, \ldots, s_{t-1}]$ satisfying $\sum_{j=1}^n s_{i,j} \bmod 2 = 0, 1 \le i \le t$ and a vector s_t satisfying $\sum_{j=1}^n s_{t,j} \bmod 2 = m$, where m is the plaintext. Secondly, pick a vector s_{t+1} satisfying $\sum_{j=1}^n s_{t+1,j} \bmod 2 = 0$. Finally, compute the ciphertext $\psi = \sum_{i=1}^\tau s_i \times p_i + s_{t+1}$. In the process of decryption, there are two steps. The one is computing $\psi' = round(\psi \times w/d)$. The other is getting the plaintext $m = \psi'(1) \bmod 2$. In addition to the parameter generation, the primary computing in the process of encryption and decryption is polynomial multiplication.

3.2 Securely Outsourcing the Extended Euclidean Algorithm

The extended Euclidean algorithm computes $xgcd(a(x), b(x))$ to get $u(x)$, $v(x)$ and $d(x)$ satisfying $u(x) \times a(x) + v(x) \times b(x) = d(x)$. For secure outsourcing, a variable substitution technique can be used to hide the coefficients and the exponents of polynomial, e.g. multiplying random polynomials and unimodular matrices. These operations make the polynomial secure. We employ the algorithm in [6] in this paper. The specific procedures are summarized as Algorithm 2. This algorithm reduced the local computational complexity to $O(nlogn)$ while ensuring security.

$$U = \begin{pmatrix} u_{11}(x) & u_{12}(x) \\ u_{21}(x) & u_{22}(x) \end{pmatrix} = \begin{pmatrix} x^2 + x + 1 & x \\ x^2 & x + 1 \end{pmatrix} \in F_2[x]^{2 \times 2} \qquad (1)$$

Algorithm 2. Secure Outsourcing of the Extended Euclidean
Input: $a(x), b(x)$
Output: $u(x), v(x), d(x)$

1: Pick a polynomial $r(x)$ and an integer α randomly
2: $f(x) \leftarrow a(\alpha x)$, $g(x) \leftarrow b(\alpha x)$, $a'(x) \leftarrow r(x) \times f(x)$ and $b'(x) \leftarrow r(x) \times g(x)$
3: $a''(x) \leftarrow u_{11}(x) \times a'(x) + u_{12}(x) \times b'(x)$ and $b''(x) \leftarrow u_{21}(x) \times a'(x) + u_{22}(x) \times b'(x)$
4: Send $a''(x)$ and $b''(x)$ to the outsourcing computational nodes
5: Get $v''(x), u''(x)$ and $d''(x)$ from the outsourcing computational nodes
6: Verify $a''(x) \times u''(x) + b''(x) \times v''(x) = d''(x)$
7: Verify $deg(u''(x)) < deg(b''(x)/d''(x))$ and $deg(v''(x)) < deg(a''(x)/d''(x))$.
8: $u(x) \leftarrow \alpha^{deg(d''(x)) - deg(r(x))}(u_{11}(\alpha^{-1}x) \times u''(\alpha^{-1}x) + u_{21}(\alpha^{-1}x) \times v''(\alpha^{-1}x))$
9: $v(x) \leftarrow \alpha^{deg(d''(x)) - deg(r(x))}(u_{12}(\alpha^{-1}x) \times u''(\alpha^{-1}x) + u_{22}(\alpha^{-1}x) \times v''(\alpha^{-1}x))$
10: $d(x) \leftarrow \alpha^{deg(d''(x)) - deg(r(x))}d''(\alpha^{-1}x)/r(\alpha^{-1}x)$

The local computation of this algorithm consists of mostly modular exponentiations and polynomial multiplications, e.g. Steps 2, 3, and 6–10. To further reduce the local computation, we propose a secure outsourcing algorithm for polynomial multiplication and extend the secure outsourcing algorithm of modular exponentiation in [9] into the blockchain-based framework.

4 The Framework of Blockchain-Based Computation Outsourcing

This section introduces a blockchain-based computation outsourcing framework, which will be followed by the specific computation tasks in the rest of this paper.

Users and computational nodes need to register before joining the network. They need to pay deposits in advance, the amount of which needs to be greater than a certain threshold or they will be rejected. The blockchain initializes the same credit value to all new nodes and users. After registration, the information of users and computational nodes are written to the blockchain.

User p posts computing tasks, data, and rewards of each task to the blockchain network ($tasks = [task_1 \ldots task_n]$, $data = [data_1, \ldots, data_n]$). If the account balance of p is insufficient for the computations, blockchain refuses this service and declines the credit of p, to prevent malicious users from attacking the blockchain by constantly sending tasks that cannot afford. After the blockchain accepts the tasks of p, the user's computing tasks are stored in the task queue and wait for the selection of computing nodes.

It is active for the computational nodes to undertake computing tasks. If multiple computational nodes select the same task at the same time, the node with the highest credit will win the task. The nodes that undertake the computing task should submit the results after completing. If any computational node cannot finish on time, it will be added to the dishonest set. If all computing nodes are able to submit the results on time, the blockchain sends the results to the user and initiates the disputing period. During this period, the user needs to verify the results locally and notify the blockchain whether the results are accepted or not. If the disputing period ends and there is no feedback from the user is received, the blockchain assumes that the computing is successful and performs the reward and charge operations. If the user does not accept the results in the feedback, the blockchain will verify the results by itself.

If the user does not accept the computing results, the blockchain will perform verification operations and find dishonest nodes or users according to the recorded data before the final transaction. The dishonest nodes and users will be put into the dishonest set.

If no one is added into the dishonest set, the user will pay the reward to all participating nodes. Otherwise, the cheating nodes will be punished and the user will be compensated. The credit value of the participating nodes that have correctly completed their tasks will increase, while the credit value of the malicious nodes will decrease. When the deposit of a node is lower than the threshold value or the credit value is reduced to 0, the system will remove it.

5 Polynomial Multiplication and Modular Exponentiation Secure Outsourcing Algorithm

5.1 Secure Outsourcing Algorithm of Polynomial Multiplication

The computational complexity of traditional polynomial multiplication is $O(n^2)$, which is reduced to $O(nlogn)$ by the FFT. The local computational complexity of the secure outsourcing algorithm in this paper is $O(n)$. The main idea of this algorithm is as follows: Firstly, securely outsource the Fourier transform of the polynomial coefficients. Secondly, locally perform correlation operation on the results of the Fourier transform. Finally, securely outsource the inverse Fourier transform of the result of the correlation operation.

Description. The input polynomials of this algorithm are $f(x) = a_0 + a_1 x + \cdots + a_n x^n$ and $g(x) = b_0 + b_1 x + \cdots + b_n x^n$. And output is $t(x) = f(x) \times g(x) = c_0 + c_1 x + \cdots + c_{2n} x^{2n}$. For convenience, replace polynomials with vectors of polynomial coefficients $(a = [a_0, a_1, \ldots, a_n], b = [b_0, b_1, \ldots, b_n], c = [c_0, c_1, \ldots, c_{2n}])$.

We use $6p$ computational nodes in this algorithm, where p is a parameter associated with the number of computational nodes. There are six steps. (a) Six parameters are picked randomly, three of which are i, j, β, s.t. $0 \leq i \leq n, 0 \leq j \leq n$ and $0 \leq \beta \leq 2n$. The other three are $k_1, k_2, k_3 \in_R Z$. We define that $L(i, k, n) : \mathbb{Z}^3 \rightarrow \mathbb{Z}^n$ can generate one n-dimensional vector in which the ith element is k, and all the other elements are 0. The user generates $r_1 = L(i, k_1, n)$, $r_2 = L(j, k_2, n)$ and $r_3 = L(\beta, k_3, 2n)$. We define that $T(v, r) : \mathbb{Z}^{n*2} \to \mathbb{Z}^{n*n}$ can generate a random matrix $W = [w_1, w_2, \ldots w_p]$ satisfying $v = \sum_{i=1}^{p} w_i + r$. Then, the user generates $V = T(a, r_1)$, $U = T(a, r_1)$, $Z = T(b, r_2)$, and $S = T(b, r_2)$. (b) The vectors in V, U, Z, and S are distributed to $4p$ computational nodes. After receiving the vectors, the computational nodes perform the Fourier transform on the received vectors and return $\{F_{v_1}, \ldots, F_{v_p}\}$, $\{F_{u_1}, \ldots, F_{u_p}\}$, $\{F_{z_1}, \ldots, F_{z_p}\}$ and $\{F_{s_1}, \ldots, F_{s_p}\}$. Meanwhile, the user computes Fourier transform of r_1, r_2, and inverse Fourier transform of r_3 locally. (c) The user verifies the Eq. (2) and (3). If they are valid, the user computes $F_a = \sum_{i=1}^{p} F_{v_i} + F_{r_1}$, $F_b = \sum_{i=1}^{p} F_{z_i} + F_{r_2}$, and $F_c = F_a \circledast F_b$. (d) The user generates $D = T(F_c, r_3)$ and $E = T(F_c, r_3)$ and distributes the vectors in D and E to $2p$ computational nodes. (e) After receiving the vectors, the computational nodes perform the inverse Fourier transform on the received vectors and return $\{F^{-1}_{d_1}, \ldots, F^{-1}_{d_p}\}$ and $\{F^{-1}_{e_1}, \ldots, F^{-1}_{e_p}\}$. (f) The user computes $c = F^{-1}_{r_3} + F^{-1}_{d_1} + \cdots + F^{-1}_{d_p}$, selects two integers $m, l \in_R \{0, \ldots, 2n\}$ randomly, and verifies the Eq. (4), (5), and (6). If they are valid, the computing is successful.

$$F_{v_1} + F_{v_2} + \cdots + F_{v_p} = F_{u_1} + F_{u_2} + \cdots + F_{u_p} \tag{2}$$

$$F_{z_1} + F_{z_2} + \cdots + F_{z_p} = F_{s_1} + F_{s_2} + \cdots + F_{s_p} \tag{3}$$

$$F^{-1}_{d_1} + F^{-1}_{d_2} + \cdots + F^{-1}_{d_p} = F^{-1}_{e_1} + F^{-1}_{e_2} + \cdots + F^{-1}_{e_p} \tag{4}$$

$$c_l = \sum_{i=0}^{l} a_i b_{l-i} \tag{5}$$

$$F_c[m] = \sum_{i=0}^{2n} W_{2n+1}^{mi} c_i \tag{6}$$

Algorithm 3. Secure outsourcing of Polynomial Multiplication
Input: a, b
Output: $c = a \times b$

1: Pick six random parameters $i, j, \beta, k_1, k_2, k_3$.
2: Generate three vectors r_1, r_2, r_3 and four matrixes V, U, Z, S.
3: Distribute vectors in V, U, Z and S to $4p$ computational nodes.
4: Get $\{F_{v_1}, \ldots, F_{v_p}\}, \{F_{u_1}, \ldots, F_{u_p}\}, \{F_{z_1}, \ldots, F_{z_p}\}, \{F_{s_1}, \ldots, F_{s_p}\}$ from computational nodes.
5: Compute F_{r_1}, F_{r_2} and the inverse Fourier transform of $F^{-1}_{r_3}$ locally.
6: Verify the Eq. (2) and (3).
7: Compute F_a, F_b and F_c and generate D and E.
8: Distribute vectors in D and E to $2p$ computational nodes.
9: Get $\{F^{-1}_{d_1}, \ldots, F^{-1}_{d_p}\}, \{F^{-1}_{e_1}, \ldots, F^{-1}_{e_p}\}$ from the computational nodes.
10: Compute c and verify the Eq. (4), (5) and (6).

In Algorithm 3, if any verification fails, the user will report cheating and the algorithm will come to an end.

Correctness Because $r_1 + v_1 + v_2 + \cdots + v_n = a$, $r_2 + z_1 + z_2 + \cdots + z_n = b$, and the Fourier transform is a linear transform, we can have $F_{r_1} + F_{v_1} + \cdots + F_{v_n} = F_a$ and $F_{r_2} + F_{z_1} + \cdots + F_{z_n} = F_b$. We get F_c by $F_a \circledast F_b$. Because of the convolution theorem, $F^{-1}_{F_c} = a \times b = c$.

Security Analysis. A participating node may be a passive or active attacker. Passive attackers will follow the scripts of the algorithm while exploiting the intermediate information to breach the privacy of polynomials. Active attackers will inject false computations into the algorithm to tamper with the whole process. In the following, we analyze the security of our algorithm against passive attackers and active attackers.

Security Against Passive Attackers. Take the attack on $f(x)$ as an example. Only if all the computational nodes collude, they can guess a set of values $[a_0', a_1', \ldots, a_n']$, in which n values are consistent with the true coefficients of $f(x)$, while one value is not. Because of r_1, they even do not know the position of the false value. They still have to make a brute-force guessing. assuming that the range of the coefficient of $f(x)$ is a field with m numbers, the attacker should traverse all possibilities by taking M different values for each coefficient. In this case, the attacker has to make m^{n+1} attempts to get $f(x)$. However, the values of polynomial coefficients are from the integer domain Z in FHEHIL. When $m \rightarrow \infty$, the attacker cannot get $f(x)$. In the same way, the inverse Fourier transform is also privacy-protected.

Security Against Active Attackers. It is easy to see that the lowest risk way for computational nodes to cheat is to tamper with only one item of the result returned to the user, while the other items are correct. There is one way to cheat in the process of securely outsourcing fourier transform. For example, the nodes of computing the DFT of $f(x)$ perform honestly, while computing the DFT of $g(x)$, one node n_j changed the ith term in F_{z_j} and another node n_k' also changed the ith term in F_{s_k}'. This way of cheating can nullify the verifications at Eq. (3) and (4), whereas the resulted error in the ith term of F_b will be propagated to every term of c, through Step 7–9. Thus, the verification at Eq. (5) can certainly detect this cheating.

There is the other way to cheat in the process of securely outsourcing inverse Fourier transform. For example, in the process of computing the inverse DFT of F_c, one node n_j changed the ith term in $F_{d_j}^{-1}$ and another node n_k' also changed the ith term in $F_{e_k}^{-1}$. This way can nullify the verification at Eq. (5), whereas the forged c' with an error in the ith term cause that the mth item in $F_{c'}$ is not equal to $F_c[m]$ for all $0 \leq m \leq 2n$. Therefore, the verification at Eq. (6) can certainly detect this way of cheating.

Algorithm Complexity. Computing Fourier transforms of r_1, r_2 and inverse Fourier transform of r_3 need $4n$ multiplications. Because of the characteristics of r_1, r_2 and r_3, only one multiplication is needed to compute each term in $F_{r_1} F_{r_2}$ and $F_{r_3} (eg : F_{r_1}[i] = F_{r_1}[i-1]W_n^{k_1})$. Computing F_a and F_b need $2pn$ addition operations. Computing F_c needs $2n$ multiplications. The verifications of Eq. (2), (3), and (4) take $6(p-1)n$ additions operations. $2pn$ additions are needed to compute c. The final verifications (Eq. (5) and (6)) need $l + 2n$ multiplication operations. To sum up, the local complexity of this algorithm is $O(n)$.

5.2 Secure Outsourcing Algorithm of Modular Exponentiation

To the secure outsourcing of modular exponentiation, we extend the algorithm of Fu et al. [6] and apply it to the blockchain. In our extension, different exponentiations are outsourced to different computational nodes, instead of one single node as in [6], aiming to protect against possible attacks on small discrete logarithms.

Description. The input of Algorithm 4 is two integers. the output is u^d where u is the base and d is the exponent.

Algorithm 4. Secure outsourcing of modular exponentiation

Input: u, d

Output: u^d

1: Pick random integers $g_1, g_2, e, k_1, k_2 \in Z$.

2: $v_1 = g_1^e, v_2 = g_2^e, w_1 = {}^u/g_1, w_2 = {}^u/g_2$.

3: $t_1 = d - k_1 e$ and $l_1 = d - k_2 t_1$.

4: Send $(k_1, v_1), (k_1, v_2), (l_1, w_1), (k_2, w_1), (l_1, w_2), (k_2, w_2)$ to six computational nodes.

5: Computational nodes compute $b_i{}^{a_i}$ after receiving (a_i, b_i).

6: Get $v_1{}^{k_1}, \ v_2{}^{k_1}, \ w_1{}^{l_1}, \ w_1{}^{k_2}, \ w_2{}^{l_1}, \ w_2{}^{k_2}$ from computational nodes.

7: Verify $v_1{}^{k_1} w_1{}^{l_1} (g_1 w_1{}^{k_2})^{t_1} = v_2{}^{k_1} w_2{}^{l_1} (g_2 w_2{}^{k_2})^{t_1}$.

8: Accept the result $u^d = v_1{}^{k_1} w_1{}^{l_1} (g_1 w_1{}^{k_2})^{t_1}$ if verification is passed.

Correctness and Security. It is easy to prove that the algorithm is correct from Eq. (7) and (8). The only way to pass the verification is that the six computational nodes perform correctly. The forged results of active attackers cannot pass the verification in step 7. The user only needs to know the results are correct or not, and the blockchain can detect the cheating nodes according to the records.

$$v_1{}^{k_1} w_1{}^{l_1} \left(g_1 w_1{}^{k_2}\right)^{t_1} = v_1{}^{k_1} g_1{}^{t_1} w_1{}^{l_1} \left(w_1{}^{k_2}\right)^{t_1} = g_1{}^d w_1{}^d = u^d \tag{7}$$

$$v_2{}^{k_1} w_2{}^{l_1} \left(g_2 w_2{}^{k_2}\right)^{t_1} = v_2{}^{k_1} g_2{}^{t_1} w_2{}^{l_1} \left(w_2{}^{k_2}\right)^{t_1} = g_2{}^d w_2{}^d = u^d \tag{8}$$

We analyze the security in the worst case, i.e., the conspiring of six computational nodes. The exponents, k_1, l_1, k_2 are visible for attackers. And the attackers cannot get t_1, d, e. For the base information, v_1, v_2, w_1, w_2 are visible and g_1, g_2, e are invisible for attackers. We discover that the privacy of u is not safe in [9], which sends six pairs to a single node in the cloud. In [9] the base and exponent are about 1000-bit, while the parameters including g_1, g_2, e, k_1, k_2 are only 64-bit long, to reduce the overhead of local computation. The shorter bit length of parameters may promote an easier attack on the small discrete logarithms. In this kind of attack, an attacker in the cloud can exhaust an x, so that $w_1{}^x * v_1 = w_2{}^x * v_2$. Then e is breached. The attacker then exhausts g satisfying $g^e = v_1$. Finally, the cloud can get u by $w_1 * g$.

We solve this attack by distributing six pairs of data to six computational nodes, which increases the difficulty of the above attack.

Algorithm Complexity. In the process of parameter generation, there are two exponentiations, two divisions, and two multiplications. Two exponentiations and six multiplications are involved during the verification. Compared with the exponentiation, the complexity of multiplication and division can be ignored. However, in the algorithm, the exponents are e and t_1 that are much smaller than the original exponent d through the transformation of $t_1 = d - k_1 e$. Therefore, local complexity will be greatly reduced.

6 Performance Evaluation

Our experiments were conducted on one machine with Intel Core i7 processors running at 2.00 GHz and 8G memory. We simulated the blockchain-based outsourcing on this machine, where all the local operations and the outsourced operations were undertaken and measured. We employ the relevant security parameters recommended in [3], i.e., $n = 1024$, $\tau = 310$, and $p = 3$. Therefore, in the public key generation stage and encryption stage, this machine as a computational node undertook about 3720 FFT and 1860 IFFT computations. It means that we should divide the FFT running time and IFFT running time of the program of computational nodes by 3720 and 1860 respectively to get an approximate computation outsourcing time, in the analysis of experimental results. Our outsourcing scheme consists of the local user's program and the computational nodes' program. The programs were written in Python3, and the smart contract based on the Ethereum platform was written in Solidity. The smart contract interacts with computational nodes program and local program by the interface provided by Web3.js.

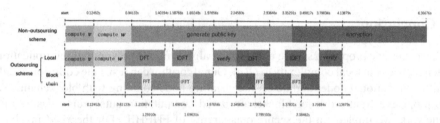

Fig. 1. Time consumptions on stages of the outsourcing and non-outsourcing schemes

Figure 1 demonstrates the running times at stages of the two schemes. This figure does not show the time-consumption of generating parameters in the FHEHIL, because that is not what we are improving. In the process of *xgcd* to obtain *w*, the efficiency is slightly improved. Our scheme saves about 2.2 s. The overall time consumption is improved by about 36.5%. (The unmarked areas in Fig. 1 are the communication time consumption for interacting with the blockchain).

We also made experiments to separately evaluate the efficiency of the secure outsourcing algorithm of polynomial multiplication. In different numbers of polynomial multiplications, we compared it with the non-outsourcing algorithm on time consumption ($n = 1024$ and $p = 3$). As demonstrated in Fig. 2, it is easy to see that when the number of polynomial multiplications is less than 60, the efficiency of the outsourcing scheme is lower than the non-outsourcing one due to the communication time consumption for interacting with the blockchain. However, when the number of polynomial multiplications increases, the efficiency of the outsourcing scheme becomes higher than the non-outsourcing scheme. When the number of polynomial multiplications is less than 300, the bottleneck of the outsourcing scheme is the time consumption on nodes' computations and interactions. When the number of polynomial multiplications becomes larger, the bottleneck is the time consumption on local computations.

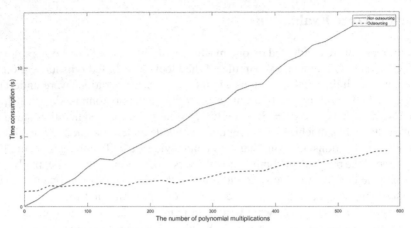

Fig. 2. Comparison of time consumption on the outsourcing and non-outsourcing scheme of polynomial multiplications

7 Conclusion

In this paper, we propose a secure outsourcing algorithm for polynomial multiplication which reduces the local complexity to $O(n)$. Our algorithm can resist the collusion attack of all computational nodes and detect the cheating node combing with blockchain. We also propose a framework for blockchain-based computation outsourcing. Using this framework, we implement the secure outsourcing of FHEHIL. For the $xgcd$ involved in FHEHIL, we improve the scheme in [6] by simplifying the local operations and outsourcing polynomial multiplication and modular exponentiation. The security analysis and experimental results show that our proposed outsourcing scheme for FHEHIL is secure and efficient.

Acknowledgement. This work was supported by the Key-Area Research and Development Program of Guangdong Province (NO. 2020B010164003), the Science and Technology Program of Guangzhou, China (No. 201904010209), and the Science and Technology Program of Guangdong Province, China (No. 2017A010101039).

References

1. Nakamoto, S.: Bitcoin: A Peer-to-Peer Electronic Cash System (2008). https://bitcoin.rog/bitcoin.pdf
2. Zhong, H., Sang, Y., Zhang, Y., Xi, Z.: Secure multi-party computation on blockchain: an overview. In: Shen, H., Sang, Y. (eds.) PAAP 2019. CCIS, vol. 1163, pp. 452–460. Springer, Singapore (2020). https://doi.org/10.1007/978-981-15-2767-8_40
3. Thomas, T., Willy, S., Zhenfei, Z.: Fully homomorphic encryption using hidden ideal lattice. IEEE Trans. Inf. Forensics Secur. **8**(12), 2127–2137 (2013)
4. Smart, N.P., Vercauteren, F.: Fully homomorphic encryption with relatively small key and ciphertext Sizes. In: Nguyen, P.Q., Pointcheval, D. (eds.) PKC 2010. LNCS, vol. 6056, pp. 420–443. Springer, Heidelberg (2010). https://doi.org/10.1007/978-3-642-13013-7_25

5. Gentry, C., Halevi, S.: Implementing Gentry's fully-homomorphic encryption scheme. In: Paterson, K.G. (ed.) EUROCRYPT 2011. LNCS, vol. 6632, pp. 129–148. Springer, Heidelberg (2011). https://doi.org/10.1007/978-3-642-20465-4_9

6. Zhou, Q., Tian, C., Zhang, H., Yu, J., Li, F.: How to securely outsource the extended Euclidean algorithm for large-scale polynomial over finite fields. Inf. Sci. **512**, 641–660 (2020)

7. Gennaro, R., Gentry, C., Parno, B.: Non-interactive verifiable computing: outsourcing computation to untrusted workers. In: Rabin, T. (ed.) CRYPTO 2010. LNCS, vol. 6223, pp. 465–482. Springer, Heidelberg (2010). https://doi.org/10.1007/978-3-642-14623-7_25

8. Fei, C., Tao, X., Yuanyuan, Y.: Privacy-preserving and verifiable protocols for scientific computation outsourcing to the cloud. J. Parallel Distrib. Comput. **74**(3), 2141–2151 (2014)

9. Anmin, F., Shuai, L., Shui, Y., Yuqing, Z., Yinxia, S.: Privacy-preserving composite modular exponentiation outsourcing with optimal checkability in single untrusted cloud server. J. Netw. Comput. Appl. **118**, 102–112 (2018)

10. Brakerski, Z.: Fully homomorphic encryption without switching from classical GapSVP. In: Safavi-Naini, R., Canetti, R. (eds.) Advances in Cryptology-CRYPTO 2012. LNCS, vol. 7417, pp. 868–886. Springer, Heidelberg (2012). https://doi.org/10.1007/978-3-642-32009-5_50

11. Gentry, C., Sahai, A., Waters, B.: Homomorphic encryption from learning with errors: conceptually-simpler, asymptotically-faster, attribute-based. In: Canetti, R., Garay, J.A. (eds.) CRYPTO 2013. LNCS, vol. 8042, pp. 75–92. Springer, Heidelberg (2013). https://doi.org/10.1007/978-3-642-40041-4_5

12. Zvika, B., Vinod, V.: Efficient fully homomorphic encryption from (standard) LEW. SIAM J. Comput. **43**(2), 831–871 (2014)

13. Hohenberger, S., Lysyanskaya, A.: How to securely outsource cryptographic computations. In: Kilian, J. (ed.) TCC 2005. LNCS, vol. 3378, pp. 264–282. Springer, Heidelberg (2005). https://doi.org/10.1007/978-3-540-30576-7_15

14. Chen, X., Li, J., Ma, J., Tang, Q., Lou, W.: New algorithms for secure outsourcing of modular exponentiations. IEEE Trans. Parallel Distrib. Syst. **25**(9), 2386–2396 (2014)

15. Ren, Y., Ding, N., Zhang, X., Lu, H., Gu, D.: Verifiable outsourcing algorithms for modular exponentiations with improved checkability. In: ASIA CCS 2016: Proceedings of the 11th ACM on Asia Conference on Computer and Communications Security, pp. 293–303. ACM (2016)

16. Zang, Y., Deng, R.H., Liu, X., et al.: Blockchain based efficient and robust fair payment for outsourcing services in cloud computing. Inf. Sci. **462**, 262–277 (2018)

17. Wang, H., Wang, X.A., Wang, W., Xiao, S.: A basic framework of blockchain-based decentralized verifiable outsourcing. In: Barolli, L., Nishino, H., Miwa, H. (eds.) Advances in Intelligent Networking and Collaborative Systems, INCoS 2019. Advances in Intelligent Systems and Computing, pp. 415–421, vol. 1035. Springer, Cham. https://doi.org/10.1007/978-3-030-29035-1_40

18. Hao, K., Xin, J., Wang, Z., Wang, G.: Outsourced data integrity verification based on blockchain in untrusted environment. World Wide Web **23**(4), 2215–2238 (2020). https://doi.org/10.1007/s11280-019-00761-2

19. Zheng, H., Shao, J., Wei, G.: Attribute-based encryption with outsourced decryption in blockchain. Peer-to-Peer Netw. Appl. **13**(5), 1643–1655 (2020). https://doi.org/10.1007/s12083-020-00918-1

Multiple Projections Learning
for Dimensional Reduction

Lin Jiang, Xiaozhao Fang$^{(\boxtimes)}$, and Na Han

School of Computer Science and Technology, Guangdong University of Technology,
Guangzhou 51006, China
isabel@mail2.gdut.edu.cn, xzhfang@126.com

Abstract. Locality Preserving Projection (LPP) is a dimensional reduction method that has been widely used in various fields. While traditional LPP only uses a single projection matrix to reduce the dimension and preserve the locality structure of data, it may cause the single matrix may not handle these two tasks well at the same time. Therefore, in this paper, we proposed relaxed sparse locality presenting projection (RSLPP) which introduces two different projection matrices to better accomplish the two tasks. The addition of another projection matrix can help the original projection matrix has more freedom to select the appropriate feature for preserving the local structure of data. The experimental results on two data sets prove the effectiveness of the method.

Keywords: Dimensional reduction · Locality preserving projection · Sparse constraint

1 Introduction

With the development of information technology and image acquisition equipment, high-dimensional data are widely acquired. While in real world application, compare with low-dimensional data, processing high-dimensional data requires higher computational complexity, calculation time and memory requirements. In order to solve this problem, dimensional reduction(DR) was proposed to find a low-dimensional subspace to map the high-dimensional data into a low dimensional form. In the past few years, many dimensionality reduction methods have been put forward and extensive used in various fields [3,5].

Locality preserving projection (LPP) [1], as a promising DR method, utilizes the graph to preserve the local structure of data optimally. However, the above-mentioned algorithms still have a shortcoming that they only use a single matrix as projection matrix. Which may cause excessive pressure on the projection

This work was supported in part by the National Natural Science Foundation of China under Grant 61772141, Grant 62006048 and Grant 61972102, in part by Science and Technology Planning Project of Guangdong Province, China, under Grant 2019B020208001, Grant 2019B110210002, and in part by the Guangzhou Science and Technology Planning Project under Grant 201903010107 and Grant 201802010042 .

© Springer Nature Switzerland AG 2021
Y. Zhang et al. (Eds.): PDCAT 2020, LNCS 12606, pp. 166–171, 2021.
https://doi.org/10.1007/978-3-030-69244-5_14

matrix to learn low-dimensional subspaces and retain data structures, thereby affecting the effect of the model.

To address this issue, in this paper, we propose a novel dimensionality reduction algorithm called relaxed sparse locality presenting projection (RSLPP). To alleviate the pressure of the single matrix, RSLPP applies two projection matrices to handle these tasks. Moreover, in order to explore the correlation between two matrices, RSLPP uses linear reconstruction to reduce the dependency between the two projection matrices.

2 Related Work

2.1 Locality Preserving Projection

LPP is a nonlinear projection method that maintains the local structure of the original data. Let matrix $X = [x_1, x_2, ..., x_n] \in \Re^{m \times n}$ as the collection of training samples in which m and n are the dimensionality and the number of training samples, respectively. The objective function is as follows

$$\min_Y \sum_{ij} (y_i - y_j)^2 w_{ij} \quad s.t. \quad y^T D y = 1 \tag{1}$$

Where $y_i = p^T x_i$ is the low-dimensional representation of x_i, and p is the projection vector, $D_{ij} = \sum_j W_{ij}$, W_{ij} is defined as

$$W_{ij} = \begin{cases} \exp\left(- \|\mathbf{x}_i - \mathbf{x}_j\|^2 / t\right), & \|\mathbf{x}_i - \mathbf{x}_j\|^2 < \varepsilon \\ 0 \text{ otherwise}, \end{cases} \tag{2}$$

Where t is the heat kernel parameter.

2.2 Graph Regularization Technique

Graph regularization technique is proposed to preserve the locality of original data in the low-dimensional subspace [6]. The similarity W_{ij} between samples x_i and x_j is defined as:

$$W_{ij} = \begin{cases} 1, & \text{if } x_i \in \aleph_K(x_j) \text{ or } x_j \in \aleph_K(x_i) \\ 0, & \text{otherwise} \end{cases} \tag{3}$$

where $\aleph_K(x_i)$ represents the set of K nearest neighbors of x_i.

3 Proposed method

On the basic of LPP, we propose to add another projection matrix $B \in \Re^{m \times d}$ to relax the work of projection matrix $A \in \Re^{m \times d}$, in which m is the dimension of original data and d is the dimension of the low-dimensional subspace. To capture the similarity between these two projection matrices, we use matrix $Z \in \Re^{c \times c}$ and projection matrix B here to reconstruct another projection matrix A. Also add a sparse constraint to the reconstruction matrix Z to maintain the similarity structure of matrices A and B. Therefore, we formulate the objective function of RSLPP as follows:

$$\min_{A,B,Z} \sum_{i}^{n} \sum_{j}^{n} \| A^T x_i - B^T x_j \|^2 w_{ij} + \lambda_1 \| A - BZ \|_F^2 + \lambda_2 \| Z \|_1 \tag{4}$$

$$s.t. \ A^T A = I$$

Where the constraint $A^T A = I$ is introduced to avoid the trivial solution of A. $\|Z\|_1$ is the ℓ_1-norm which is the sum of absolutes of all entries. $\lambda_1 \geq 0$ and $\lambda_2 \geq 0$ are two trade-off parameters. The definition of graph Laplacian is $L = D - W$, where D is a diagonal matrix and $D_{ij} = \sum_j W_{ij}$. The weight w_{ij} of graph is defined as:

$$w_{ij} = \begin{cases} e^{-\frac{\|x_i - x_j\|^2}{\sigma}}, & \text{if } x_i \text{ and } x_j \text{ are } k \text{ nearest neighbors} \\ 0, & \text{otherwise} \end{cases} \tag{5}$$

3.1 Optimization

We adopt an iterative optimization algorithm to solve problem (5). For calculation convenience, we introduce an auxiliary variable J to replace Z. We use the augmented lagrange multiplier (ALM) [4] method to solve the problem. We reformulate (5) into the following ALM function:

$$L(A, B, Z, J, Y_1, Y_2) = Tr(A^T X D X^T A) + Tr(B^T X D X^T B) - 2Tr(A^T X W X^T B)$$

$$+ \lambda_1 \|A - BZ\|_F^2 + \lambda_2 \|J\|_1 + Tr(Y_1^T(Z - J)) + Tr(Y_2^T(A^T A - I)) \tag{6}$$

Where D is a diagonal matrix $(D_{ij} = \sum_j W_{ij})$ and the graph Laplacian matrix L is defined as $L = D - W$. $\mu > 0$ is a penalty parameter, and Y_1, Y_2 are Lagrange multipliers. Parameters A, B, Z, J can be minimized respectively. Through uncomplicated calculation, the optimization results are as follows.

A can be updated by the following formula:

$$(XDX^T + \lambda_1)A + AY_2 - XWX^T B - \lambda_1 BZ = 0 \tag{7}$$

A is essentially updated by solving Sylvester equation.

B can be updated by the following formula:

$$XDX^T B + \lambda_1 BZZ^T - XW^T X^T A - \lambda_1 AZ^T = 0 \tag{8}$$

B is also update by utilizing Sylvester equation.

Z can be updated by the following formula:

$$Z = (2\lambda_1 B^T B + \mu I)^{-1}(\mu(J + \frac{Y_1}{\mu} + 2\lambda_1 A^T B)) \tag{9}$$

J can be updated by the following formula:

$$J_{K+1} = max\{\Theta_{\frac{\lambda_2}{\mu}}(\mu(Z_k - J_k + \frac{Y_1^k}{\mu})), 0\} \tag{10}$$

Where Θ is the ℓ_1 shrinkage minimization operator [2].

Lagrange multipliers Y_1, Y_2 and penalty parameter μ are updated by using:

$$\begin{cases} Y_1 = Y_1 + \mu(Z - J) \\ Y_2 = Y_2 + \mu(A^T A - I) \\ \mu = \min\{\rho\mu, \mu_{max}\} \end{cases} \tag{11}$$

The algorithm framework of solving problem (9) summarizes in Algorithm 1.

Algorithm 1. RSLPP

---
procedure
Input: Training samples matrix X; Weight matrix W; Parameters λ_1, λ_2 and μ.
Initialization: $A^* = \arg\min_A \text{Tr}(A^T(-\Sigma)A)$; where Σ is the data covariance;
$A = B$; $Z = \mathbf{1}_{c \times c}$; $J = \mathbf{1}_{c \times c}$; $Y_1 = \mathbf{1}_{c \times c}$; $Y_2 = \mathbf{1}_{c \times c}$; $\mu = 1e - 6$; $\rho = 1.01$
while not converged do
1. Update A by solving (7);
2. Update B by solving (8);
3. Update Z by solving (9);
4. Update J by solving (10);
5. Update Y_1, Y_2, μ by (11)
End while
Output: The projection matrix A and structure similarity matrix Z.
end procedure

---

4   Experiment

In this section, we evaluate the effectiveness of our method on two public of data sets with different types: 1) AR; 2) COIL20. In our experiments, we use the nearest neighbor classifier with the Euclidean distance to classify sample.

4.1   Data Sets and Experiment Setup

AR: The database contains 126 people and a total of more than 4000 images. Each classes contains 26 frontal face images with different facial expressions, lighting conditions, and facial occlusions. In this experiment, we randomly select 8, 10, 12, 15, 20 images for training, and the remaining images for testing. Parameters $\lambda_1 = 1e2, \lambda_2 = 1e1, dim = 400$ are used in this experiment.

Table 1. Classification accuracies (%) of different methods on AR data set

#Tr	NN	PCA	NPE	SPP	LPP	LPP-ℓ_1	OLPP	**RSLPP**
8	73.74	74.33	80.47	88.73	67.53	90.62	75.67	**93.36**
10	78.64	78.60	85.09	92.43	73.56	93.23	77.06	**94.97**
12	81.79	82.26	90.02	94.67	78.95	95.39	82.20	**96.48**
15	86.40	86.48	87.70	96.70	84.73	97.21	85.46	**97.67**
20	91.44	90.60	93.17	98.09	91.31	98.47	91.25	**98.92**

Table 2. Classification accuracies (%) of different methods on COIL20 data set

#Tr	NN	PCA	NPE	SPP	LPP	LPP-ℓ_1	OLPP	RSLPP
8	87.41	87.89	88.04	87.34	78.59	86.82	87.47	**88.39**
10	89.45	89.97	90.27	89.48	80.41	89.54	89.91	**91.34**
15	93.37	93.15	92.95	91.94	84.55	93.05	92.94	**94.60**
20	94.96	95.25	94.79	94.17	86.73	94.68	95.32	**96.56**
25	96.84	96.21	95.30	95.74	88.85	96.43	96.51	**98.06**

COIL20: The COIL20 data set consists of 1440 images of 20 categories. These images are rotated 360 degrees for each object, and one image is taken every 5 degrees. Each object provides 72 images. We randomly select 8, 10, 15, 20, and 25 images for each object for training, and the rest are tested separately. Parameters $\lambda_1 = 1e2, \lambda_2 = 1e-1, dim = 75$ are used in this experiment.

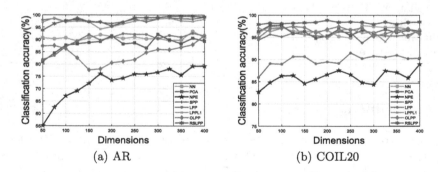

(a) AR   (b) COIL20

Fig. 1. The classification accuracy(%) versus different dimensions on (a) AR (b) COIL20 data sets.

4.2 Experiment Result and Analysis

The experimental results on AR and COIL20 data sets are shown in Tables 1 and Tables 2, respectively. $\#Tr$ represents the number of training samples per class. Each algorithm runs 20 times and records the average classification accuracy.

In Table 1, our proposed obtains the best classification results, and the classification accuracy of our method is as high as 93% with small base sample. This shows that our method has strong generalization ability. Figure1(a) shows the classification accuracy versus different dimensions on AR data set, it can be seen that our algorithm achieves ideal experimental results in each dimension of AR dataset, which shows that our algorithm has strong generalization ability and can achieve good results in a certain range of dimensions.

From Table 2, we can easily find that the experimental results of our method achieves the best results among all algorithms. Although the images in this data set have some rotations, our method is also robust to such rotation. Figure1(b) shows the classification accuracy versus different dimensions on COIL20 data set, it can bee seen that RSLPP achieve the best classification accuracy in all dimensions. Moreover, the classification accuracy of the RSLPP changes smoothly with the change of the dimension, and is more stable than other algorithms, indicating that the algorithm has strong robustness and is less affected by the dimension.

5 Conclusions

This paper proposes a novel dimensional reduction method called relaxed sparse locality presenting projection (RSLPP). RSLPP uses two different projection matrices to project the original data into the low dimensional subspace. This allows the task of the projection matrix to be shared by the two matrices so that the matrices have more freedom to preserve the local structure and data representation of the data. The promising experimental results on two datasets show that the performance of RSLPP outperforms the state-of-the art methods.

References

1. He, X., Niyogi, P.: Locality preserving projections. Adv. Neural Inf. Process. Syst. **16**(16), 153–160 (2003)
2. Lin, Z., Chen, M., Ma, Y.: The augmented lagrange multiplier method for exact recovery of corrupted low-rank matrices. CoRR abs/1009.5055 (2010)
3. Xu, H., Caramanis, C., Mannor, S.: Outlier-robust PCA: the high-dimensional case. IEEE Trans. Inf. Theory **59**(1), 546–572 (2013)
4. Xu, Y., Fang, X., Wu, J.: Discriminative transfer subspace learning via low-rank and sparse representation. IEEE Trans. Image Process. **25**(2), 850–863 (2016)
5. Yan, S., Xu, D., Zhang, B., Zhang, H.: Graph embedding: a general framework for dimensionality reduction. In: 2005 IEEE Computer Society Conference on Computer Vision and Pattern Recognition (CVPR 2005), pp. 830–837 (2005)
6. Yi, S., Liang, Y., He, Z., Li, Y., Cheung, Y.: Dual pursuit for subspace learning. IEEE Trans. Multi. **21**(6), 1399–1411 (2019)

Preventing DDoS Attacks on Bitcoin Memory Pool by the Dynamic Fee Threshold Mechanism

Shunchao Luo, Yingpeng Sang$^{(\boxtimes)}$ (ID), Mingyang Song, and Yuying Zeng

School of Data and Computer Science, Sun Yat-sen University, Guangzhou, China
{luoshch5,songmy5,zengyy26}@mail2.sysu.edu.cn, sangyp@mail.sysu.edu.cn

Abstract. Blockchain is a well-known distributed technology which can be regarded as a decentralized database and combines the peer-to-peer architecture, cryptography and consensus mechanism. As the first and most famous application of blockchain and cryptocurrencies, Bitcoin also suffers from different kinds of attacks. Several studies have been done to analyze and solve these attacks. In this paper, we research the Distributed Denial-of-Service (DDoS) on Bitcoin's memory pool (Mempool). We analyze the feasibility of this kind of attack based on the current mining and relay progress of Bitcoin, and propose the possible effects caused by this attack from some novel aspects of Bitcoin, including the confirmation time, market price and legitimate user. We further present the Dynamic Fee Threshold Mechanism to counter this attack. Our method uses several characteristics to distinguish the normal transactions in Bitcoin from the malicious spam transactions. We also make an analysis and experiments on the method, to demonstrate its effectiveness and accuracy of detecting the malicious transactions.

Keywords: Blockchain · Memory pool · DDoS attack · Dynamic Fee Threshold

1 Introduction

Blockchain is one of the most famous information technology and structures in recent years. Due to its features like decentralization, transparency, immutability, it is widely used in several fields, especially in cryptocurrencies like Bitcoin (BTC) [2], Ethereum [3], etc.

Despite some advantages that blockchain provides to those applications, it meanwhile raises some risks. Blockchain is based on peer-to-peer (p2p) architecture that allows all the participants to exchange data without a central node. Participants in blockchain transfer the transactions and blocks to all the directly connected nodes within a few seconds [1], which is called "flooding". This p2p system is the main reason why blockchain has decentralized trust, however it is also vulnerable to attacks. Distributed Denial-of-Service (DDoS) is a classic attack towards traditional p2p system which can also aim at blockchain.

© Springer Nature Switzerland AG 2021
Y. Zhang et al. (Eds.): PDCAT 2020, LNCS 12606, pp. 172–184, 2021.
https://doi.org/10.1007/978-3-030-69244-5_15

According to statistics of Bitcoin we can know each block contains about 2200 transactions on average, and each block costs about 10 min to be mined and generated. Such a low transaction throughput of cryptocurrencies (e.g., Bitcoin with 4–6 transactions per second and Ethereum with about 15 transactions per second) leads to the opportunity for DDoS attack.

The first kind of DDoS attack uses the limit of block size in Bitcoin, in which the attacker generates the spam transactions with low value to waste the space of block and decrease the rate of legitimate transactions' verification. One of the methods to prevent this kind of attack is that the miner can locally compute priority of transactions and decide which transaction to be mined first.

To store all the unconfirmed transactions, each Bitcoin node has to maintain a local memory space called "Memory Pool" (Mempool). So even though the spam transactions cannot be included in blocks, they still can occupy the space of Mempool. As mentioned above, the transaction throughput of Bitcoin is so low that if the input rate is higher than the transaction throughput, the Mempool size will keep growing, and finally the memory pressure will force nodes to drop some transactions which may include some legitimate transactions. Therefore, DDoS attack on Mempool affects the honest users on cryptocurrencies like Bitcoin. Saad et al. [4] proposed the countermeasures based on fee and age of transactions to filter out malicious transactions, while their measures were not dynamically adjusted according to the condition of Mempool size. Jung W et al. [5] presented a dynamic transaction limit volume approach to limit the transaction a node can create during a period time. However, since the attacker only holds a small part of nodes in Bitcoin, in their method the normal nodes can still get limited, and some urgent transactions cannot be processed in time.

Confronted with this attack, our goal is to maintain the Mempool size in an appropriate range, and to exhaust the attackers' budget. Thus, we propose Dynamic Fee Threshold Mechanism to counter this kind of DDoS attack on Mempool. Our contributions can be summarized as follows: 1) We analyze the feasibility of the DDoS attack on Mempool based on the current Bitcoin core. 2) We propose a series of effects that may be caused by this attack from some novel aspects of Bitcoin, including the confirmation time, market price and legitimate user. 3) We propose a novel countermeasure using dynamic fee threshold, and conduct several experiments to prove its effectiveness and illustrate its advantages compared with related work.

The rest of this paper is organized as follows. We introduce some background of Bitcoin and the related work in Sect. 2. In Sect. 3, we discuss and analyze the DDoS attack on Mempool. In Sect. 4, we present a countermeasure. In Sect. 5, we conduct some experiments and analyze the performance. The paper is concluded in the Sect. 6.

2   Background

2.1   Bitcoin

Transaction. In Bitcoin, a transaction is generated by the user to transfer value between wallets which contains at least one input and one output. Each input of the transaction spends the value that is paid to an output of previous transaction. Then the output becomes an Unspent Transaction Output (UTXO) if the transaction gets confirmed and mined, until another transaction spends it. Each participant's balance is made up of all its UTXOs. A transaction's input can be the output of an unconfirmed old transaction, but the new one cannot be confirmed until the old one gets confirmed.

Transaction Fee. Transaction fee equals the value of all outputs minus all inputs, which is an incentive reward for the miner who confirms and includes that transaction into a block [6]. There is a minimum transaction fee a transaction must pay (unless it is a high-priority transaction) for being relayed to other nodes and included into the Mempool called the relay fee. Each node can choose its own minimum relay fee (per byte).

Mining. Miners demonstrate the proof of work, verify the transactions, and create the new blocks, then they get Coinbase transactions as rewards. Since the rewards decrease as time goes by, transaction fee becomes the main incentive for miners. There are two ways for a transaction to be confirmed and included into a block. Firstly, there is a section of block for "high-priority transaction". The priority of a transaction with n inputs is defined by

$$\sum_{i=1}^{n}(value_i \times age_i)/transaction\ size \tag{1}$$

where the value and age are respectively defined as an input's value and difference of height between the input's block and the latest block. Each block would reserve 50KB space for this kind of high-priority transactions before Bitcoin Core 0.12, but now the default value of this section is set to 0 KB [6]. Secondly, behind the former section, the transactions are prioritized by their transaction fee per byte. The higher proportion a transaction occupies, the earlier it gets included in the block. Those remaining transactions keep staying at Mempool until they satisfy the above conditions.

Mempool. In Bitcoin or some other cryptocurrencies, Memory pool (Mempool) is a local non-persistent memory to store all the unconfirmed transactions. If the node shuts down or reboots, the Mempool is lost and all the unconfirmed transactions gradually disappear from this p2p network and never get chances to be mined.

2.2   Related Work

Vasek et al. [7] discussed the DDoS attacks on the Bitcoin system including the mining pool and currency exchange. Johnson et al. [8] then used a series of game-theoretic models to analyze the DDoS attack which targets the Bitcoin mining pool. Then a spam attack occurred on Bitcoin in 2015, which caused more than 10% nodes were reduced from Bitcoin network. Then Bitcoin raised the default relay fee to 5 times the old one to counter the attack. Baqer et al. [9] conducted an analysis of this kind of spam attack, used the k-means clustering to identify and differentiate spam from non-spam transactions, and measured the impacts on Bitcoin.

Jung W et al. [5] defined the DDoS attack on Mempool as an overflood attack, and presented a dynamic transaction limit volume approach to limit the transaction a node can create during a period of time. Although it can effectively control the Mempool size, it also has negative impacts on legitimate users as we know that the attacker only holds a small part of nodes in Bitcoin. Thus, the normal nodes get limited by this method. Some urgent transactions with high transaction fees also cannot be processed in time. A severe DDoS attack was launched against Bitcoin Gold on November 12, 2017 [10].

Saad et al. [4] presented the attack procedure and threat model of DDoS attack on Mempool, and proposed the countermeasures based on fee and age. They also took the impacts of legitimate users into consideration and used the average age of parent transactions to filter out the transactions. However, they do not dynamically adjust the countermeasure according to the condition of Mempool size. Therefore, it may reject some unsatisfied but legitimate transactions like fast transactions with low fee even when the Mempool pressure is not that high. Later they presented a size-based method [11], but also pointed out that increasing the block size may have multiple disadvantages [12].

3   DDoS Attack on Bitcoin Memory Pool

The procedure of attack can be summarized as two phases [4]: In the first "distribution phase", the attacker observes and estimates the relay fee of the network for the follow-up attack, then distributes its budget ("UTXOs") into several transactions and transfers them to the sybil accounts controlled by the attacker. Then in the second phase, which is called "attack phase", the sybils will generate the dust transactions [6] from the balance received in the first phase. Then the sybils relay those spam transactions on the network. The inputs of those dust transactions are the transactions of distribution phase which are not confirmed. The exchange rate will be higher than the transaction throughput which leads to transactions backlog. Thus, the Mempool size grows. The attacker's goal is to fill the Mempool with this kind of malicious dust transactions, and meanwhile try not to let their dust transactions get mined in order to reduce their cost and launch more attacks.

In the following part of this section, we analyze the feasibility of this kind of DDoS attack, and propose its notable effects and potential risks in different aspects.

3.1   Feasibility

Firstly, from the angle of mining procedure, as mentioned in the Sect. 2.1, a transaction with a transaction fee that satisfies the relay fee (or a high-priority transaction) can be propagated in Bitcoin network and accepted by Mempool. Then there are two ways for a transaction to be confirmed and mined, i.e., high-priority and high transaction fee per byte. However, because the default space for high-priority transactions had been set as 0 KB since Bitcoin Core 0.12, transaction fee becomes the dominant factor in most cases. Thus, the malicious transactions that include only the minimum relay fee have few chances to get mined and can stay in the Mempool for a long time as the attacker wishes.

Secondly, from the angle of the attacker, since the output of an unconfirmed transaction can be spent to generate a new transaction before the confirmation, the attacker can continually generate and distribute the transactions among sybils as long as the transactions satisfy the minimum relay fee. The cost for attacker only happens when these malicious transactions get mined, while in these cases the attacker just loses a small amount of relay fee and can keep launching this kind of DDoS attack.

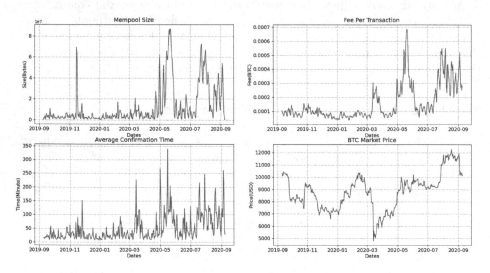

Fig. 1. The Mempool size, Average confirmation time, Fee per transaction and BTC Market price in the past year.

3.2   The Effects of Attack

Though it is obvious that the DDoS attack on Mempool results in the transactions backlog and growth of Mempool size, there still lacks a quantitative analysis. For this aim, we used the data provided by Blockchain.com [11] from September 2019 to September 2020, calculated the relations among Mempool size, average confirmation time, fees per transaction (in BTC) and BTC market price (in USD) using the Pearson Correlation Coefficient. Then we analyzed and came up with the following effects from different aspects.

Average Confirmation Time. As mentioned above, miners have two ways to select and mine the transactions. In the first way they need to calculate the priority of each transaction, and in the second way they need to sort all the transactions by their transaction fee per byte. Apparently, the time that both operations cost depends on the amount of unconfirmed transactions in Mempool, so when Mempool suffers from a DDoS attack, and its size grows, the node uses more time and resources to choose the transactions. Due to the transaction backlog, the average confirmation time of a transaction definitely increases. In Fig. 1, we show the fluctuation of average confirmation time in one whole year.

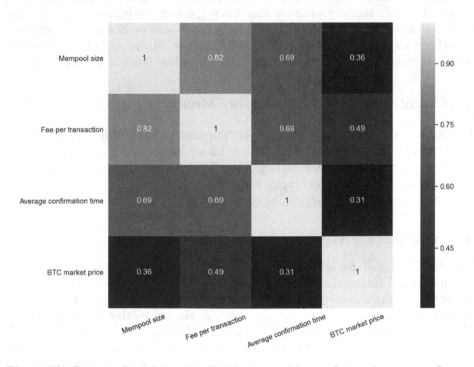

Fig. 2. The Pearson Correlation Coefficient among Mempool size, Average confirmation time, Fees per transaction and BTC market price.

As shown in Fig. 2, by calculating the Pearson Correlation Coefficient we found a high correlation between the Mempool size and average confirmation time with the value of 0.69, which supports our analysis above.

Fee Per Transactions and Market Price. Since the miners prioritize the transactions by the transaction fee per byte, and Mempool size is a public information. Thus, when users observe the growth of Mempool size, they tend to pay more transactions fees in order to guarantee the confirmation of their transactions. From Fig. 2, by calculation we also found a high correlation of 0.82 between the fee per transaction (BTC) and Mempool size.

Users attain the same service by spending more BTC which may cause the variation of BTC market price, as we can observe from Fig. 2 that BTC market price is partially associated with the fee per transaction with the Pearson Correlation Coefficient of 0.49. Attacker might gain some profits based on this observation, by increasing the Mempool size in a DDoS attack.

Legitimate Users. If the DDoS attack is severe and the Mempool size keeps growing, the node may suffer the memory pressure. Except for the unconfirmed transactions, node has to maintain the UTXO set to easily verify the received transactions [9]. Memory pressure may force the node to remove some unconfirmed transactions which include both legitimate and malicious transactions. Attacker does not suffer a loss and can continue the attack, while for legitimate users, transaction failure or delay may make them lose trust in Bitcoin.

4   Countering DDoS Attack on Mempool

In this section, we propose Dynamic Fee Threshold Mechanism to counter the DDoS attack. As mentioned in the Sect. 3, Mempool size is the dominant factor for those effects, so our mechanism has the following two goals: 1) maintaining the Mempool size in an appropriate range, 2) reducing the impact towards the legitimate users and transactions as much as possible.

4.1   Dynamic Fee Threshold Mechanism

To achieve our goal, we plan to filter the transactions at network level and refuse the unqualified transactions to occupy the space of Mempool. From the analysis in Sect. 3, the malicious transactions generated by the attacker pay the relay fee in order to be propagated without being mined. So, we use this characteristic to filter the transactions and maintain the Mempool size.

Since each node has a different capability to store and confirm the transactions, our countermeasure allows nodes to set the *active_size* of themselves. The countermeasure activates if the Mempool size exceeds this parameter. There is also a parameter *maximum_size* which is the node's maximum memory space for Mempool. To dynamically adjust the fee threshold, we set the memory level based on the space that Mempool occupies, and the fee threshold increases as the memory level rises. Assume the total number of memory levels is N, we define the current memory level CML and the $fee\_threshold$ as:

$$CML = \lfloor \frac{current\_size - active\_size}{maximum\_size - active\_size} * N \rfloor \qquad (2)$$

$$fee\_threshold = relay\_fee + f(base\_fee, CML) \qquad (3)$$

The Fig. 3 shows a situation of Mempool in which the countermeasure is activated. The $f(base\_fee, CML)$ in Eq. (3) is a monotonic increasing function such as linear function: $f(base\_fee, CML) = base\_fee * CML$, logarithmic function:$f(base\_fee, CML) = base\_fee * ln(CML + 1)$ and exponential function:$f(base\_fee, CML) = base\_fee^{CML} - 1$. Thus, the $fee\_threshold$ is positively correlative with the CML. When the Mempool size gets higher, the fee threshold gets higher, thus the mechanism gets stricter and filters out more unqualified transactions. If the incoming transaction has the transaction fee per byte that satisfies the condition, it gets into the Mempool and can be propagated to the adjacent nodes, otherwise it gets refused but can also be propagated to neighbors. Each node can adjust how strict he filters the transactions by setting the $base\_fee$ and level numbers N according to their own conditions.

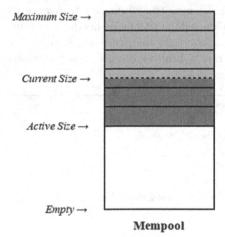

Fig. 3. The Layout of Mempool. The Mempool is divided into several levels to dynamically adjust the fee threshold. As shown in the above fig, the number of levels N is set as 6, and according to $current\_size$, the current memory level CML is 3.

Simply using the fee threshold filters not only the malicious transactions but also some legitimate transactions with low transaction fees. To achieve the second goal, we need to recognize as many legitimate transactions as possible to reduce the impact on legitimate users. We notice that when the attacker generates spam transactions, he uses the unconfirmed malicious transactions from the distribution phase as inputs. So, if all inputs of a transaction can be found in the UTXO set, that means it uses the outputs of confirmed transactions, then we accept it into the Mempool even though it does not satisfy the fee threshold.

Each time a node receives the transactions, if the current Mempool size exceeds the $active\_size$, the Dynamic Fee Threshold Mechanism starts. For each transaction, if it already exists in the Mempool, the mechanism continues to check the next one. For a new transaction, the mechanism firstly calculates the CML and the corresponding $fee\_threshold$, then checks if it pays a transaction fee per byte higher than $fee\_threshold$. The qualified transactions get accepted into the Mempool. For the unqualified transaction, the mechanism then checks if all of its inputs are the outputs of confirmed transactions. If so, it has a high probability to be a legitimate transaction according to the above analysis. If not, it gets rejected. Finally, the mechanism updates the Mempool $current\_size$ for the follow-up calculation and deal with the next incoming transaction. The pseudocode is shown in Algorithm 1.

4.2 Analysis of Our Method

With a limited budget, the attacker can only generate limited transactions. Firstly, he launches this kind of DDoS attack and part of spam transactions get into the Mempool. Then the Mempool size keeps grow up and the Dynamic Fee Threshold Mechanism is activated when it exceeds the $active\_size$. The $fee\_threshold$ gradually increases. Therefore, the rest of spam transactions cannot satisfy the threshold. Additionally, those spam transactions generated by sybil accounts use the unconfirmed transactions generated by the attacker as inputs. So, they also fail to pass the second check in line 10 of Algorithm 1 and finally get rejected. Thus, the Mempool size temporarily stays in a stable range.

To continually launch the attack, the attacker has two options: 1) pays more transaction fees for each spam transaction, 2) uses the confirmed transactions as the input of the follow-up spam transaction, which means he needs the miner to confirm the transactions he generates in the first distribution phase. Both options will bring cost to him. Since he has a limited budget, the malicious

---

Algorithm 1. Dynamic Fee Threshold Mechanism

Condition: $current\_size$ exceeds $active\_size$
Input: $active\_size$, $maximum\_size$, $current\_size$, N, $base\_fee$

```
 1: for transaction T of incoming transactions do
 2:     if T is in the Mempool then
 3:         Continue to check the next transaction
 4:     if current_size + T's size is bigger than maximum_size then
 5:         Reject the transaction
 6:     calculate CML by Eq. (2)
 7:     calculate fee_threshold by Eq. (3)
 8:     if T's transaction fee per byte ≥ fee_threshold then
 9:         Accept the transaction into Mempool
10:         Update the Mempool current_size
11:     else
12:         if All inputs of the T can be found in the UTXO set then
13:             Accept the transaction into Mempool
14:             Update the Mempool current_size
15:         else
16:             Reject the transaction
17: return Mempool Size Maintained and Malicious transactions rejected
```
---

transactions he can generate in next attack will decrease. Each node can adjust their *active_size* and *base_fee* to resist different levels of attacks.

Compared with the normal situation, while suffering from the DDoS attack, the Mempool with the Dynamic Fee Threshold Mechanism can reject malicious transactions, and force the attacker to pay more for each spam transactions and generate fewer malicious transactions because of his limited budget. Thus, the Mempool size will be usually smaller than the Mempool without a defensive mechanism. More importantly, by setting the parameters, a node can prevent its Mempool size from exceeding the *maximum_size* that the space it saves for Mempool. Therefore, the node does not have to remove the transactions from Mempool. For legitimate users, as most of them tend to use the confirmed transactions as inputs except for some fast trade, their transactions can successfully get into the Mempool even when the mechanism is activated.

5 Experiments

In this section, we describe our experiments and analyze the results. We simulated the DDoS attack on Mempool and applied our proposed method to show its effectiveness.

5.1 Experiment Setting

We developed a blockchain by implementing the basic components in Bitcoin including mining, Mempool management, network relay, etc. To simulate the environment of Bitcoin, we set the block generation time, legitimate transactions generation rate based on the current Bitcoin statistics [11].

For the parameters setting, we used the data of Mempool size, Fee Per Transaction from Bitcoin.com [11] to set the parameters of our Dynamic Fee Threshold Mechanism, including *maximum_size*, *active_size*, *base_fee*. In the simulated attack, we initially set *current_size* = *active_size*.

The procedure of the experiment is as follows. Firstly, we collected the latest unconfirmed transactions from Bitcoin.com [11] and used them as the incoming legitimate transactions for Mempool. Then we allocated a budget for the attacker to simulate the attack based on the attack procedure in Sect. 3 and generated the spam transactions that only pay the relay fee. In each experiment step we proportionally generated the legitimate transactions and malicious transactions based on the statistics and the former attacks. We carried out the experiments with and without the Dynamic Fee Threshold Mechanism in the blockchain and evaluated the results respectively.

Since our main goal is to maintain the Mempool size in an appropriate range, the Mempool size would be an evaluation metric. Our proposed mechanism can also be regarded as a classifier that filters out the malicious transactions, we also used the evaluation metrics in machine learning including accuracy: $\frac{TP+TN}{TP+FP+TN+FN}$ and precision: $\frac{TP}{TP+FP}$. We used the definitions from [4]. True positive (TP) and false positive (FP) respectively denote the legitimate and malicious transactions that pass the mechanism and get into the Mempool. True negative (TN) and false negative (FN) respectively denote the malicious and legitimate transactions that get rejected by the mechanism. We calculated the metrics for three different $f(base\_fee, CML)$ that mentioned in Sect. 4.

5.2 Results Analysis

In Fig. 4, the Mempool size increases when transactions are generated, and decreases when the block is generated. From Fig. 4. We can observe that when Mempool suffers from DDoS attack without any defensive mechanism, its size grows rapidly. However, with the Dynamic Fee Threshold Mechanism, when Mempool size gets larger, the fee threshold gets higher and the attacker can only generate fewer spam transactions, thus the growth of it correspondingly becomes slower. Finally, it stays in a stable range (In Fig. 4 it stays in the memory level 3).

We calculated the accuracy and precision, our results and some best results of other methods with their parameters in brackets are shown in Table 1. In the fee-based and age-based methods of [4], their main goal is to limit the malicious transactions to get into Mempool as much as possible, so they have high precision especially when they set the strict parameter. From the results we can observe that their methods have better precision, but they also reject huge part of legitimate transactions which leads to only 60% accuracy at most. However, since our main goal is not to reject all the malicious transactions, but to prevent Mempool size from continuously growing, we allow some malicious transactions to get into Mempool when the memory level is low, so our method affects fewer

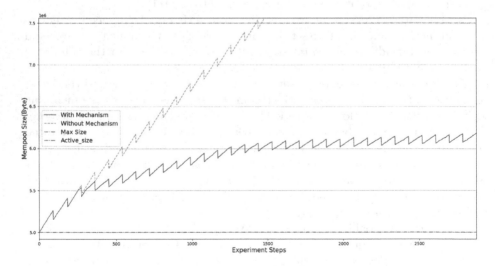

Fig. 4. The Increasing of Mempool size while suffering from DDoS with and without the Dynamic Fee Threshold Mechanism. $maximum\_size$ is 7500000, $active\_size$ is 5000000, and the number of memory levels is 5. Here we used linear function as f.

legitimate transactions at the same time. Our proposed method with 3 different functions has nearly 70% accuracy and 73% precision. Compared with the fee-based and age-based methods, our method achieves higher accuracy without trading off too much precision, which shows that our method can detect most of the malicious transactions and rarely affect the legitimate user.

Table 1. Experiment result with corresponding parameters

Countermeasure	Accuracy	Precision
proposed(linear)	69.3%	73.4%
propose(logarithmic)	69.4%	74.2%
propose(exponential)	69.1%	73%
fee-based(13)	60%	78%
age-based(50)	58%	75%
age-based(200)	58%	98%

Therefore, with our proposed method, the Mempool size maintains in an appropriate range and hardly exceeds the $maximum\_size$, thus the node can be free from memory pressure. The legitimate transactions also have great chances to pass the mechanism. Thus, the impacts mentioned in Sect. 3 are reduced as much as possible, and the attacker likely gives up the attack while he knows that it cannot raise the expected damage.

6   Conclusions

In this paper, we analyze the feasibility of DDoS attack on Bitcoin Mempool and the relation between Mempool and some statistics of Bitcoin. Then we propose some potential effects that can be caused by this kind of DDoS attack. To counter the DDoS attack, we propose the Dynamic Fee Threshold Mechanism to filter out the malicious spam transactions and maintain the Mempool size. Finally, we carry out some experiments to show the effectiveness of our proposed method and compare our methods with others.

In the future, we will further analyze the relation between the attacker's cost and benefit in this kind of DDoS attack. Meanwhile, we will explore more characteristics to precisely distinguish between the spam transaction and legitimate transaction. Then we will try to improve our method and test its scalability.

Acknowledgements. This work was supported by the Key-Area Research and Development Program of Guangdong Province (No. 2020B010164003), the Science and Technology Program of Guangzhou, China (No. 201904010209), and the Science and Technology Program of Guangdong Province, China (No. 2017A010101039).

References

1. Decker, C., Wattenhofer, R.: Information propagation in the bitcoin network. In: IEEE P2P Proceedings, pp. 1–10. IEEE (2013)
2. Bitcoin, N.S.: A peer-to-peer electronic cash system. Manubot (2019)
3. Wood, G.: Ethereum: a secure decentralised generalised transaction ledger. Ethereum Project Yellow Paper **2014**(151), 1–32 (2014)
4. Saad, M., Njilla, L., Kamhoua, C., et al.: Mempool optimization for defending against DDoS attacks in PoW-based blockchain systems. In: 2019 IEEE International Conference on Blockchain and Cryptocurrency (ICBC). IEEE (2019)
5. Jung, W., Park, S.: Preventing DDoS attack in blockchain system using dynamic transaction limit volume. Int. J. Control Autom. **10**(12), 131–138 (2017)
6. Community, B.: Developer's Guide, Confirmation Score, Transaction Fee and Miner Fee, Minimum Relay Fee, UTXO, Memory Pool, Child Pays for Parent, Raw Transactions (2018). https://developer.bitcoin.org/reference/
7. Vasek, M., Thornton, M., Moore, T.: Empirical analysis of denial-of-service attacks in the bitcoin ecosystem. In: Böhme, R., Brenner, M., Moore, T., Smith, M. (eds.) FC 2014. LNCS, vol. 8438, pp. 57–71. Springer, Heidelberg (2014). https://doi. org/10.1007/978-3-662-44774-1_5
8. Johnson, B., Laszka, A., Grossklags, J., Vasek, M., Moore, T.: Game-theoretic analysis of DDoS attacks against bitcoin mining pools. In: Böhme, R., Brenner, M., Moore, T., Smith, M. (eds.) FC 2014. LNCS, vol. 8438, pp. 72–86. Springer, Heidelberg (2014). https://doi.org/10.1007/978-3-662-44774-1_6
9. Baqer, K., Huang, D.Y., McCoy, D., Weaver, N.: Stressing out: bitcoin "Stress Testing". In: Clark, J., Meiklejohn, S., Ryan, P.Y.A., Wallach, D., Brenner, M., Rohloff, K. (eds.) FC 2016. LNCS, vol. 9604, pp. 3–18. Springer, Heidelberg (2016). https://doi.org/10.1007/978-3-662-53357-4_1
10. Memoria, F.: 700 Million Stuck in 115, 000 Unconfirmed Bitcoin Transactions. (2017). https://goo.gl/mYX14V. Accessed Mar 2017
11. B. Community: Bitcoin Data from Blockchain. info (2020). https://www. blockchain.com/charts
12. Zamani, M., Movahedi, M., Raykova, M.: Rapidchain: Scaling blockchain via full sharding. In: Proceedings of the ACM SIGSAC Conference on Computer and Communications Security, pp. 931–948 (2018)

The Compiler of DFC: A Source Code Converter that Transform the Dataflow Code to the Multi-threaded C Code

Zheng Du[1,2], Jing Zhang[2], Jinrong Li[2], Haixin Du[2], Jiwu Shu[1], and Qiuming Luo[2(✉)]

[1] Department of Computer Science and Technology, Tsinghua University, Beijing, China
[2] NHPCC/Guangdong Key Laboratory of Popular HPC, College of Computer Science and Software Engineering, Shenzhen University, Shenzhen, China
lqm@szu.edu.cn

Abstract. The working principle of DFC compiler is introduced in this article. DFC is a grammatical extension of standard C language, with special DF function which describe the dependence of computing DAG. DFC compiler, dfcc, is used to convert the DFC source codes to standard C codes with the assistance of multi-threaded library. The lexical rules and grammatical rules are used to setup the AST of DFC codes, and then it is converted to the AST of standard C without DF function nodes. The derived AST is printed into a text file, a normal C file, and finally is processed by GCC to obtain the executable file. The experiment gives some demonstration of that converting details, and the memory footprint is studied to show an ideal scalability.

Keywords: Dataflow · Compiler · AST · Grammatical rules

1 Introduction

Facing the challenge of exascale computing, breakthroughs will be made in four aspects [1]: execution model, memory access mode, parallel algorithm and programming interface. As the execution model is concerned, dataflow execution model dramatically increases both scalability and efficiency to drive future computing across the exascale performance region, even possibly to zetaflops. So, as the number of cores in the host system increases (1–2 orders of magnitude higher than the current one), exascale-level computing will gradually migrate to the data flow execution mode.

Dataflow is a parallel execution model originally developed in the 1970s but explored and enhanced as the basis for non-von Neumann computing architecture and techniques. The dataflow model of computation presents a natural choice for achieving concurrency, synchronization, and speculations. In the basic form, activities in a dataflow model are enabled when they receive all the necessary inputs; no other triggers are needed. Thus, all enabled activities can be executed concurrently if functional units are available. And the only synchronization among computations is the flow of data.

The recent research related to dataflow systems can be classified into three categories: (1) One of these dataflow systems is based on dedicated customized hardware.

© Springer Nature Switzerland AG 2021
Y. Zhang et al. (Eds.): PDCAT 2020, LNCS 12606, pp. 185–197, 2021.
https://doi.org/10.1007/978-3-030-69244-5_16

These dedicated hardware, such as Maxeler [2], TRIPS [3, 4] can achieve fine-grain dataflow tasks and excellent performance. (2) Another type is called the macro dataflow (hybrid dataflow/controlflow) system. These systems are based on off-shelf platform or customized hardware, and the presentive macro dataflow system is Teraflux [5, 6]. (3) The third one is application-related system. These systems are based on underlying dataflow execution software engine. The well-known application-related dataflow system includes Labview [7], Orcc [8] and Tensorflow [9], etc.

Besides those dataflow systems, some research focus on lightweight dataflow solution, who provides a new programming language or execution library. DSPatch is a C++ library for dataflow framework. StreaMIT is a language for streaming application based on dataflow execution. Programmer of StreaMIT should construct the dataflow graph explicitly in natural textual syntax. These solutions are more likely to be prevalent, because they are general-purpose and capable to run on the commercial hardware/software platforms on the market.

Among such lightweight solutions, DFC is with a tiny extension of C language to obtain the capacity of describing the dataflow tasks, without any special underlying hardware [10]. Because it is based on C, it is much easier for programmer. And this make it highly coupled with Unix-like OSs, which means we can build other upper system software layers in DFC environment.

The compiler of DFC is implemented as a source code converter which use GNU compiler collection as its backend. We will discuss how DFC compiler perform the following operations: preprocessing, lexical analysis, parsing, semantic analysis (syntax-directed translation), conversion of input programs to an intermediate representation and code generation.

This paper is organized as follows. In Sect. 2, we give an overview of DFC briefly. In Sect. 3, the principle of DFC compiling is demonstrated in detail. Some basic verification and memory footprint analysis is presented in Sect. 4. Conclusion is drawn in Sect. 5.

2 Brief of DataFlow C

DataFlow C, short for DFC, is designed by NHPCC of Shenzhen University. It extended the C language to enable the capable of describe the dataflow by enhanced C functions. Because of implementing in C language, DFC might be used to support the programs written in higher level languages, such as StarSs and OpenStream, by source-to-source translating to DFC.

By extending C syntax, a new type of function is defined to describe the nodes and edges in a dataflow graph. The computing nodes are presented by the DFC functions' body explicitly, and the edges are derived implicitly from the Input and Output elements of preceding and following nodes. After building the AST of C source file, the DAG of dataflow graph is setup.

There are two type of function coexisting in DFC program, the typical C function and DF function. The main function is constructed in the form of typical C function, where it invokes the DF runtime functions to trig the dataflow computing defined by DAG of DF functions constructed by compiler.

2.1 DF Function

The DF function is defined by a statement, as Fig. 1 shows.

> *Dataflow_function_name*
> *(arg_in_1,arg_in2,...,arg_inK; arg_out_1,arg_out_2, ...*
> *, arg_outL)*
> *{*
>
>   *...*
>   *body*
>   *...*
>
> *}*

Fig. 1. Declaring a DF function

All the nodes in dataflow graph should be declared as DF functions. The input and output ports of one edge are represented by the preceding DF function's output argument and the input argument of a succeeding DF function. The DF function's two argument lists of input ports and output ports are separated by a semicolon. The argument of input/output lists are separated by commas.

As a dataflow graph shown by Fig. 2 is under consideration, there needs four DF functions.

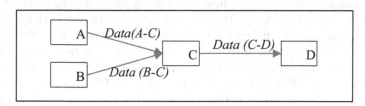

Fig. 2. A demo of dataflow graph

The corresponding DFC source code can be shown as Fig. 3. Form Fig. 3, we can easily identify four DF functions, which corresponding the four computing nodes (A, B, C and D) in the graph of Fig. 2. But the directed edges should be derived from the conjunction of DF functions, by matching the input and output arguments between the two DF functions. By matching symbol names of the input and output arguments, the compiler can reconstruct the DAG of the whole dataflow computing.

The DFC compilation flow has been implemented as a frontend extension to GCC 4.8.5.

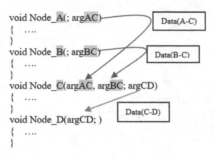

Fig. 3. A demo of DFC source code

2.2 DFC Runtime

The DFC program is started by main function in C language's traditional way, then invokes DF_run() to enable the dataflow computing. There are seven DF runtime functions to help manipulating the dataflow computing.

Two DF runtime functions are called by user. "void DF_Run()" is called by user to start the proceeding of dataflow graph. "void DF_Source_Stop(DF_TFL *table, int item_index)" is called to stop the execution when the preset conditions are met.

The rest five of that seven DF runtime functions are inserted into the source code in the compiling phase of source-to-source converting. "void DF_Init(DF_TFL* table, int InputFNNum,...)" is responsible for initiating the dataflow environment. "void DF_ADInit (DF_AD* AD, int persize, int FanOut)" is inserted to initiate the driven data, here we called them Active Data (AD). When the compiler identifies one input /output argument in DF function, that argument will be consider as one AD and initiated by DS_ADInit(). Then it will insert "void DF_FNInit1(DF_FN* FN, void*FunAddress, char *Name, int InPutADNum,...)" and "void DF_FNInit2(DF_FN *FN, int OutPutADNum,...)" to bind the AD with DF function's input port and output port. "DF_AD_GetData(DF_FN * FN, outdata address, int persize)" and "DF_AD_UpData(int DF_count, DF_TFL, DF_FN* FN, DF_AD* AD, int persize)" are inserted into the related DF function body to get the AD (driven data) and produce the AD (driven data) respectively.

3   The Compiler for DFC

A compiler is needed to convert the DFC code into multi-threaded code, as there is no direct support for dataflow execution on computing platform of contemporary era. We chose Linux on X86 as the target system, as it is the most popular system. The details of DFC compiler are discussed in this section.

3.1   The Source and Target of Compiling

DFC compiler program translates between the high-level languages, which are DFC and standard C, so it is usually called the source-to-source compiler or trans-compiler. The gap and conversion between these two languages are show as Fig. 4. DFC runtime library provides the kernel environment supporting dataflow execution. It fills the gap by performing the following operations: 1) managing the computing DAG of the dataflow graph; 2) invoking the dataflow computing; 3) tagging the distinct data of various passes.

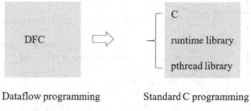

Dataflow programming   Standard C programming

Fig. 4. The gap between DFC and C

Besides the kernel function provided by runtime, a deliberated C code framework should be figured out, which is responsible for preparing the environment and initiating the DAG execution by invoking DFC runtime functions.

The underlying multi-threaded parallel execution is fulfilled by pthread library. DFC runtime will setup a thread pool to perform the ready tasks of DAG.

3.2 Working Principle

The DFC codes must be converted to normal C codes, so that it can be compiled to executable program by GCC to run with the DFC runtime library. To achieve that, the user can use the command "dfcc -k -v -o Myfile Myfile.dfc", where dfcc is the compiler of DFC.

The working flow of DFC compiler driver is shown as Fig. 4. DFC compiler is to perform the following operations: preprocessing, lexical analysis, parsing, semantic analysis (syntax-directed translation), conversion of input programs to an intermediate representation and code generation.

Firstly, during pre-processing, three lines of **Code. 1** will be inserted to the very beginning of the source DFC codes (Myfile.dfc, as shown in Fig. 5) to include the necessary headerfiles. The C preprocessor of GCC then extends Myfile.dfc to Myfile.p. And then, the lexical analyzer/scanner and parser will then process the Myfile.p with the guided of "scanner.l" and "parser.y" which describe the lexical and grammatical rules of DFC.

DFC compiler adopts AST (abstract syntax tree) as the intermediate representation of the parser output, which can keep hierarchical grammatical structure of the source codes and from which the codes can be easily gotten conversely. After building DFC AST, DFC compiler will transform the DFC AST into AST of standard C, so that the transformed program implemented in C code can be derived. Once the C code version of the program is available, the generation of the executable program is handed over to GCC. As the previous description, the DFC compiler is a pre-compiler or source code converter, and only responsible for transforming the DFC codes to standard C codes. The DFC compiler reused part of the opensource OMPi compiler's source code for lexical/grammatical analysis and the framework of AST.

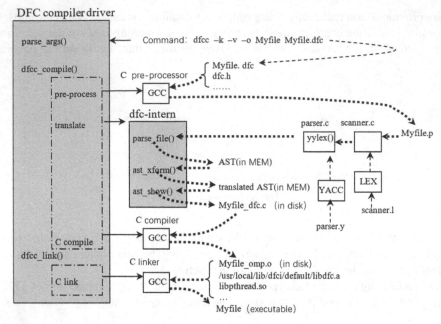

Fig. 5. The working flow of DFC compiler driver

```
1.    #include <pthread.h>
2.    #include "threadpool.h"
3.    #include "dfc.h"
```

Code. 1 The necessary header files

3.3 Lexical Analysis

As DFS defines no new lexical item, it shared the same lexical scanner as standard C language. Theoretically, any C scanner will work. DFC reused the open source OMPi compiler's lexical rules file "scanner.l", which is text file in GCC Flex format with nearly 600 lines.

3.4 Parsing

DFC extends syntax of C language by adding DF function which is slightly different from C normal function. **Code. 2** shows how the parser describes the syntax of the declarations of DF functions corresponding to the normal DF function with both input arguments and output arguments, the Source DF function without input arguments, and the Sink DF function without output arguments. The general DF function is with two "parameter list" separated by a semicolon. **Code. 2** also instructs the semantic actions for GCC Bison parser. It invokes DfcFuncDecl to make AST node for DF function, which invokes DfcMakeVars to handle the DFC argument list.

Code. 3 presents the data structures of DF function declaration node of AST. The input arguments and output arguments are stored in dfcvar_ which used for DF function transformation. For example, to handle the normal DF function's declaration, the input argument list('$3') and the output argument list('$5') are both as the input arguments of DfcMakeVars to derive the dfcvar_. And then, according to '$1' and the dfcvar_, DfcFuncDecl creates the declaration node of AST. Figure 6 shows the AST node of DF function, and DF function transformation will be talked about in the next section.

```
1.   direct_declarator:
2.       /* DF-C function declaration. */
3.       direct_declarator '(' parameter_list ';' parameter_list ')'
4.       {
5.           __has_dfc = 1;
6.           $$ = DfcFuncDecl($1, DfcMakeVars($3, $5));
7.       }
8.     | direct_declarator '(' ';' parameter_list ')'
9.       {
10.          __has_dfc = 1;
11.          $$ = DfcFuncDecl($1, DfcMakeVars(NULL, $4));
12.      }
13.    | direct_declarator '(' parameter_list ';' ')'
14.      {
15.          __has_dfc = 1;
16.          $$ = DfcFuncDecl($1, DfcMakeVars($3, NULL));
17.      }
18.
19.  astdecl DfcFuncDecl(astdecl decl, dfcvar var)
20.  {
21.      astdecl d = Decl(DDFCFUNC, 0, decl, NULL);
22.      d->u.dfcvars = var;
23.      return (d);
24.  }
25.
26.  dfcvar DfcMakeVars(astdecl inParams, astdecl outParams)
27.  {
28.      dfcvar d = smalloc(sizeof(struct dfcvar_));
29.      d->inParams = inParams;
30.      d->outParams = outParams;
31.      return (d);
32.  }
```

Code. 2 Grammatical rules for AST node of DF function

```
1.   struct astdecl_
2.   {
3.       enum decltype type;
4.       int        subtype;
5.       astdecl    decl;     /* For initlist,initializer,declarator */
6.       astspec    spec;     /* For pointer declarator */
7.       union
8.       {
9.          symbol   id;      /* Identifiers */
10.         astexpr  expr;    /* For initializer/bitdeclarator */
11.         astdecl  next;    /* For lists */
12.         astdecl  params;  /* For funcs */
13.         dfcvar   dfcvars; /* For DF-C */
14.      } u;
15.      int     l, c;        /* Location in file (line, column) */
16.      symbol file;
17.  };
18.
19.  struct dfcvar_
20.  {
21.      astdecl inParams;
22.      astdecl outParams;
23.  };
```

Code. 3 AST node for DF function

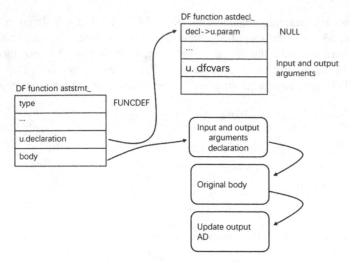

Fig. 6. A transformed AST node of DF function

3.5 AST Transformation

AST transformation is the critical step from DFC codes to C codes, so we can use GCC as backend of DFC compiling. AST transformation mainly includes three aspects: 1) DF function transformation; 2) inserting global variables for active data (or driven data); 3) forging the framework of main function. It will be detailed below.

DF Function Transformation. DFC executable file is finally built by GCC. However, GCC can't recognize DF function, so it's necessary to turn the DF function to normal C function. **Code. 4** presents a demo of normal DF function to normal C function. It can be seen that the DF function has been transformed to a C standard function that has no input arguments and returns void. Input and output arguments are declared as local variables in function. And DF function fetching data and outputting data are respectively implemented by DF_AD_GetData() and DF_AD_UpData(). This transformation happens during parsing the declaration of DF function.

To declare the input/output argument as local variable in the function, just go through the dfcvars of the astdecl's union and build declaration statements, cause all the input and output arguments are recorded in dfcvars.

After declaring the local variable, "DF_persize_%s = sizeof(%s);" will be inserted to update the value of "DF_persize_%s" which records the data size and is declared as global variable. "%s" is the symbol name obtained from the argument list by previous steps.

And then DF_AD_GetData() is inserted to load input data from global variable to local variable. Finally, the compiler insert DF_AD_UpData() to output data after the original body of DF function. It should be noted that there is a little difference in transforming Source DF function. It's easily to distinguish Source DF function from other DF function by checking the number of input arguments. To Source DF function, "int DF_count" is inserted in the beginning of the function body and the function getting data is replaced by DF_SOURCE_Get_And_Update(). Actually, the DF function transformation happens during parsing.

```
1.    /* original DF function */
2.    void FUNS(int A, int B; int C) {
3.    ...
4.    }
5.    /* transformed DF function */
6.    void FUNS(/* DF-C function */)
7.    {
8.    int B;
9.    DF_persize_B = sizeof(B);
10.
11.   int A;
12.   DF_persize_A = sizeof(A);
13.
14.   int C;
15.   DF_persize_C = sizeof(C);
16.   int DF_count=DF_AD_GetData(&DF_FN_FUNS, &B, DF_persize_DF_Source_B, &A, DF_persize_A);
17.   {
18.   ...
19.   }
20.   DF_AD_UpData(DF_count,&DF_TFL_TABLE, &DF_FN_FUNS, &C, DF_persize_DF_C);
21.   }
```

Code. 4 A demo of code transformation

Inserting Global Variables for Active Data (or Driven Data). In DFC, all ADs (active data) are declared as global variables. For convenience to get data and output data in DF function, "DF_persize_%s" (variable recording the data size), "DF_fanout_%s" (variable recording the number of DF function inputs this AD) are all needed. "DF_fanout_%s" is calculated by matching the symbol names of DF input arguments and output arguments. What's more, "DF_FN_%s" (node of DAG) and "DF_TFL_TABLE" (structure recording all information of DAG) are declared as global variables as well. All these declarations are inserted before the first DF function.

Forging the Framework of Main Function. The original main() function is renamed to __original_main() during parsing and a new main() is inserted to initialize the DAG and call __original_main(). The reason why renaming the original and inserting new main() is that during parsing, all information of DAG isn't available so that inserting codes initializing DAG in main() is not workable. The framework of new main() is shown as **Code. 5**. The new main() is mainly responsible for initialization and executing the DAG. DF_ADInit() is called to allocate memory for AD. The adjacent nodes are connected by calling DF_FNInit1() and DF_FNInit2(). Because of the particularity of Source DF function as the start of the DAG, DF_SourceInit() is called to exert extra initialization to Source DF function. Now the DAG is built, and then all information of the DAG are recorded in the DF_TFL for convenience to trig the nodes of the DAG. After initialization, the __original_main() is called to ensure the semantics of the original program unchanged.

```
1.   int __original_main()
2.   {
3.     DF_Run(&DF_TFL_TABLE, DF_COUNT);
4.     return (0);
5.   }
6.   /* DF-C-generated main() */
7.   int main(int argc, char **argv)
8.   {
9.     DF_ADInit(DF_AD * AD, int persize, int InPutADNum);
10.    DF_FNInit1(DF_FN* FN,void*FunAddress ,char *Name, int InPutADNum,...);
11.    DF_FNInit2((DF_FN *FN, int OutPutADNum, ...));
12.    DF_SourceInit(DF_TFL, int SourceDFNum, DF_FN* FN,...);
13.    DF_Init(DF_TFL* DF_TFL_TABLE, int DFNum,DF_FN* FN,...);
14.    int DF_original_main_ret = (int) __original_main();
15.    return(DF_original_main_ret);
16.  }
```

Code. 5 Framework of new main()

4   Analysis and Evaluation

4.1   Experimental Environment

We compare the performance of DFC and DSPatch on machines with various number of cores. The first one is based on Intel® Xeon® Gold 6130 CPU with 32 cores and 64 hardware threads. And the operating system on it is Ubuntu with kernel version 4.15.0, and the version of GCC is 7.4.0. The second one is based on Intel® Xeon® E5–2620 v4 with 16 cores and 32 hardware threads. Its operating system is Ubuntu with kernel version5.4.0, and the version of GCC is 7.5.0.

The DFC code for analysis is the one that build a graph of two binary trees connected back-to-back as Fig. 7 shown. The first node randomly generate an integer array, and the nodes of the second layer to the $(n-1)$th layer simply divide the input array equally into two arrays average. The nodes of the n-th layer will bubble sort the input array, and the other nodes will merge the two input orderly arrays.

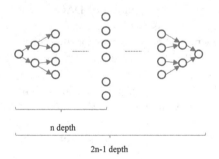

n depth

2n-1 depth

Fig. 7. Binary trees connected back-to-back

The DFC source code files of various tree depth are compiled by DFC compiler to demonstrate the memory consumption and time to compile.

4.2 Memory Footprint Analysis

The massif tool of Vargind is used to trace the memory consumption. Figure 8 recodes the memory that DFC compiler, dfcc, used during the compiling the program of depth of 12. As dfcc setting up the AST and then transform the AST, the memory demand is increase steady.

As the depth of binary tree increase, the amount of code lines number increases exponentially. Subsequently, the memory to store the AST for that DFC code increase exponentially. The curve of memory consumption for various code sizes is drawn as Fig. 9. The amount of code lines is drawn in Fig. 9 as well. It shows that the memory consumption increases faster than the amount of code lines, slightly, which means the optimization margin for the memory management of dfcc.

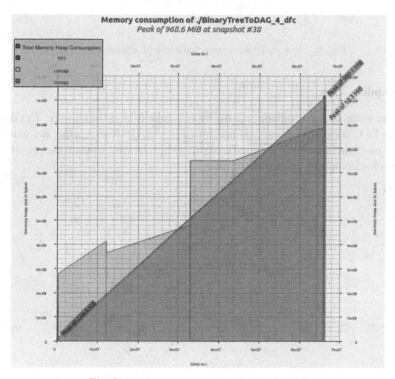

Fig. 8. Memory consumption for depth of 12

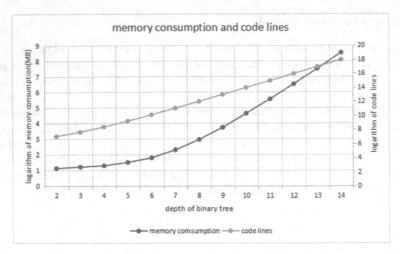

Fig. 9. Memory consumption for tree with different depth

4.3 Compiling Time

Besides the memory consumption, the time to compile is analyzed briefly. The time for compiling DFC code of various sizes are drawn in Fig. 10. It shows that the compiling time increases faster than the amount of code lines slightly, which means the optimization margin for the compiling procedure of dfcc.

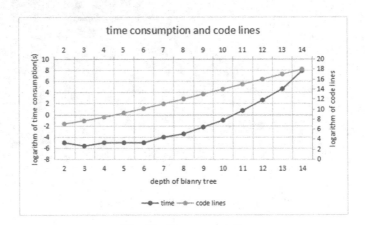

Fig. 10. Time to compile

5   Conclusion and Future Works

The proposed compiler of DFC, dfcc, is implemented as a source code converter which use GNU compiler collection as its backend.

DFC compiler is designed to perform the following operations: preprocessing, lexical analysis, parsing, semantic analysis (syntax-directed translation), conversion of input programs to an intermediate representation and code generation. This compiler can convert the DFC code into multi-threaded code to support data flow execution, successfully. Besides the functional verification, a preliminary performance analysis is carried out. The memory consumption of dfcc is traced by massif tool of Valgrind, which shows a good scalability with some margin for optimization. It shows an ideal scalability of the time for compiling as the code size increase, too.

The following jobs of the next step will be included in our future work. The first one is the optimization to keep the memory consumption and the time to compiling to increase at the same speed as the increasing of code size. The second one is to instrument the dfcc with profiling capability to obtain the in-depth performance factors.

References

1. Sterling, T., Brodowicz, M., Anderson, M.: High Performance Computing: Modern Systems and Practices, pp. 616–618. Morgan Kaufmann, Cambridge (2018)
2. Pell, O., Averbukh, V.: Maximum performance computing with dataflow engines. Comput. Sci. Eng. **14**(4), 98–103 (2012). https://doi.org/10.1109/MCSE.2012.78
3. Burger, D., et al.: The TRIPS team, scaling to the end of silicon with EDGE architectures. IEEE Comput **37**(7), 44–55 (2004). https://doi.org/10.1109/MC.2004.65
4. Gebhart, M., et al.: An evaluation of the TRIPS computer system. SIGPLAN Not. **44**(3), 1 2 (2009). https://doi.org/10.1145/1508284.1508246
5. Giorgi, R.: Teraflux: exploiting dataflow parallelism in teradevices. In: Proceedings of the 36th Annual IEEE/ACM International Symposium on Microarchitecture, CF 2012, ACM, 2012. New York, NY, USA (2012). https://doi.acm.org/10.1145/2212908.2212959.
6. Portero, A., Yu, Z., Giorgi, R.: Teraflux: exploiting tera-device computing challenges. Procedia CS **7**, 146–147 (2011)
7. Ursutiu,D., Samoila, C., Jinga, V.: Creative developments in LabVIEW student training: (Creativity laboratory — LabVIEW academy). In: 2017 4th Experiment@International Conference (exp.at 2017), Faro, pp. 309–312 (2017). https://doi.org/10.1109/EXPAT.2017.798 4399
8. Chavarrias, M., Pescador, F., Juárez, E., Garrido, M.J.: An automatic tool for the static distribution of actors in RVC-CAL based multicore designs, Design of Circuits and Integrated Systems, Madrid, pp. 1–6 (2014)
9. Lin, H., Lin, Z., Diaz, J.M., Li, M., An, H., Gao, G.R.: swFLOW: a dataflow deep learning framework on Sunway TaihuLight Supercomputer. In: 2019 IEEE 21st International Conference on High Performance Computing and Communications; IEEE 17th International Conference on Smart City; IEEE 5th International Conference on Data Science and Systems (HPCC/SmartCity/DSS), Zhangjiajie, China, pp. 2467–2475 (2019)
10. Du, Z., Zhang, J., Sha, S., Luo, Q.: Implementing the matrix multiplication with DFC on Kunlun small scale computer. In: 2019 20th International Conference on Parallel and Distributed Computing, Applications and Technologies (PDCAT), Gold Coast, Australia, pp. 115–120 (2019)

Online Learning-Based Co-task Dispatching with Function Configuration in Edge Computing

Wanli Cao[1], Haisheng Tan[1(✉)], Zhenhua Han[2], Shuokang Han[1], Mingxia Li[1], and Xiang-Yang Li[1]

[1] LINKE Lab, University of Science and Technology of China, Hefei, China
hstan@ustc.edu.cn
[2] Microsoft Research Asia, Shanghai, China

Abstract. Edge computing is a promising cloud computing paradigm that reduces computing latency by deploying edge servers near data sources and users, which is of great importance to implement delay-sensitive applications like AR, Cloud Gaming and Auto Driving. Due to the limited resources of edge servers, task dispatching and function configuration are the key to fully utilize edge servers. Moreover, a typical task request in edge computing (called a co-task) is consisted of a set of subtasks, where the task completion time is determined by the latest completed subtask. In this work, we propose a scheme named OnDisco, which combines reinforcement learning and heuristic methods to minimize the average completion time of co-tasks. Compared with heuristic algorithm, deep reinforcement learning can learn the inherent characteristics of the environment without any prior knowledge, and OnDisco is therefore well adapted to varying environments. Simulations on Alibaba traces shows that OnDisco reduces the average task completion time by 58% and 76% compared with the heuristic and random algorithm, respectively. Moreover, OnDisco outperforms the baselines consistently in various data environments and parameter settings.

1 Introduction

The rapid development of IoT has led to the emergence of a large number of intelligent terminals. IDC predicts that by 2025, the global connected IoT devices will reach 41.6 billion, generating 79.4 ZB data. Although, computation of edge-generated data can be done in remote cloud data centers, however, the long propagation delay, limited Internet bandwidth, and unstable networking environments makes it hard to meet many latency-critical applications, e.g., autonomous driving and augmented reality.

Edge computing [4] has been proposed to extend the capabilities of the cloud to the edge of the Internet. Typically, edge computing provides networking computing, and storage services by deploying relatively small-scale servers (a.k.a,

---

Part of the first two authors' work was done when visiting at PCL, Shenzhen, China.

© Springer Nature Switzerland AG 2021
Y. Zhang et al. (Eds.): PDCAT 2020, LNCS 12606, pp. 198–209, 2021.
https://doi.org/10.1007/978-3-030-69244-5_17

edge servers) that are close to terminals or data sources (e.g., wireless access points). As a complement to cloud computing, this paradigm greatly reduces the massive data transmission and avoids the long latency to remote cloud data centers thus can potentially solve the above -mentioned issues.

However, it is challenging to shift cloud-native applications to analyze data at the edge, because of the limited computation and network resources. Data analytic frameworks like Spark [14], MapReduce [3] involves. A typical task, called *co-task*, in these frameworks has multiple subtasks. A co-task is considered completed only after all its subtasks have been completed. In edge computing scenarios, most requests from users also contain a set of subtasks. Due to latency sensitivity and limited bandwidth from the edge server to the cloud server, the cloud data center cannot meet al.l the requests that transfer massive data from different locations. So running functions on edge servers to serve these requests is an appropriate solution. However, there are three challenges to deploy the service at the edge servers:

- **Task dispatching:** When a co-task request arrives at an edge server with its data, we can execute its subtasks after all the existing subtasks in the server's execution queue have been executed, or dispatch its subtasks to another edge/cloud server, suffering the latency between the two servers. We should develop a task dispatch policy to take online decision on the edge server, without the knowledge of the task execution time.
- **Function configuration:** A user-generated request requires specific functions to serve. Due to limited resources of edge servers, we can not configure all the functions at each edge server. To adapt to the varying environment, we should reconfigure the existing functions on the edge servers.
- **Diverse and time-varying environment:** Massive user-generated data is distributed in different physical locations, which may be very far away from each other. And the user's favorite applications vary according to the user's location. Even at different times of the day, users prefer to use different applications, such as watching news in the morning and playing games in the evening. The diverse and time-varying environments mentioned above make it hard for a simple heuristic algorithm to adapt.

To solve the above challenges, we propose an online algorithm, named OnDisco, which combines deep reinforcement learning and a heuristic approach, to minimize the average request completion time, without requiring the execution time of the tasks at their arrivals. Our main contributions can be highlighted as:

- We considered task dispatching and function configuration jointly and formulated the online decision problem in edge computing to minimize the average task completion time. Specifically, we consider co-tasks, each of which is consisted of a set of subtasks with the completion time determined by the latest completed subtask.
- We propose an Online algorithm for Co-task Dispatching with Function Configuration, called OnDisco, that leverages deep reinforcement learning to

be adaptive to various edge environments without any prior knowledge, e.g., the task execution time, by using neural networks to extract the intrinsic of the environment.

– Based on the real production traces from Alibaba consisted of more than 3 million complicated co-tasks, we conduct extensive simulations to evaluate the performance of OnDisco. OnDisco reduces the average co-task completion time by 58% and 76% respectively when compared with the heuristic and random baselines.

The rest of this paper is organized as follows. In Sect. 2, we present the model and define the problem formulation. In Sect. 3, we develop OnDisco. Section 4 shows the simulator and training method. In Sect. 5 we discuss the evaluation results. And Sect. 7 concludes our work.

2   System Model and Problem Formulation

2.1   System Model

Edge System: We consider an edge-cloud system including an edge cluster with $K - 1$ heterogeneous edge servers and one remote cloud server, denoted by $\{s_1, s_2, ..., s_K\}$. Each edge server $s_i (i = 1, 2, ...K - 1)$ has limited resources, denoted by $u_i (i = 1, 2, ..., K - 1)$. Compared with edge servers, the remote cloud server s_K is usually assumed with unlimited resources. There will be a communication latency l_{ij} between server s_i and server s_j.

Application and Function: Consider there are N different types of functions $\{f_1, f_2, ..., f_N\}$. Each function f_i runs on a server and consumes 1 unit of the server. Each edge server may initially be configured with some functions. Since the remote cloud server has unlimited resources, we assume it has all functions configured. There are M types of applications in the edge system, denoted by $\{a_1, a_2, .., a_M\}$. Each application contains a subset of functions Note that one application may contains several identical functions.

Co-task and Subtask: When a specific application is invoked, we call each of its request as a co-task, denoted by $\{r_1, r_2, ..., r_M\}$. Each co-task contains a number of subtasks. Our scheduler will dispatch its subtasks to the servers. A co-task is considered completed only if all subtasks corresponding to the co-task are completed. Besides, functions are implemented at the subtask level. Let $\{v_1, v_2, ..v_N\}$ be the set of different types of subtasks. Without loss of generality, we assume that each function f_i exactly maps to subtask v_i, and the execution time of the subtask v_i is uncertain and varies depending on the state of the server. Typically, when a server s_j without the function f_i receives the subtask v_i, the server s_j has to bypass the subtask v_i to the remote cloud server. The remote cloud server will return the result after a constant time cost $time_{cloud}$, which is dominated by the WAN latency.

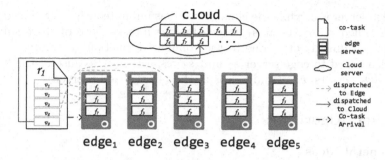

Fig. 1. An example of the cloud-edge system model, where co-task r_1 arrives at $edge_1$, which consists of 5 subtasks from v_1 to v_5. v_1 and v_2 are served at the local server $edge_1$. v_3 and v_4 are dispatched to $edge_2$ and $edge_3$, respectively. v_5 is dispatched to the cloud server.

2.2 Problem Formulation

We consider a series of online co-tasks arriving at arbitrary time, denoted by $\mathcal{J} = \{r_{j_1}, r_{j_2}, ... | 1 \leq j_i \leq M\}$. Each co-task r_{j_i} initially arrives at a specific edge server at time t_i^s. The probability of co-task arrival on each server satisfies a distribution. The scheduler determines where all subtasks of a co-task are executed. The finish time of co-task is denoted by t_i^f. If the initial arrival server of the co-task is different from the execution server of the subtask, the subtask will suffer from an extra overhead of the latency between the two servers. The completion time of co-tasks r_i is defined as the difference between the last subtask's completion time and the arrival time of r_i.

The input of our model is as follows:

- The status of the environment, including the function configuration status on each server, the arrival frequency of co-tasks, the latency between each server, and the remaining resources on each server.
- The static information of the system, including all the correspondence between co-tasks and subtasks, the number of the servers, and the total resources of each server.

Note that our model does not require the execution time of subtasks, which is difficult to obtain in the real production environment. Our optimization objective is to minimize the average co-task completion time, i.e., $\frac{1}{|\mathcal{J}|}\sum_{r_{j_i} \in \mathcal{J}} t_i^f - t_i^s$. The system must meet the following constraints:

- Each subtask $v_i \in \mathcal{J}$ must be dispatched to a specific server that is already configured with function f_i.
- The number of functions configured on each edge servers cannot exceed the limits of the server resources u_i.

A simple example of our model is illustrated in Fig. 1. There is an edge computing system that contains 5 edge servers and a cloud server. The capacity of the edge server is 3, while the cloud server has unlimited capacity. Specifically, a co-task called r_1 arrives at s_1, which has 5 subtasks. Two of these subtasks can be served directly at s_1 due to the configured function f_1. Subtask v_2 and v_3 are dispatched to edge server s_2 and s_3 respectively. Because of the lacking of f_8 at edge servers, v_8 should be dispatched to the cloud server.

3   Online Algorithms

3.1   General Ideas

The model mentioned above is a continuous online decision problem. The characteristics of the environment, such as the proportion of different co-tasks and the real-time delays, may change over time. Therefore, heuristic algorithms can not adapt to various environmental conditions. We propose a two-layer algorithm that combines heuristic approach and reinforcement learning. At the function configuration layer, every few hundred milliseconds, we use a neural network to extract the features of the environment and tasks, and the following function configuration on each server is generated by a reinforcement learning model, as shown in Fig. 2. At the task dispatching layer, which will happen every few milliseconds. Since the edge server has limited resources and real-time sensitivity is required, we use a heuristic algorithm to dispatch all the requests arriving at the edge server.

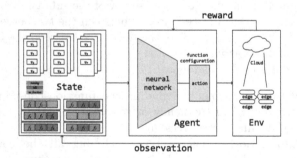

Fig. 2. Reinforcement Learning framework in the function configuration layer: the agent observes the state of the edge system, generates the function configuration action, and receives the reward according to the average execution time.

3.2   Function Configuration

Feature Engineering: The input of our model is divided into two layers: environmental information and task information, corresponding to environmental feature and task feature extracted by neural network respectively. We denote as $\mathcal{F}^t = \{F_i^t | 1 \leq i \leq N\}$ and $\mathcal{L}^t = \{l_{ij}^t | 1 \leq i, j \leq K\}$ the number of function configuration on each server and the latency between each server as the environmental information at time step t, respectively. In addition, we take the number of arrival tasks between the previous step and the current step as the task information, which is the input vector of the neural network, noted by $\mathcal{C}^t = \{C_i^t | 1 \leq i \leq N\}$. Unlike the execution time of the tasks, the number of arrival tasks is easy to obtain in the real production environment.

Reward: OnDisco will calculate the reward of steps spanning $\mathcal{T}$ milliseconds based on the environmental information, which is used for training reinforcement learning agent. The reward in step t is defined as: $reward_t = \sum_{r_t} -c_{r_t}$, where c_{r_t} is the completion time of co-task r_t and r_t completed between step $t-1$ and step t. The optimization objective of the reinforcement learning is to minimize the time-average of rewards: $\mathbb{E}[\sum_{t=1}^{T} \frac{1}{|\mathcal{J}|} \sum_{r_t} c_{r_t}]$, where T is the total time steps. This objective exactly minimizes the average co-task completion time.

Algorithm 1: function configuration

---

1 **for** *every $\mathcal{T}$ milliseconds* **do**
2   Calculate $\mathcal{F}^t, \mathcal{L}^t, \mathcal{C}^t$
3     $f^t = $ RL_inference$(\mathcal{F}^t, \mathcal{L}^t, \mathcal{C}^t)$
4   Update the estimated execution time e_i of each task v_i $(i = 1, ..., N)$
5   Configure f^t to the server that can earliest start serving
      tasks according to the estimated task execution time e_i

---

Action: OnDisco's action is used to generate the function configuration (1). The action of OnDisco is divided into two stages. The first stage (Line 2–3) is to determine the total number of function configuration for the entire edge system. Here, we use reinforcement learning approach to obtain the result. The second stage (Line 4–5) is to decide where each function is placed. In order to get a better placement of functions, we calculate the average execution time of different types of tasks on the server based on historical information. In the second stage, the chosen function is configured to the server, so that this function can start serving tasks at the earliest, according to the estimated execution time of the queuing tasks. When the system could not configure more functions, OnDisco will select a function with the least remaining execution time to evict. When the function finishes all remaining tasks, it will be replaced by a new function using the reinforcement learning. Note that, instead of a complete configuration scheme for all functions, the reinforcement learning agent at the first stage just suggests the function that is urgently needed. Then, we will increase the number of this function by 1 so that the future tasks can be served timely.

Network Updating: We have two neural networks, the policy network and the value network, that partially share weights. Both networks take $\mathcal{F}^t, \mathcal{L}^t, \mathcal{C}^t$ as the inputs, where the output of the policy network is a probability distribution of all possible actions, denoted by $\pi_\theta(action_t | \mathcal{F}^t, \mathcal{L}^t, \mathcal{C}^t)$ and the output of the value network is the evaluation of the current state, represented as the estimation of the sum of expected future rewards, denoted as $V_w(\mathcal{F}^t, \mathcal{L}^t, \mathcal{C}^t)$. The agent updates the parameters θ and w using actor-critic algorithm [7]:

$$\theta \leftarrow \theta + \alpha \nabla_\theta \log \pi_\theta(\mathcal{F}^t, \mathcal{L}^t, \mathcal{C}^t)(reward_t + \gamma V_w(\mathcal{F}^{t+1}, \mathcal{L}^{t+1}, \mathcal{C}^{t+1}) - V_w(\mathcal{F}^t, \mathcal{L}^t, \mathcal{C}^t)),$$

$$w \leftarrow w + \beta \nabla_w (reward_t + \gamma V_w(\mathcal{F}^{t+1}, \mathcal{L}^{t+1}, \mathcal{C}^{t+1}) - V_w(\mathcal{F}^t, \mathcal{L}^t, \mathcal{C}^t))^2$$

Here, α and β are the learning rates. Intuitively, $reward_t + V_w(\mathcal{F}^{t+1}, \mathcal{L}^{t+1}, \mathcal{C}^{t+1}) - V_w(\mathcal{F}^t, \mathcal{L}^t, \mathcal{C}^t)$ estimates whether current action performs better than the average case. $\nabla_\theta \log \pi_\theta(\mathcal{F}^t, \mathcal{L}^t, \mathcal{C}^t)$ is the direction of increasing the probability to choose a_t in the parameter space. The value network will be updated to narrow the gap between the evaluated value and the sum of weighted rewards.

3.3   Task Dispatching

Algorithm 2: task dispachcing

---

1 **for** *every v_k of the arrival co-task r* **do**
2 　　Calculate the average execution time e_j of all the task v_j $(j = 1, ..., N)$ according to the completed historical tasks
3 　　**for** *i in range K-1* **do**
4 　　　　$wait\_time_i = \text{sum}([e_j \text{ for } v_j \text{ in } queue_i^k]) / num_i^k$
　　　　/* $queue_i^k$ is the execution queue of function f_k in edge server s_i and num_i^k is the number of configured function f_k in edge server s_i */
5 　　**if** $min(wait\_time_i) + e_k < time_{cloud}$ **then**
6 　　　　$id = \text{argmin}(wait\_time_i)$ dispatch v_k to edge server s_{id}
7 　　**else**
8 　　　　dispatch v_k to cloud server

At the task dispatching layer, the millisecond-level reinforcement learning inference model is not suitable as a task dispatching algorithm, because those tasks with short execution time can be executed in a few milliseconds. The model inference progress will greatly increase the overhead of the system. We adopt a heuristic algorithm illustrated as Algorithm 2. When a co-task arrives at an edge server, we will trigger Algorithm 2 to decide where to dispatch the subtasks. For every v_k of the arrival co-task r, Algorithm 2 will calculate the average execution time e_i of all the subtasks according to the completed historical subtasks. If there is no subtask completed in the early stage, then we assume its average execution time is 0 (Line 2). And we will calculated the waiting time $wait\_time_i$ if we dispatch v_k to s_i (Line 3–4). If the estimated completion time of v_k is less than $time_{cloud}$, we will dispatch v_k to an edge server that can serve v_k earliest, or we will dispatch v_k to the cloud server (Line 5–8).

4   Implementation

Simulator: Training reinforcement learning models require many interaction traces between agents and the environment. Because of the fixed-scale servers in a real production environment, we cannot obtain enough traces for training in a short time or train models in a large-scale distributed environment. Therefore, we build a simulator based on the real production environment, which contains all the servers and can simulate the interaction between agent and environment. Compared with training in a real environment, a simulator can simulate the task's execution time instead of waiting for its real completion.

For training instances that contains agents and environment, there will be at most 10 functions in the environment. Moreover, we have 10 servers, one of which is the cloud server, and the rest are edge servers. Due to the heterogeneity of the server environment, each edge server will have its own function configuration limit, ranging from 2 to 8. Relatively cloud server can deploy unlimited functions, and the constant completion time of cloud is $time_{cloud} = 1000$.

Training the Neural Network: In the early stage of training, we distribute evenly half of the total number of functions that can be contained in the system to each function, which guides the agent to a more optimized state. In the later stage of training, we will gradually reduce this ratio, until finally, there is no initial function in the edge system, so that the agent can explore more possible states.

Distributed Training: To further accelerate the training and make the it more stable, we set up a total of 30 training instances in parallel, as well as a central parameter server that interacts with all agents in every step. At the beginning of a step, the parameter server will send its weights to all agents, which interacts with the environment in calculating the gradients. After that, the parameter server collects all the gradients and updates the weights of the model.

5 Performance Evaluation

5.1 Evaluation Settings

Dataset: We use the real production traces from Alibaba [1] as our workload. After removing some unnecessary attributes from the original data traces, each piece of data has the following attributes: *request id, task id, and execution time*. We put all the request traces into a request pool, and when the simulator needs a request, we randomly select one from the request pool. To better reflect the characteristics of edge computing scenarios, we scale all the processing time of each task to the millisecond level.

Heuristic Baselines: We adopt the following two heuristic algorithms as the baseline of OnDisco, which is

- **Weighted load balance**: In the function configuration layer, the weighted load-balancing algorithm counts the number of all the subtasks, and configure the functions according to the proportion of subtasks. In the task dispatching layer, this baseline dispatches tasks to corresponding functions in a balanced manner.
- **Random:** Random algorithms configure function and dispatch subtasks randomly to the edge servers, and also dispatch subtasks randomly.

5.2   Large-Scale Simulation Results

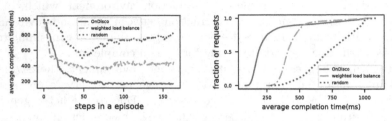

(a) Average co-task completion time   (b) CDF of co-task completion time
in an episode

Fig. 3. Performance overview of co-task's average completion time.

Overview Performance: Figure 3(a) demonstrates the average co-task completion time of the algorithms in an episode. After more than 50 steps, the function configuration status of the edge system is close to saturation and the average completion time tends to be stable. OnDisco reduces the average co-task completion time by 58% and 76% compared with the weighted load-balancing and the random baselines, respectively. At the beginning of an episode (less than 20 steps), OnDisco performs slightly worse than the weighted load-balancing algorithm. Since the weighted load balance algorithm is heuristic, it moves directly towards the fixed target. However, our approach will explore various possible states in the early stages due to the non-deterministic policy to find better state. The random algorithm cannot adapt well to an online decision-making problem in the changing environment, thus lead to worse performance after the 50-th step.

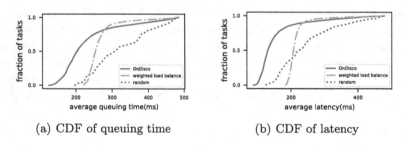

(a) CDF of queuing time   (b) CDF of latency

Fig. 4. CDF of queuing time and latency. OnDisco gains less completion time in both queuing time and latency, because the edge servers' resources are used more efficiently.

CDF Analysis: To understand why OnDisco performs better than the baselines, we further analyze the distribution of co-task queuing time and latency. Figure 3(b) illustrates the proportion of subtasks with different completion time over 15 episodes. Compared with load-balancing algorithms, OnDisco gains lower average co-task completion time before the 91st quantile. Furthermore, most of the co-tasks between the 91st and 100th quantile arrive at the early stage of the episode, when OnDisco is exploring the states of the system. The completion time of a subtask is composed of queuing time, latency between servers, and execution time, in which queuing time and latency vary greatly with the scheduling policy. Figure 4(a) and Fig. 4(b) illustrate the proportion of subtasks with respect to queuing time and latency respectively. OnDisco performs better before the 82nd quantile and before the 89th quantile respectively with respect to queuing time and latency.

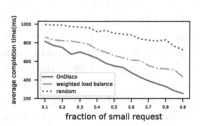

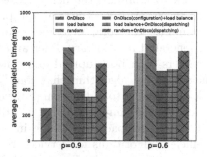

Fig. 5. Task completion time under different small request fractions

Fig. 6. Average completion time on two different datasets, where p is the fraction of small request

Impact of Varying Environment: In the real environment, at different times of the day, the the proportion of requests arriving for different types of requests vary greatly. In order to analyze the performance of the algorithms under various request arrival distributions. We divide all requests of the Alibaba traces into two categories: large requests (the number of subtasks are more than 9) and small requests (the number of subtasks are less than or equal to 9). Figure 5 demonstrates the performance of each algorithm in different data sets, with the fraction of small requests changing from 0.1 to 0.9.

Performance in Each Layer: OnDisco consists of two layers and adopts different methods. In order to reflect the improvement of OnDisco in function configuration and task dispatching respectively, we only use OnDisco in one layer and the baseline policy in the other layer, as illustrated in Fig. 6. We evaluated all the algorithms in two different datasets (fraction of small request is equal to 0.9 and 0.6 respectively). No matter in which layer using OnDisco alone, the task completion time will be improved to a certain extent compared with baselines.

6   Related Work

The bloom of cloud computing and edge computing has promoted the research of scheduling multiple concurrent jobs on clusters. Zhao et al. [15] has proved that the edge system of heterogeneous cloud can increase the probability of satisfying the delay demands by 40%, demonstrating the potential of this research field. Here we survey the mostly closed work to our online learning-based co-task dispatching problem in edge computing.

6.1   Task Dispatching and Function Configuration

Considering the correlation of jobs, Isard et al. [6] maps the scheduling problem to a graph structure, and a scheduling scheme is given by a central cost model. Chen et al. [2] proposed a task dispatching scheme based on software defined network to minimize the delay while saving the battery life of user devices in ultra-dense networks. Tan et al. [11] proposed a general model for task scheduling to minimize the total weighted response time over all the tasks, where tasks are generated in arbitrary order and times at the mobile devices and offloaded to servers with both upload and download delays. Xu et al. [12] proposed an efficient online algorithm that jointly optimizes dynamic service caching and task dispatching to reduce computation latency for end users while maintaining low energy consumption. Lei et al. [13] studied the joint optimization of service placement and load dispatching in mobile cloud systems, and developed a set of efficient algorithms for service providers to achieve various trade-offs among the average request latency of mobile users and the cost of service providers. However, few previous works studied online dispatching of co-related tasks in the scenarios of edge computing. Paper [8] studied DAG dispatching and scheduling in edge computing with online greedy algorithms. In this work, we investigate co-task dispatching problem with function configuration in edge computing using online learning techniques.

6.2   Scheduling with Reinforcement Learning

The development of hardware, like GPU and CPU, triggered the bloom of machine learning, including deep reinforcement learning (RL). RL is a natural way to handle the cases where scheduler does not have any prior knowledge of the arrival jobs. Considering the dependencies among tasks, Decima, proposed in [9], abstracts the tasks as DAGs, which is described by a graph neural network, to minimize the average completion time. Similar to Decima, Hu et al. devised a learning-based algorithm called Spear [5], which also considers the scheduling of DAGs consist ed of related jobs. Choosing different jobs on different branches in a DAG will lead to different states, so inspired by AlphaGo [10], they applied Monte Carlo tree search to their system, using deep reinforcement learning as a guide to avoid invalid searches. In this work, we adopt RL to guide our online algorithm design.

7   Conclusion

In this work, we study online task dispatching and function configuration in edge computing without knowing the task execution time in advance. We propose an online learning-based algorithm to minimize the average task completion time. Extensive simulations on Alibaba traces show that `OnDisco` consistently performs better than baselines in multiple different data environments and various parameter settings. An interesting and challenging future work is to model each co-task as a directed acyclic graph, i.e., taking the precedence relationship among subtasks into account.

Acknowledgements. This work is supported partly by NSFC Grants 61772489, 61751211, Key Research Program of Frontier Sciences (CAS) No. QYZDY-SSW-JSC002, and the project of "FANet: PCL Future Greater-Bay Area Network Facilities for Large-scale Experiments and Applications (No. LZC0019)".

References

1. Alibaba trace (2018). https://github.com/alibaba/clusterdata
2. Chen, M., Hao, Y.: Task offloading for mobile edge computing in software defined ultra-dense network. IEEE J. Sel. Areas Commun. **36**(3), 587–597 (2018)
3. Dean, J., Ghemawat, S.: MapReduce: simplified data processing on large clusters. Commun. ACM **51**, 107–113 (2008)
4. Garcia Lopez, P., et al.: Edge-centric computing: vision and challenges. SIGCOMM Comput. Commun. Rev. (2015)
5. Hu, Z., Tu, J., Li, B.: Spear: optimized dependency-aware task scheduling with deep reinforcement learning. In: IEEE ICDCS (2019)
6. Isard, M., Prabhakaran, V., Currey, J., Wieder, U., Talwar, K., Goldberg, A.: Quincy: fair scheduling for distributed computing clusters. In: ACM SOSP (2009)
7. Konda, V.R., Tsitsiklis, J.N.: Actor-critic algorithms. In: NIPS (2000)
8. Liu, L., Huang, H., Tan, H., Cao, W., Yang, P., Li, X.Y.: Online DAG scheduling with on-demand function configuration in edge computing. In: WASA (2019)
9. Mao, H., Schwarzkopf, M., Venkatakrishnan, S.B., Meng, Z., Alizadeh, M.: Learning scheduling algorithms for data processing clusters. In: SIGCOMM (2019)
10. Silver, D., et al.: Mastering the game of go without human knowledge. Nature **550**, 354–359 (2017)
11. Tan, H., Han, Z., Li, X., Lau, F.C.M.: Online job dispatching and scheduling in edge-clouds. In: IEEE INFOCOM (2017)
12. Xu, J., Chen, L., Zhou, P.: Joint service caching and task offloading for mobile edge computing in dense networks. In: IEEE INFOCOM (2018)
13. Yang, L., Cao, J., Liang, G., Han, X.: Cost aware service placement and load dispatching in mobile cloud systems. IEEE Trans. Comput. **65**(5), 1440–1452 (2016)
14. Zaharia, M., Chowdhury, M., Franklin, M.J., Shenker, S., Stoica, I., et al.: Spark: cluster computing with working sets. HotCloud (2010)
15. Zhao, T., Zhou, S., Guo, X., Niu, Z.: Tasks scheduling and resource allocation in heterogeneous cloud for delay-bounded mobile edge computing. In: IEEE ICC (2017)

System-Level FPGA Routing for Logic Verification with Time-Division Multiplexing

Long Sun[1], Longkun Guo[1,2], and Peihuang Huang[3(✉)]

[1] College of Mathematics and Computer Science, Fuzhou University,
Fuzhou 350116, People's Republic of China
longsun100@foxmail.com, longkun.guo@gmail.com
[2] School of Computer Science, Qilu University of Technology (Shandong Academy
of Sciences), Jinan 250353, People's Republic of China
[3] College of Mathematics and Data Science, Minjiang University,
Fuzhou 350108, China
peihuang.huang@foxmail.com

Abstract. Multi-FPGA prototype design is widely used to verify modern VLSI circuits, but the limited number of connections between FPGAs in a multi-FPGA system may cause routing failure. Therefore, using time-division multiplexing (TDM) technology, multiple signals are transmitted through the same routing channel to improve utilization. However, the performance of this type of system depends on the routing quality within the FPGAs due to the signal delay between FPGA pairs. In this paper, we propose a system-level routing method based on TDM to minimize the maximum TDM ratio that satisfies the strict ratio constraint. Firstly, we weight the edges and use two methods to build approximate minimum Steiner trees (MST) to route the nets. Then we propose a ratio assignment method based on edge-demand which satisfy the TDM ratio constraint. We tested our method with the benchmarks provided by 2019 CAD Contest at ICCAD and compared it with the top two. The experimental results shows that our method not only solves all problems but also achieves a good TDM ratio.

Keywords: FPGA · Logic verification · Approximate Steiner Tree · Routing

1 Introduction

Nowadays, the geometric size of integrated circuit (IC) technology is rapidly shrinking. By increasing the number of components on a single chip and reducing the feature size of the chip, the density and integration of the chip has been increased. With the level of development and complexity of very large scale integrated circuits (VLSI) [5], the design and manufacturing costs of VLSI are gradually increasing [8]. So it is necessary to find an effective method to verify the effectiveness and efficiency of the VLSI in industry.

© Springer Nature Switzerland AG 2021
Y. Zhang et al. (Eds.): PDCAT 2020, LNCS 12606, pp. 210–218, 2021.
https://doi.org/10.1007/978-3-030-69244-5_18

There are several ways of logical verification. One is software logic simulation [2]. It provides visibility and debugging capabilities with a considerable cost. However, it must simulate each logic gate one by one but the circuit size is always very large, it consumes a lot of running time. Another way is hardware emulation [7,15]. It greatly reduces the running time, but the cost of implementation is high. The third method of logic verification is to use an FPGA prototype system. The FPGA prototype verifies the circuit through a configurable FPGA system and it achieve better trade-off between cost and runtime. Therefore, the FPGA prototype system has been widely used in industry [14].

In order to adapt to the design of the FPGA prototype system, a large VLSI circuit must be divided into multiple sub-circuits [9], each of which corresponds to an FPGA. Since the number of I/O pins in FPGA is fixed and limited, routing signals usually exceed the number of I/O pins. Bobb et al. [3] introduced a time division multiplexing (TDM) technique that can transmit multiple routing signals in one system clock cycle. The technique increases the signal capability in one FPGA and the routability for the prototyping system. However, this technique also slows down the inter-FPGA signal delay [3].

There are several works targeting on the optimization of inter-FPGA connections for logic verification. The work [12] presented an integer linear programming (ILP) based method to select I/O signals to achieves higher frequency of 2-FPGA systems. However, due to the sharp increase in the size of the internal signal of the FPGA, it is difficult for the method based on ILP to generate a promising solution within a reasonable runtime. To optimize the TDM ratio, Pui et al. [16] presented an analytical framework for multi-FPGA based system. However, in this architecture, only cross-FPGA networks with the same direction and TDM ratio can be allocated to the same line. In addition, the above work does not consider the TDM ratio to be an even number, which is an impractical hardware implementation of multiple channels [13]. Therefore, it is desirable to develop an effective and efficient TDM based FPGA routing method.

In this paper we proposed an efficient TDM based FPGA routing method. The major contributions of our work are summarized as follows:

- We have solved the system-level FPGA routing problem while considering the TDM constraints and the maximum TDM ratio.
- We transformed the routing problem into an MST construction problem, and proposed an method to effectively solve it.
- We proposed a ratio assignment method based on edge-demand and it works well.
- We compared the results of the top two in the 2019 CAD Contest at ICCAD, and experimental results show that our algorithm not only solves all TDM constraints but also have a good performance.

The rest of this paper is organized as follows. Section 2 introduces the calculation of the TDM ratio and formulates the system-level FPGA routing problem. Section 3 introduces our TDM based FPGA routing method. Section 4 provides the experimental results. Finally, conclusions are drawn in Sect. 5.

2   Preliminaries

The system-level FGPA routing problem can be modeled as an undirected graph $G(V, E)$, where vertices $V = \{v_1, v_2, ..., v_i\}$ are the set of FPGAs and edges $E = \{e_1, e_2, ..., e_j\}$ are the connections between FPGAs. There also defines a set of nets $\mathcal{N} = \{N_1, N_2, ..., N_k\}$ and a set of groups $\mathcal{P} = \{P_1, P_2, ..., P_l\}$ for the problem. A net is a set of vertices in graph G, and a net-group includes several nets with same power consumption or similar attributes. In addition, a vertex can be connected by one or multiple nets and a net belongs to one or multiple net-groups.

In this section, we first introduce the TMD ratio and its computation, and then give the TMD constraints. Finally, we will formulate the system-level FPGA routing problem based on the TMD ratio constraints.

2.1   TDM Raito

The challenge of system-level FPGA routing lies in the side-effect of TDM. That is the FPGA routing can always complete, but the inter-FPGA signal delay is increased by TDM ratio [10]. The TDM ratio of each edge e can be defined as:

$$r_e = \frac{demand_e}{capacity_e} \tag{1}$$

where $demand_e$ is the demand of edge e and $capacity_e$ is the capacity of edge e.

In fact, the actual capacity of each edge is 1 in the TDM based system-level FPGA routing and the TDM ratio is defined as an even number due to multiplexing hardware implementation. The TDM ratio of each net is defined as the sum of ratios of all edges it routes, and the TDM ratio of each group is defined as the sum of ratios of all the nets it contains. Which can be defined as:

$$r_N = \sum_{e \in N} r_e \\ r_P = \sum_{N \in P} r_N \tag{2}$$

where r_N is total TDM ratio of net N and r_P is total TDM ratio of group P.

2.2   Problem Statement

The remodeled problem can be described as follows:

Given a multi-FPGA platform with timing division multiplexing (TDM) wire between each FPGA connection pair, a netlist, a net-groups list, route the nets and assign TMD ratio for each connection in net, so that the maximum total TMD ratio in all net-groups are minimized, and the crucial TDM constraints are satisfied.

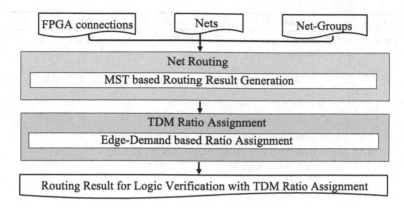

Fig. 1. The algorithm flow

3 Our Algorithm

Figure 1 show our system-level FPGA routing method with crucial ratio constraints. It basically consists of two steps: (1) MST based net connection; (2) TDM ratio assignment.

In the MST based net connection step, we firstly weight the edge in a linear-growth way and formulate the connection task as the structure of a minimum Steiner tree(MST) [6] while updating the edge cost. In the TDM ratio assignment step, we propose an edge-demand based TDM ratio assignment method to assign the TDM ratio for every groups to satisfied ratio constraints. In the following subsections we will explain the details of each technique for the two major steps.

3.1 Edge Weight Setting and Problem Reformulation

For a net with n points, it can be proved that the tree is optimal [4]. For each inter-FPGA connections, it should be routed by as few nets as possible, so we must increase its routing cost when an inter-FPGA connection is routed. In our algorithm, we initialize the initial wight of each edge to 2, and once this edge is routed by a net, the weight increases by 2. In addition, considering the potential impact of routing order on TDM ratio assignment results, we route the nets by the size of the net-group, and the net with more vertexes will be routed first in same net-group. The size of net is defined as the number of FPGAs it contains and the size of net-group is the sum of the net size it contains.

After setting the weight for edges, we remodel the problem as a minimum Steiner tree (MST) problem: Given an undirected and weighted graph $G(V, E, W)$. The goal is to find a sub-graph $G' = (V', E', W')$ of G to connect the vertices sets V' which satisfies min $\sum_{e \in E'(G')} W'(e)$.

Algorithm 1. Shortest path Based Approximate MST Algorithm

Input: An undirected and weighted graph $G(V, E, W)$, a vertex set $V' \subseteq V$.
Output: A set of connecting edges E'.

1. $U \leftarrow \forall v' \in V'$;
2. $U' \leftarrow V' - U$;
3. add vertex s and t
4. **While** $U' \neq \emptyset$ **do**
5. $E_s :=$ add edges from s to vertices in U with weight 0;
6. $E_t :=$ add edges from vertices in U' to t with weight 0;
7. $path \leftarrow$ shortest path from s to t;
8. $path \leftarrow path - E_s - E_t$;
9. $u \leftarrow$ vertices in $path$ without s, t;
10. $E' \leftarrow E' + path$;
11. $U \leftarrow U + u$;
12. $U' \leftarrow U' - u$;
13. **End while**
14. **Return** E'.

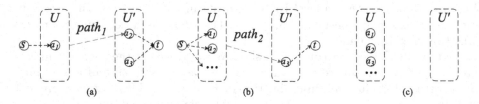

        (a)   (b)   (c)

Fig. 2. The routing procession for a 3-FPGAs net.

3.2   MST-Based Net Routing

In this section, we develop an efficient method of generating MST to route the net. For the net with small size we used an exact MST algorithm which time complexity is $O(3^{|V'|}|V| + 2^{|V'|}(|V|log|V| + |E|))$ to generate an optimal solution, for the net with large size we propose an approximate MST algorithm to reach a reasonable runtime.

Algorithm 1 shows the details of the our approximate MST algorithm. In lines 1–2, we first construct a set U that represents the set of routed vertexes and U' that represents the set of remaining vertexes that have not been routed. For each loop, we add the auxiliary edge from the source vertex s to vertexes in U and also add the auxiliary edge from vertexes in U' to sink vertex t, and then use the Dijkstra algorithm to solve the shortest path from s to t (line 5–8); In lines 9–12, we have updated U, U' and E'.

Figure 2 shows the routing process for a 3-FPGAs net using our approximate MST algorithm.

In addition, in the previous algorithm we used Fibonacci heap [1] to optimize the path solving algorithm.

Algorithm 2. Routing Result Generation Algorithm

Input: An undirected and weighted graph $G(V, E, W)$, a list of net-groups $\mathcal{G}$, a list of
       nets $\mathcal{N}$.

Output: The routing result R.

1. pre-route all nets in group with largest size by algorithm 1;
2. reinitialize the weight of each edge;
3. **For** each group G in $\mathcal{G}$ ordered by their size **do**
4.   **For** each net N in $\mathcal{N}$ ordered by their size **do**
5.   **If** N is not routed **do**
6.   Route N by algorithm 1;
7.   Update R;
8.   **End For**
9. **End For**
10. **Return** R.

3.3 Routing Result Generation

In subsection, we propose an efficient routing method based on edge prediction.
First, we use Algorithm 1 to pre-route all nets in the group with largest size,
and use n_e to record the number of times each edge is routed in pre-routing and
we set the initial weight of each edge $w_e = ne$. Then we route all the net-groups
ordered by group size and we re-record the number of times each edge is routed
once a net is routed, If one edge does not exceed n_e, then the weight of edge e is
not updated, else the weight of edge $+2$ before routing the next net. Algorithm 2
shows the process of our method.

3.4 Ratio Assignment

After the nets connection phase, all nets are routed by our routing method.
In the TDM ratio assignment step, the task is assign TDM ratio to each net
and minimize the maximum TDM ratio in net-groups while satisfying the ratio
constraints. The optimization problem can be formulated as:

$$
\begin{aligned}
\min \max \quad & \sum_{N \in P} \sum_{e \in N} r_e \quad \forall P \in \mathcal{P} \\
s.t. \quad & R_e(r_e) = \sum \frac{1}{r_e} \leq 1 \ \forall e \in E \\
& mod(r_e, 2) = 0 \quad \forall e \in E \\
& r_e \geq 2 \quad\quad\quad \forall e \in E
\end{aligned}
\tag{3}
$$

where r_e is the TDM ratio of edge e in net N, and R_e represents the reciprocal
of the TDM ratio of the same edge in all nets, whose sum must not exceed 1.

    Due to the huge scale of modern circuit design, it is difficult for ILP-based
algorithms such as that in paper [11] to solve complex assignment problems
within a reasonable runtime. Therefore, we propose an efficient TDM ratio
assignment method to the routing signals. Its process is as follows:

– We use $\alpha(e)$ to represent the routed times of edge e.

Table 1. Statistics about Benchmarks

Benchmark	#Vertexes	#Edges	#Nets	#Groups	Avg. #vert	Avg. #nets
synopsys01	43	214	68456	40552	2	27
synopsys02	56	157	35155	56308	10	51
synopsys03	114	350	302956	334652	11	36
synopsys04	229	1087	551956	464867	11	37
synopsys05	301	2153	881480	879145	11	37
synopsys06	410	1852	785539	910739	48	54
hidden01	73	289	54310	50417	17	54
hidden02	157	803	610675	501594	31	54
hidden03	487	2720	720520	886720	50	54

Table 2. The comparison of maximum TDM ratio

Benchmark	2nd place	1st place	Ours
synopsys01	40190	40498	40436
synopsys02	32093856	32006748	31630660
synopsys03	129290206	128206488	127132204
synopsys04	6334724	7049316	6419814
synopsys05	4613110	5190132	4672398
synopsys06	15759302934	15751028904	15743366622
hidden01	409891196	409300116	408575164
hidden02	45952594974	45942030192	45933427024
hidden03	4872933150	4867325852	4860390872
Normalized	1.001	1.026	1.000

- If $\alpha(e)$ is even, we assign $\alpha(e)$ as the TDM ratio of edge e is each net; and if $\alpha(e)$ is odd, we assign the TDM ratio $\alpha(e) + 1$ for $\lceil \frac{\alpha(e)}{2} \rceil$ routing signals and $\alpha(e) - 1$ for the remaining $\lfloor \frac{\alpha(e)}{2} \rfloor$ nets.

4   Experimental Result

To evaluate our system-level FPGA routing algorithm, we implemented our algorithm in C++ programming language and tested it on benchmarks provided by the 2019 CAD contest at ICCAD. Table 1 lists the statistics of benchmarks, where "#Vertexes", "#Edges", "#Nets" and "#Groups" are denoted as vertexes, edges, nets and net-groups, and "Average Vertexes" and "Average Nets" are denoted as the average number of vertexes per net and the average number of nets per group.

The experimental results of the top 2 teams of the 2019 CAD Contest at ICCAD and ours are listed in Table 2. Our method was run on a Linux workstation and the results of the top 2 teams of the contest are provided by the contest organizer. Our method has better performance in most benchmarks, especially in large-scale cases. Large-scale FPGA routing design is the main challenge of contest.

5   Conclusion

In modern logic verification, multi-FPGA platforms with time-division multiplexing (TDM) I/Os have become more and more popular, and it also brings many challenges in system-level FPGA routing. In this paper, we have presented an effective method to solve the system-level FPGA routing problem satisfying the strict TDM constraints. First, we divide nets into two sets according to their sizes, and then we use two different MST construction methods to route them. Then we assign the TDM ratio for each net-group according to the number of times each edge is used. Finally, we have evaluated our method on the 2019 CAD Contest at ICCAD benchmark suites and we achieved good performance on them.

Acknowledgements. The first and the second author are supported by Natural Science Foundation of China (No. 61772005). The third author is supported by Natural Science Foundation of Fujian Province (No. 2020J01845) and Educational Research Project for Young and Middle-aged Teachers of Fujian Provincial Department of Education (No. JAT190613).

References

1. Ahuja, R.K., Mehlhorn, K., Orlin, J., Tarjan, R.E.: Faster algorithms for the shortest path problem. J. ACM (JACM) **37**(2), 213–223 (1990)
2. Asaad, S., et al.: A cycle-accurate, cycle-reproducible multi-FPGA system for accelerating multi-core processor simulation. In Proceedings of the ACM/SIGDA International Symposium on Field Programmable Gate Arrays, pp. 153–162 (2012)
3. Babb, J., Tessier, R., Agarwal, A.: Virtual wires: overcoming pin limitations in FPGA-based logic emulators. In: 1993 Proceedings IEEE Workshop on FPGAs for Custom Computing Machines, pp. 142–151. IEEE (1993)
4. Chen, G., Young, E.F.Y.: Salt: provably good routing topology by a novel steiner shallow-light tree algorithm. IEEE Trans. Comput. Aided Des. Integr. Circuits Syst. **39**(6), 1217–1230 (2019)
5. Constantinescu, C.: Trends and challenges in VLSI circuit reliability. IEEE Micro **23**(4), 14–19 (2003)
6. de Vincente, J., Lanchares, J., Hermida, R.: RSR: a new rectilinear steiner minimum tree approximation for FPGA placement and global routing. In Proceedings. 24th EUROMICRO Conference (Cat. No. 98EX204), vol. 1, pp. 192–195. IEEE (1998)
7. Graham, P.S.: Logical hardware debuggers for FPGA-based systems. PhD thesis, Citeseer (2001)

8. Hung, W.N.N., Sun, R.: Challenges in large FPGA-based logic emulation systems. In Proceedings of the 2018 International Symposium on Physical Design, pp. 26–33 (2018)
9. Inagi, M., Takashima, Y., Nakamura, Y.: Globally optimal time-multiplexing in inter-FPGA connections for accelerating multi-FPGA systems. In: 2009 International Conference on Field Programmable Logic and Applications (2009)
10. Inagi, M., Takashima, Y., Nakamura, Y.: Globally optimal time-multiplexing of inter-FPGA connections for multi-FPGA prototyping systems. IPSJ Trans. Syst. LSI Des. Methodol. **3**, 81–90 (2010)
11. Inagi, M., Takashima, Y., Nakamura, Y., Takahashi, A.: ILP-based optimization of time-multiplexed i/o assignment for multi-FPGA systems. In: 2008 IEEE International Symposium on Circuits and Systems, pp. 1800–1803. IEEE (2008)
12. Inagi, M., Takashima, Y., Nakamura, Y., Takahashi, A.: Optimal time-multiplexing in inter-FPGA connections for accelerating multi-FPGA prototyping systems. IEICE Trans. Fund. Electron. Commun. Comput. Sci. **91**(12), 3539–3547 (2008)
13. Pui, C.-W., Wu, G., Mang, F.Y.C., Young, E.F.Y.: An analytical approach for time-division multiplexing optimization in multi-FPGA based systems. In: 2019 ACM/IEEE International Workshop on System Level Interconnect Prediction (SLIP), pp. 1–8. IEEE (2019)
14. Schelle, G., et al.: Intel nehalem processor core made FPGA synthesizable. In Proceedings of the 18th Annual ACM/SIGDA International Symposium on Field Programmable Gate Arrays, pp. 3–12 (2010)
15. Wang, W., Shen, Z., Dinavahi, V.: Physics-based device-level power electronic circuit hardware emulation on FPGA. IEEE Trans. Industr. Inf. **10**(4), 2166–2179 (2014)
16. Lai, H.-H., Su, Y.-H., Huang, E., Zhao, Y.-C.: 2019 CAD contest at ICCAD on system-level FPGA routing with timing division multiplexing technique (2019)

Protein Interresidue Contact Prediction Based on Deep Learning and Massive Features from Multi-sequence Alignment

Huiling Zhang[1,2], Hao Wu[2], Hing-Fung Ting[3], and Yanjie Wei[1,2(✉)]

[1] University of Chinese Academy of Sciences, No.19(A) Yuquan Road, Shijingshan District, Beijing 100049, China
[2] Shenzhen Institutes of Advanced Technology, Centre for High Performance Computing, Chinese Academy of Sciences, Shenzhen 518055, Guangdong, China
yj.wei@siat.ac.cn
[3] Department of Computer Science, The University of Hong Kong, Hong Kong 999077, China

Abstract. Predicting the corresponding 3D structure from the protein's sequence is one of the most challenging tasks in computational biology, and a confident interresdiue contact map serves as the main driver towards *ab* initio protein structure prediction. Benefiting from the ever-increasing sequence databases, residue contact prediction has been revolutionized recently by the introduction of direct coupling analysis and deep learning techniques. However, existing deep learning contact prediction methods often rely on a number of external programs and are therefore computationally expensive. Here, we introduce a novel contact prediction method based on fully convolutional neural networks and extensively extracted evolutionary features from multi-sequence alignment. The results show that our deep learning model based on a highly optimized feature extraction mechanism is very effective in interresidue contact prediction.

Keywords: Residue contact prediction · Protein structure prediction · Multi-sequence alignment · Deep learning

1 Introduction

Protein accounts for 18% of the total weight of the human body and performs a variety of life functions. The unavailability of 3D structures for many protein sequences remains a daunting problem, which creates bottlenecks in functional annotation and many other applications. Since Anfinsen suggested that the advanced spatial structure of a protein is determined by its amino acid sequence [1], it has been a "holy grail" for the computational biology community to develop an algorithm that can accurately predict a protein's structure from its amino acid sequence. For protein structure prediction, ab initio methods are gaining importance because the well-established traditional method of template-based modeling is limited by the number of structural templates available in the Protein Data Bank. In the absence of homologous structural templates,

© Springer Nature Switzerland AG 2021
Y. Zhang et al. (Eds.): PDCAT 2020, LNCS 12606, pp. 219–228, 2021.
https://doi.org/10.1007/978-3-030-69244-5_19

a key input for successful ab initio protein structure prediction is the residue-residue contact map.

Residue-residue contacts refer to the residue pairs that are close within a specific distance threshold in a three-dimensional protein structure, as shown in Fig. 1. Protein contact maps are "simplified" 2D representations of the 3D protein structure yet, they retain most of the important features for protein folding. Simulation techniques can be used to predict protein structures using predicted contacts as restraints. Contact prediction has been considered an avenue for the prediction of protein structures. There are many 3D structure prediction methods [2–6] that enable the prediction of 3D structures from residue constraints or the combination of residue constraints with other evolutionary information. The applications of residue-residue contacts have been extended to the protein topology prediction [7], potential 3D model scoring and filtering [8, 9], protein-protein interaction prediction [10]. Residue contacts can also be used as the distance restraints to accelerate the procedure of molecular dynamics simulations [11, 12] and predict the binding affinity docking simulations [13].

Accurate residue contact prediction and its corresponding application have been one of the most challenging and promising problems in structural bioinformatics. The last two decades have witnessed significant progress in the algorithm development of residue contact prediction. Early contact prediction methods are mainly based on correlated mutation analysis (CMA)/mutual information (MI), mathematical optimization techniques, and machine learning algorithms. When evaluated with the contact definition of $C\beta - C\beta < 8\text{Å}$, CMA/MI-based methods such as Pollock et al. [14], MIp [15], MIc [16] generally achieved prediction precision of about 20%–30% with a relatively low recall of about 20%. Rajgaria et al. [17, 18] used integer linear programming techniques for residue contact prediction and achieved prediction precision of 50%–60% when evaluated based on a threshold of 12 Å between Ca-Ca atoms. The ab initio contact prediction methods do not rely on detectable homology to experimentally solved structures but still defect with low coverage. Traditional machine learning (ML) methods like SVMcon [19], NNcon [20] and SVMSEQ [21] are trained on a limited number of proteins and make a prediction for each residue pair without considering the status of other pairs. Therefore, these methods still show low recall and unsatisfactory prediction, especially for long-range contacts. Methods combining integer linear programming together with machine learning techniques [22, 23] can improve contact prediction performance to some extent.

One landmark work in contact map prediction is direct coupling analysis (DCA) [24] which combines covariance analysis with global inference analysis to capture the correlated pattern of coevolved residues. DCA emphasizes the importance of distinguishing between directly and indirectly correlated residues. Representative methods developed or promoted using the ideas of DCA include EVFold-Contact (mfDCA) [25] using mean-field approximation to accelerate the computation of DCA; GDCA [26] using continuous Gaussian random variables to replace the discrete amino acid variables; PSI-COV [27] using an estimate of the inverse covariance matrix to assign a score to residue pairs; PlmDCA [28], GREMLIN [29], CCMpred [30] learning the direct couplings as parameters of a Probabilistic Graphical Model (Markov random field) by maximizing its pseudo-likelihood. DCA-based methods assume that contacting residue pairs should present correlated mutations in the long-term evolutions reflected in multi-sequence alignment (MSA). Although DCA-based methods show higher accuracy compared to the MI methods and traditional machine learning methods, however, frequently become powerless for proteins with a limited number of effective sequences in MSA. To further increase accuracy and recall, meta-predictors like PconsC [31], MetaPSICOV [32] and NeBcon [33] combine the output of different DCA-based or ML-based contact predictors to create consensus predictions.

Benefiting from the ever-increasing sequence databases and powerful GPU computing capability, deep learning methods open a new chapter in residue contact prediction by combining sequence profile features together with the widespread adoption of evolutionary couplings. Many state-of-the-art methods for contact prediction rely on many different sources of information in order to predict residue–residue contacts, such as solvent accessibility, predicted secondary structure, and scores from other contact prediction methods. Methods in this category include RaptorX-Contact [34], DeepConPred2 [35], DNCON2 [36], DEEPCON [37] and MapPred [38]. On the contrary, deep learning methods like DeepCOV [39] use pure MSA as input, assuming that contacting residue pairs should present correlated mutations in the long-term evolutions reflected in the MSA.

Here, we would like to further study how to mine the effective information in MSA as much as possible to achieve the best performance of the prediction model. We show that using deep neural network models, extensively extracted statistics from the alignment (with enough effective sequences) contain sufficient information (such as self-information, partial entropy, covariance information, mutual information, normalized mutual information and cross-entropy) can achieve state-of-the-art precisions.

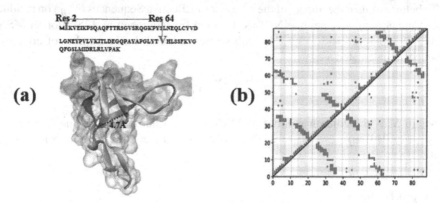

Fig. 1 (a) A cartoon illustrates an example of two residues that are far from each other in sequence but close (4.7 Å) in 3D structure. (b) A typical residue contact map. The upper left triangle shows all the native residue contact pairs and the lower right triangle contains the non-local (sequence separation > 6) native residue contacts.

2 Materials and Methods

2.1 Contact Definition

Two residues can be defined in contact if the distance between side-chain and backbone heavy atoms is less than 5.5 Å, or the distance between any two atoms from a residue pair is less than the sum of their van der Waals radii plus a threshold of 0.6 Å [40]. Rajgaria et al. [17, 18] used a maximum prediction width of 12 Å between Ca-Ca atoms to define a contact for globular proteins. Duarte et al. [41] pointed out that the cut-off distances from 9 to 11 Å, especially 11 Å around the Cb atoms generated the lowest RMSD structures when tested on a set of 60 proteins.

In this study, the definition of contacts is directly taken from the CASP experiments. A pair of residues in the experimental structure is considered to be in contact if the distance between their Cβ atoms (Cα for Gly) is less than or equal to 8Å. Depending on the separation of two residues along the sequence (seq_sep), the contacts are classified into four classes: all-range (seq_sep > 6), short-range (6 ≤ seq_sep < 1 2), medium-range (12 ≤ seq_sep < 24) and long-range (seq_sep > 24).

2.2 Number of Effective Sequences

The majority of recent approaches for contact prediction methods all start with the analysis of correlated mutations in multiple sequence alignments (MSA). The idea behind these approaches is based on the fact that within a protein structure, interacting residue pairs are under evolutionary pressure to maintain the structure. So generating high-quality sequence alignments is the first step for may contact prediction methods.

To better evaluate the impact of the Number of Effective Sequences (N_{eff}) on residue contact prediction, we calculate Neff as depicted by Morcos et al. [25]. For an MSA $=\{(A_1^a, A_2^a \ldots, A_L^a)|a = 1, 2 \ldots, M\}$ with sequence length L, the number of effective sequences is calculated as:

$$N_{eff} = \sum_{a=1}^{M} 1/m^a \tag{1}$$

where M is the column number of MSA, m^a is a number determined by $m^a = |b \in \{1, 2 \ldots M\}|seq_{identity(A^a, A^b)} > Threshold|$. In this work, the *Threshold* is defined as 68%.

2.3 Input Features

For a multi-sequence alignment with m columns and n rows, each row represents a different homologous sequence and each column a set of equivalent amino acids across the evolutionary tree, with gaps considered as an additional amino acid type. We consider 21 amino acid types in this study.

We have 6 types of features extracted from the alignment: self-information, partial entropy, covariance information, mutual information, normalized mutual information and cross-entropy. The first 2 categories are 1D features, and the others are 2D features.

1) Self-information for amino acid a in column i is calculated as:

$$I_i^a = log_2(p_i^a/\langle p_a \rangle) \qquad (2)$$

For amino acids, we compute p_i^a is the probability of each amino acid calculated through the average frequency $\langle p_a \rangle$ as the observed frequency of the amino acid on the whole dataset.

2) Partial entropy for amino acid a in column I is calculated as:

$$S_i^a = p_i^a log_2(p_i^a/\langle p_a \rangle) \qquad (3)$$

3) Covariance feature:

$$C_{ij}^{ab} = \frac{1}{n}\sum_{k=1}^n \left(x_i^{ak} - \overline{x}_i^a\right)\left(x_j^{bk} - \overline{x}_j^b\right) \qquad (4)$$

where x_i^{ak} is a binary variable indicating the presence or absence of amino acid type a at column i in row k and x_j^{bk} is the equivalent variable for observing residue type b at column j in row k.

We have 441 covariance features which is also the type of feature used by DeepCOV.

4) Mutual information:

$$\text{MI(i, j)} = \sum_{i,j} p(i,j)\log(p(i,j)/p(i)p(j)) \qquad (5)$$

5) Normalized mutual information is the Pearson correlation coefficient. The entropies of the columns play the role of the variances:

$$\text{NMI(i, j)} = MI(i,j)/MI(i,j) \qquad (6)$$

6) Cross-entropy is calculated as an additive normalization of mutual information:

$$\text{H(i, j)} = S(i) + S(j) - \text{MI(i, j)} \qquad (7)$$

Finally, the 1D features are converted to 2D features by concatenating 1D features of two residues. In total, we have 528 features extracted from the multi-sequence alignment.

2.4 Data Sets and Model Training

The proposed method is trained on a set of 3269 independent protein chains with a maximum sequence identity of 20%, minimum resolution of 2.0 Å, maximum R-factor of 0.3. Since the proposed method depends only on MSA as input, we consider only protein chains with more than 5*L effective sequences in MSA as our test target. The test set is 103 independent protein chains with the same criteria mentioned above culled against the training set of the proposed method and the methods used for comparison.

At the core of our method is architecture. Our networks involve an input layer, a Maxout layer to reduce the number of input feature channels from 528 to 64, one or more 2D convolutional layers, and a final sigmoid output layer, each with batch normalization applied. The model is trained for 100 epochs (full passes over the training data) using the Adam optimizer at an initial learning rate of 0.001. The learning rate is decreased by a factor of 0.5 if the S-score training loss did not decrease over the last five epochs. The training dataset is shuffled after each epoch. The final model was selected by taking the epoch with minimum S-score loss on the validation dataset.

2.5 Criteria for Evaluation

The predicted residue contact map is a matrix of probability estimates. We analyze the performance of predictors on reduced lists of contacts (sorted by the probability estimates) selected by either the probability threshold or the top L criteria. The prediction performance is assessed using precision (accuracy in some references), coverage (recall in some references) and Matthew's Correlation Coefficient (MCC). The standard deviation (STD) is used in this study to evaluate the dispersion of precision, coverage and MCC.

3 Results and Analysis

We compare the proposed method against SVMSEQ, CCMpred, MetaPSICOV and DeepCOV as representative methods based on traditional machine learning, direct-coupling analysis, meta-method and deep learning, respectively.

3.1 Performance on Model Reliability

The probability (score) of the predicted class can be viewed as a method's prediction reliability. Rather than evaluating all the methods with the same probability standard, this section focuses on the performance trend of each method as the probability threshold changes.

Figure 2 illustrates the prediction performance in terms of different metrics with the increasing probability threshold given by the prediction methods. It is shown that the prediction coverages for all methods decrease gradually with the increase of the probability threshold. The prediction precisions of meta-method MetaPSICOV and two DL methods(DeepCOV and the proposed method) increase monotonically with the probability threshold. Nevertheless, this trend is not monotonous for traditional ML method SVMSEQ and DCA method CCMpred; as the threshold increases, their accuracy curves go down at some probability value. Traditional ML methods and ECA methods also show much larger STDs on precisions and relatively lower coverages/MCCs when compared with DL methods. "Shallow" neural network-based method MetaPSICOV shows very low coverage and MCC, which results in fewer and fewer proteins with predictions returned as the probability threshold increases. On the contrary, DL methods can still maintain a larger number of predicted proteins and higher MCCs and coverages. Notably, as the threshold increases, the proposed method can retain predictions for more protein than DeepCOV.

3.2 Performance on Precisions of Top L Predictions

To further validate our method, we also compared the proposed method against SVM-SEQ, CCMpred, MetaPSICOV and DeepCOV on precisions of top L predictions (Fig. 3).

It is clear that the proposed method is significantly better than SVMSEQ/CCMpred /MetaPSIOV for all different sequence separations, outperforming these methods

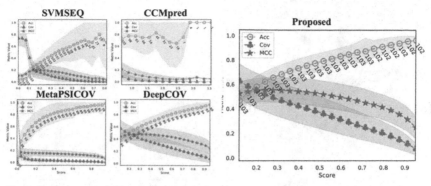

Fig. 2. Performance of different methods in terms of precision, coverage, MCC and the corresponding standard deviation (the shaded area around the curves) with the increasing probability (score) threshold given by the predictors. The numbers under the precision curve (blue) are the numbers of proteins with predictions returned using the corresponding probability threshold on the x-axis.

by 6%/11%/3%, 13%/13%/7%, 40%/16%/27% and 43%/24%/33% for short-range, medium-range, long-range and all-range contacts, respectively. Although the proposed method shows only 2% and 3% improvement on short- and medium-range contact predictions, it surpasses DeepCOV with 7% and 6% for long- and all-range contacts. The improvement is meaningful since long-range contacts are thought to be the most informative for tertiary structure modeling.

A demonstrative example (PDB ID: 1OI0D, which is the a Jab1/MPN domain protein from Archaeoglobus fulgidus with a sequence length of 124) shows the contact prediction with disulfide bonds for top L predictions by the proposed method (Fig. 4). The precisions are 85% and 100% for contact prediction and disulfide bond prediction. Although contact prediction methods are not designed specifically for chemical bond prediction, Fig. 4 underscores broader applications for residue contact prediction.

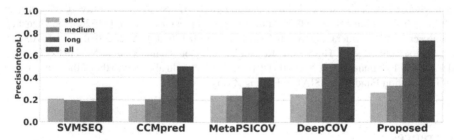

Fig. 3. Precision comparison between different methods

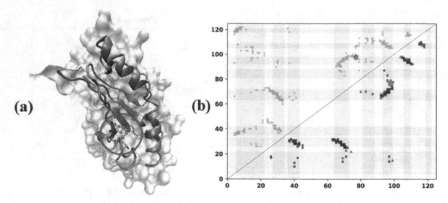

Fig. 4. Case study: residue contact map with disufide bond of a Jab1/MPN domain protein from Archaeoglobus fulgidus. (a) The cartoon illustrates the disulfide bond in the 3D structure. (b)The predicted contact map with a disulfide bond. The gray square/blue square/red triangle and yellow star stand for the native/true positive/false positive contact and the disulfide bond, respectively. (Color figure online)

4 Conclusions

Artificial intelligence techniques are proving increasingly useful for a wide range of challenging applications, including protein contact prediction. In this work, we developed a contact prediction method based on fully convolutional network models with input features derived from a multiple sequence alignment only. Our results suggest that extensively extracted features (self-information, partial entropy, covariance information, mutual information, normalized mutual information and cross-entropy) from MSA contain more sufficient information than covariance feature only for predicting contact. Our method is very effective in terms of prediction precision and model reliability compared with state-of-the-art methods with similar input data types.

Acknowledgment. This work is partly supported by Strategic Priority CAS Project XDB38000000, National Science Foundation of China under Grant No. U1813203, the Shenzhen Basic Research Fund under Grant No. JCYJ20180507182818013, JCYJ20200109114818703 and JCYJ20170413093358429, Hong Kong Research Grant Council under Grant No. GRF-17208019 and CAS Key Lab under Grant No. 2011DP173015. We would also like to thank the Outstanding Youth Innovation Fund (CAS-SIAT to Huiling Zhang).

References

1. Anfinsen, C.B.: Principles that govern the folding of protein chains. Science **181**(4096), 223–230 (1973)
2. Marks, D.S., Hopf, T.A., Sander, C.: Protein structure prediction from sequence variation. Nat. Biotechnol. **30**(11), 1072 (2012)
3. Adhikari, B., et al.: CONFOLD: residue-residue contact-guided ab initio protein folding. Proteins Structure Function Bioinf. **83**(8), 1436–1449 (2015)

4. Xu, J.: Distance-based protein folding powered by deep learning. Proc. Natl. Acad. Sci. **116**(34), 16856–16865 (2019)
5. Senior, A.W., et al.: Improved protein structure prediction using potentials from deep learning. Nature **577**, 706–710 (2020)
6. Yang, J., et al.: Improved protein structure prediction using predicted interresidue orientations. In: Proceedings of the National Academy of Sciences, p. 201914677 (2020)
7. Taylor, W.R., Jones, D.T., Sadowski, M.I.: Protein topology from predicted residue contacts. Protein Sci. **21**(2), 299–305 (2012)
8. Miyazawa, S., Jernigan, R.L.: Residue–residue potentials with a favorable contact pair term and an unfavorable high packing density term, for simulation and threading. J. Mol. Biol. **256**(3), 623–644 (1996)
9. Zhu, J., et al.: Protein threading using residue co-variation and deep learning. Bioinf. **34**(13), i263–i273 (2018)
10. Cong, Q., et al.: Protein interaction networks revealed by proteome coevolution. Science **365**(6449), 185–189 (2019)
11. Raval, A., et al.: Assessment of the utility of contact-based restraints in accelerating the prediction of protein structure using molecular dynamics simulations. Protein Sci. **25**(1), 19–29 (2016)
12. Lubecka, E.A., Liwo, A.: Introduction of a bounded penalty function in contact-assisted simulations of protein structures to omit false restraints. J. Comput. Chem. **40**(25), 2164–2178 (2019)
13. Dago, A.E., et al.: Structural basis of histidine kinase autophosphorylation deduced by integrating genomics, molecular dynamics, and mutagenesis. Proc. Natl. Acad. Sci. **109**(26), E1733–E1742 (2012)
14. Pollock, D.D., Taylor, W.R.: Effectiveness of correlation analysis in identifying protein residues undergoing correlated evolution. Protein Eng. Des. Sel. **10**(6), 647–657 (1997)
15. Dunn, S.D., Wahl, L.M., Gloor, G.B.: Mutual information without the influence of phylogeny or entropy dramatically improves residue contact prediction. Bioinf. **24**(3), 333–340 (2007)
16. Lee, B.-C., Kim, D.: A new method for revealing correlated mutations under the structural and functional constraints in proteins. Bioinf. **25**(19), 2506–2513 (2009)
17. Rajgaria, R., McAllister, S., Floudas, C.: Towards accurate residue–residue hydrophobic contact prediction for α helical proteins via integer linear optimization. Proteins **74**(4), 929–947 (2009)
18. Rajgaria, R., Wei, Y., Floudas, C.: Contact prediction for beta and alpha-beta proteins using integer linear optimization and its impact on the first principles 3D structure prediction method ASTRO-FOLD. Proteins **78**(8), 1825–1846 (2010)
19. Pierre, B., Cheng, J.: Improved residue contact prediction using support vector machines and a large feature set. BMC Bioinf. **8**(1), 113–113 (2007)
20. Tegge, A.N., et al.: NNcon: improved protein contact map prediction using 2D-recursive neural networks. Nucleic Acids Res. **37**, 515–518 (2009)
21. Wu, S., Zhang, Y.: A comprehensive assessment of sequence-based and template-based methods for protein contact prediction. Bioinf. **24**(7), 924–931 (2008)
22. Wang, Z., Xu, J.: Predicting protein contact map using evolutionary and physical constraints by integer programming. Bioinf. **29**(13), i266–i273 (2013)
23. Zhang, H., et al.: COMSAT: Residue contact prediction of transmembrane proteins based on support vector machines and mixed integer linear programming. Proteins Structure Function Bioinf. **84**(3), 332–348 (2016)
24. Weigt, M., et al.: Identification of direct residue contacts in protein–protein interaction by message passing. Proc. Natl. Acad. Sci. **106**(1), 67–72 (2009)
25. Morcos, F., et al.: Direct-coupling analysis of residue coevolution captures native contacts across many protein families. Proc. Natl. Acad. Sci. **108**(49), E1293–E1301 (2011)

26. Baldassi, C., et al.: Fast and accurate multivariate Gaussian modeling of protein families: predicting residue contacts and protein-interaction partners. PLoS ONE **9**(3), e92721 (2014)
27. Jones, D.T., et al.: PSICOV: Precise structural contact prediction using sparse inverse covariance estimation on large multiple sequence alignments. Bioinf. **28**(2), 184–190 (2012)
28. Ekeberg, M., et al.: Improved contact prediction in proteins: using pseudolikelihoods to infer Potts models. Phys. Rev. E **87**(1), 012707 (2013)
29. Kamisetty, H., Ovchinnikov, S., Baker, D.: Assessing the utility of coevolution-based residue–residue contact predictions in a sequence-and structure-rich era. Proc. Natl. Acad. Sci. **110**(39), 15674–15679 (2013)
30. Seemayer, S., Gruber, M., Söding, J.: CCMpred—fast and precise prediction of protein residue–residue contacts from correlated mutations. Bioinf. **30**(21), 3128–3130 (2014)
31. Skwark, M.J., Abdel-Rehim, A., Elofsson, A.: PconsC: combination of direct information methods and alignments improves contact prediction. Bioinf. **29**(14), 1815–1816 (2013)
32. Jones, D.T., et al.: MetaPSICOV: combining coevolution methods for accurate prediction of contacts and long range hydrogen bonding in proteins. Bioinf. **31**(7), 999–1006 (2015)
33. He, B., et al.: NeBcon: Protein contact map prediction using neural network training coupled with naïve Bayes classifiers. Bioinf. **33**(15), 2296–2306 (2017)
34. Wang, S., et al.: Accurate de novo prediction of protein contact map by ultra-deep learning model. PLoS Comput. Biol. **13**(1), e1005324 (2017)
35. Ding, W., et al.: DeepConPred2: an improved method for the prediction of protein residue contacts. Comput. Struct. Biotechnol. J. **16**, 503–510 (2018)
36. Adhikari, B., Hou, J., Cheng, J.: DNCON2: improved protein contact prediction using two-level deep convolutional neural networks. Bioinf. **34**(9), 1466–1472 (2017)
37. Adhikari, B.: DEEPCON: Protein Contact Prediction using Dilated Convolutional Neural Networks with Dropout. Bioinf. **36**(2), 470–477 (2019)
38. Wu, Q., et al.: Protein contact prediction using metagenome sequence data and residual neural networks. Bioinf. **36**(1), 41–48 (2020)
39. Jones, D.T., Kandathil, S.M.: High precision in protein contact prediction using fully convolutional neural networks and minimal sequence features. Bioinf. **34**(19), 3308–3315 (2018)
40. Nugent, T., Jones, D.T.: Predicting transmembrane helix packing arrangements using residue contacts and a force-directed algorithm. PLoS Comput. Biol. **6**(3), e1000714 (2010)
41. van Giessen, A.E., Straub, J.E.: Monte Carlo simulations of polyalanine using a reduced model and statistics-based interaction potentials. J. Chem. Phys. **122**(2), 024904 (2005)

See Fine Color from the Rough Black-and-White

Jingjing Wu[1], Li Ning[2(✉)], and Chan Zhou[2]

[1] Huawei Technologies, Hangzhou, China
wujingjing19@huawei.com
[2] Shenzhen Institutes of Advanced Technology, Chinese Academy of Sciences,
Shenzhen, China
{li.ning,chan.zhou}@siat.ac.cn

Abstract. Image super-resolution and colorization are two important research fields in computer vision. In previous studies, they have been considered separately as two unrelated tasks. However, for the task of restoring gray video to high-definition color video, when the network learns to abstract features from low-resolution images and maps them to high-resolution images, the abstract understanding of images by the network is also useful for colorization task. Treating them as two unrelated tasks have to construct two different models, which needs more time and resources. In this paper, we propose a framework to combine the tasks of image super-resolution and colorization together. We design a new network model to directly map low-resolution gray images into high-resolution color images. Moreover, this model can obtain motion information of objects in the video by predicting surrounding frames with the current frame. Thus, video super-resolution and colorization can be realized. To support studying super-resolution and colorization together, we build a video dataset containing three scenes. As far as we know, this is the first dataset for such kinds of tasks combination.

1 Introduction

With the development of science and technology, many aspects of our lives are changing. From the original black and white to color, and to the current 3D movies, people are keep on pursuing more high-definition, more colorful and true visual experience. In the early stage, television and photographs have very low pixels and no color. In order to enhance the audience's visual feeling, an interesting method is to increase the image resolution and involve color information. Another motivation is from the medical image processing. Most of the images generated in medical devices (CT, MR, etc.) are gray images. If they are properly colored, it will help doctors to identify the pathological features of patients correctly and let patients know their situation better. Besides, the super-resolution and colorization play an important role in the field of image compression, remote sensing imaging, etc.

Super-resolution and colorization of images are two important research directions in the field of image processing. The main challenges to handle these two

© Springer Nature Switzerland AG 2021
Y. Zhang et al. (Eds.): PDCAT 2020, LNCS 12606, pp. 229–240, 2021.
https://doi.org/10.1007/978-3-030-69244-5_20

tasks is finding proper mappings from low resolution to high resolution and from gray to color. However, such mappings are not clear yet. During these years, deep learning has shown excellent performance in many fields. Dong et al. [3] applied deep learning to super-resolution for the first time, they surpassed the traditional algorithm with only three layers of CNN. Implementing deep learning on colorization can be done automatically without the need for manual annotation. However, for tasks that require both super-resolution and colorization for images, if the two tasks are handled separately, we need to train two network models, which needs more time and resources. Another bad thing is, no matter we first consider super-resolution or colorization, the error from the former task will propagate to the latter and the prediction is very difficult to handle, in this way, recovering a high-resolution color images is quite mission impossible. This paper aims to propose a framework which can jointly optimize the two tasks of video super-resolution and colorization. For example, as shown in Fig. 1, the task is output high-resolution color image from the input low-resolution gray image. The abstract understanding of the image by our end-to-end network (Fig. 1(a)) is equally useful for coloring tasks when the network learns abstract features from low-resolution images and maps to high-resolution images. Compared with training two models to predict separately (Fig. 1(b)), the single model saves more resources and time. When optimizing the two tasks together, the end-to-end network (Fig. 1(a)) can optimize all the parameters in one model without biasing one sub-model.

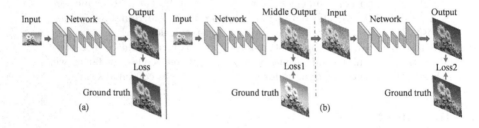

Fig. 1. Flow chart comparison of the end-to-end model(our) and two stages model.

The main methods for analyzing video motion information include RNN [1,2] and 3-dimensional convolution [4]. RNN can process sequence data of any length, which has significant performance in speech recognition, language modeling, translation and image caption. The idea behind RNN is handling the current frame by using the information of the previous frame. However, it can not establish a direct connection between adjacent multi-frames. This problem can be avoided by using 3-dimensional convolution to generate future frames, but the number of parameters in 3-dimensional convolution is too large and it is easy to over-fitting. In addition, the running time of RNN and 3-dimensional convolution calculation are quite large. In this paper, we propose a video processing method for predicting the upper and lower frames through the current frame, which has the same parameter quantity and the running time is similar to 2-dimensional Convolutional Neural Network, thus, the video motion information can be obtained effectively.

As far as we know, there is no public video dataset for super-resolution and colorization together. In our work, we have produced a new video dataset for super-resolution and colorization. The dataset includes three scenes: flowers, birds, and forests. There are 940 videos and 113115 pictures in total. The dataset includes high definition color images, gray images, gray plus noise image with 4 times lower resolution, color weakened plus noise image with 4 times lower resolution, gray plus noise images. This dataset will be update and published later.

In this paper, our main contributions include:

- We propose a video dataset consisting of three scenes for super-resolution and colorization tasks.
- We propose a new network model for super-resolution and colorization.
- We propose a new video processing method to capture motion information of objects in video effectively.

This paper is structured as follows. We introduce the related work on super-resolution and coloring in Sect. 2, propose our method in Sect. 3, the dataset and experimental results are shown in Sect. 4. Finally, we give the concluding remark in Sect. 5.

2   Related Work

The task of super-resolution can be split into the groups of single image super-resolution and multi-frame or video super-resolution methods [5]. Since 2016, Dong et al. [3] first applied deep learning to single image super-resolution, more neural network structures were proposed, which achieved significantly higher PSNR on different datasets [4,7,11]. Video and multi-frame super-resolution approaches combine information from multiple LR frames to reconstruct details that are missing in individual frames which can lead to higher quality results [13]. [14] propose a bidirectional recurrent convolutional network for efficient multi-frame SR, they used bidirectional connection between video frames with three types of convolutions: the feedforward convolution, the recurrent convolution, and the conditional convolution respectively for the purpose of solving spatial dependency, long-term temporal dependency, long-term contextual information. Recent research [12] improves temporal coherence by using multiple low-resolution frames as input to generate a high-resolution frame. Sajjadi et al. [13] propose an end-to-end trainable frame-recurrent video super-resolution framework that uses the previously inferred HR estimate to super-resolve the subsequent frame. Since 2014, Generative Adversarial Networks [6] has been proposed, researchers have found that an adversarial loss significantly helps in obtaining realistic high-frequency details, Since maximizing for PSNR leads to generally blurry images. [15] combine perceptual, adversarial and texture synthesis loss terms to produce sharper images with hallucinated details.

Previous colorization methods broadly fall into three categories: scribble based, transfer, and automatic direct prediction [17]. The fully automatic colorization can be done without human intervention. Cheng et al. [16] used neural networks to extract features from gray images, and L2 loss to train neural networks, combined bilateral filtering to improve the results, However, the training data used in this paper is less, and the network structure is relatively simple, which greatly limits the applicable types and coloring effect of images. Larsson et al. [17] used convolutional neural networks to extract low-dimensional features and semantic information from gray images and predicted the color distribution of each pixel using the super-column model. Iizuka et al. [18] used the dual-stream structure to extract the global features of the image on the one hand, and the low-dimensional local features of the image on the other hand, and fused the global features and local features to predict the color information of each pixel. Zhang et al. [19] used VGG16 convolution neural network to extract features from gray images, and predicted the color distribution histogram of each pixel to color gray images.

Whether it's recovering from low-resolution images to high-resolution images or coloring gray images is a very difficult task. For the task of recovering high-resolution color images from low-resolution gray images or low-resolution color degraded images, the previous methods of video super-resolution and video staining were processed successively, and two errors would be generated. We propose an end-to-end network structure to directly predict high-resolution color images and propose a video processing method to predict the surrounding frames through the current frame. Next, we will introduce our work.

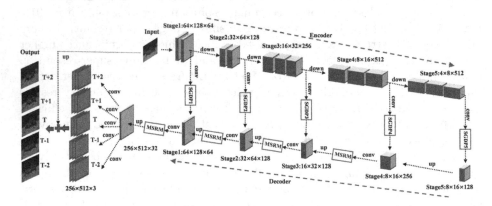

Fig. 2. Model structure. "up" and "down" represent the up-sample and down-sample respectively of the feature graph. After each "stage" is the size of the current feature map and the number of channels

3   Method

As Fig. 2 shows, the video super-resolution and colorization model proposed in this paper are mainly divided into two parts: encoder, decoder. The encoder consists of VGG's 16 convolution layers, that consists of 5 stages, the decoder consists of multiple layers of residual modules and also contains 5 stages. Through 5 stages of the encoder, the rich multi-scale feature will be extracted from input images. The decode leverage the feature to generate residual maps. Finally, an up-sample module maps the addition of residual maps and input images to high-quality color images. Considering the semantic gap between the encoder and decoder corresponding stages, we have added 4 Semantic Gap Destroy Path (SGDP) consisting of multiple residual modules. At the same time, in order to further extract more context information in the image, we introduce a Multi-scale Residual Module (MSRM) module to be applied to our network structure. In addition, we invented a novel method to model the motion information in videos by predicting the surrounding frames without increasing many parameters and could be easily integrated into 2D fully convolutional networks (FCN). In the following paragraphs, we will be detailed explain the encoder, the multi-scale residual module, discriminator we used to calculate adversarial loss and the configuration of the other loss functions in our model.

3.1   Encoder and Decoder

The encoder is a VGG16 network that is pre-trained by ImageNet. It receives an image as input and outputs five scales feature maps with different resolution and rich semantic information at its five stages. The decoder consists of five multi-scale residual modules and four up-sample modules. Starting from the feature map with the smallest resolution, the decoder gradually fuses the feature map with higher resolution than the current one to recover spatial details. At each stage, before the feature map is fused, Semantic Gap Destroy Path (SGDP) is used to eliminate the semantic gap between two groups of feature maps to be fused. The SGDP of five stages includes 5, 4, 3, 2 and 1 MSRMs respectively.

3.2   Muti-scale Residual Module

As Fig. 3(a) shows, the Multi-scale Residual Module (MSRM) mainly consists of four branches and one short connection. Each branch includes different numbers of convolution layers in series, and the quantity of convolution layer with 3×3 kernel of each branch is 1, 2, 3, and 4 respectively. MSRM accepts the feature maps with C channels, these feature maps will be associated with different scale context information by flaw through different lengths of branches, and C/4 channel feature maps will be produced at each branch. Then, these feature maps at four branches are concatenated to generate a feature map with C channels. Finally, these feature maps are added to the input feature maps by the short connection to produce the output. Short connections allow our network to avoid the gradient vanishing problem like ResNet, while by stacking 4 different lengths of branches, MSRM also could capture context information at different scales.

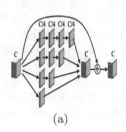
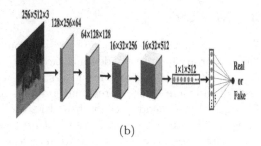

(a)   (b)

Fig. 3. (a) Muti-scale residual module. (b) Architecture of discriminator.

3.3 Discriminator

Inspired by DCGAN [21], the discriminator network we used contains four convolution layers, a global pooling and a fully connected layer as shown in Fig. 3(b). It receives an image with the resolution is $256 \times 512 \times 3$ and output the probability of the input image is real. The kernel size of four convolution layers is 4×4, while their strides are 2, 2, 4 and 1.

3.4 Loss Function

Our model can be formulated as follows Eq. 1, 2 and 3:

$$(\hat{y}_{hc}^{t-2}, \hat{y}_{hc}^{t-1}, \hat{y}_{hc}^{t}, \hat{y}_{hc}^{t+1}, \hat{y}_{hc}^{t+2}) = Model(x_{lgn}) \tag{1}$$

$$P_{\hat{y}_{hc}} = Discriminator(\hat{y}_{hc}^{t}) \tag{2}$$

$$P_{y_{hc}} = Discriminator(y_{hc}^{t}) \tag{3}$$

Where x_{lgn} represents a low-resolution noise-added input image, $\hat{y}_{hc}^{i}$ represents the high resolution color image of the model output and the image of the surrounding frames, where $i \in \{t-2, t-1, t+1, t+2\}$. The objective function of our model and the objective function of the Discriminator can be expressed as Formulation 4 and 5 respectively.

$$L_{Model} = L_{cur} + w_{adv} \times L_{adv} + w_{mot} \times L_{mot} \tag{4}$$

$$L_{Discriminator} = log(P_y) + log(1 - P_{\hat{y}_{hc}}) \tag{5}$$

The model's optimized objective function L_{Model} has three parts: the difference between the current frame and its ground truth (L_{cur}) in pixel level, the difference between the current frame and its ground truth in distribution level(L_{adv}), and the difference between the surrounding frames and their ground truth (L_{mot}). The w_{adv} and w_{mot} that are used in our model both are 10^{-6}. Their detailed definition is shown in Eq. 6, 7 and 8.

$$L_{cur} = (\hat{y}_{hc}^{t} - y_{hc}^{t})^2 + |\hat{y}_{hc}^{t} - y_{hc}^{t}| \tag{6}$$

$$L_{adv} = log(P_{\hat{y}_{hc}}) \tag{7}$$

$$L_{mot} = \sum_{i \in (t-2, t-1, t+1, t+2)} [(\hat{y}_{hc}^{i} - y^{i})_{hc}^{2} + |\hat{y}_{hc}^{i} - y_{hc}^{i}|] \tag{8}$$

4   Experiment

4.1   Dataset

To train our proposed model, we manually build a video jointly super-resolution and colorization dataset which is named VSRC. It has 940 videos with 113,115 images and provides two types of label videos: 1. high-definition color videos(HC), 2. High-resolution gray videos(HG), and three types of input videos: 1. Gray videos with 4 times lower resolution and plus noise (LGN), 2. Color partly missed videos with 4 times lower resolution and plus noise (LCN), 3. High resolution gray videos with noise (HGN).

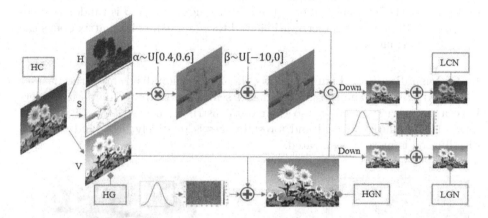

Fig. 4. Flow chart of generating every single image's LGN, LCN, HGN, HG, and HC images of the videos in video jointly super-resolution and colorization (VSRC) dataset. "HC" means high definition color videos, "HG" means high-resolution gray videos, "LGN" gray videos with 4 times lower resolution and plus noise, "LCN" means color partly missed videos with 4 times lower resolution and plus noise, "HGN" means high-resolution gray videos with noise.

The Video Collection. We use the keywords: "flower", "bird" and "forest" search video on the Internet, and select the videos with diverse colors and rich textures. Then, each video is divided into many sub-video sequences that have continuous moving objects. Finally, 940 sub-videos that contains 113,115 images are selected, and they are divided into three scenes: "flower", "bird" and "forest". Under the scene of flower: 170 videos of the training set, 60 videos of validation set and 47 videos of the test set. In the bird scene: 268 videos of the training set, 86 videos of validation set and 47 videos of the test set. In the forest scene: 162 videos of the training set, 60 videos of validation set and 30 videos of the test set.

The Procession of Producing Training Samples. Figure 4 shows the procession of producing each single image sample pair of videos in VSRC dataset. Our proposed VSRC dataset includes two types of label images: 1. high-definition color images(HC), 2. high-resolution gray images(HG), and three types of input images: 1. gray image with 4 times lower resolution and plus noise (LGN), 2. color partly missed image with 4 times lower resolution and plus noise (LCN), 3. high-resolution gray images with noise images(HGN). Among them, HC images and HG as ground truth, without adding any changes. The reason that we reduce the resolution of gray images by 4 times of original images is that it could avoid images resolution are too low to make the network unpredictable and matins our dataset is challenging. The noise addition process of three types of input images is shown in Fig. 4. The multiplicative noise scalar α is randomly distributed uniformly in the interval $[0.4, 0.6]$, the additive noise scalar β is randomly sampled from $[-10, 0]$ uniform distribution, and the Gaussian noise matrix obeys the $N(0, 10)$ distribution.

Data Visualization. Figure 5 shows the distribution diagram of the dataset visualized by t-SNE and the sample images selected from VSRC dataset. Flowers have a rich variety of colors, so their data distribution is scattered, while the intra-class similarity of bird and forest images is relatively high, and the data distribution is more concentrated.

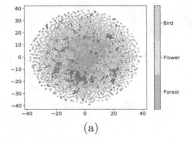
(a)

(b)

Fig. 5. Visualize dataset distribution and picture samples.

4.2 Experimental Results Analysis

Table 1. Different scenes.

Metrics	Train data	Flower	Bird	Forest
MSE	Single	0.0110	0.0060	0.0049
	All	0.0194	0.0073	0.0065
PSNR	Single	32.61	29.77	30.52
	All	31.15	29.46	29.82

Table 2. Different loss

Method	L_{cur}	$L_{cur} + L_{mot}$	$L_{cur} + L_{mot} + L_{adv}$
MSE	0.0049	0.0051	0.0049
PSNR	30.37	30.38	30.52

Table 3. Different structure.

Method	No SGDP	3D conv	ResNet	Our
MSE	0.0051	0.0049	0.0050	0.0049
PSNR	30.43	30.24	30.43	30.52

Table 4. Different inputs and two-stages.

Method	Our LGN	Our LCN	Two-stages
MSE	0.0049	0.0035	0.0102
PSNR	30.52	31.35	29.11

Quantitative Analysis. Our method was tested on VSRC dataset and several comparative experiments also were carried out, the results are shown in the Table 1, 2, 3 and 4 that respectively indicate: 1. Set the VSRC dataset benchmark and verify the versatility of our model in three scenes. 2. Verify the validity of the loss function we designed. 3. Verify the rationality and effectiveness of our model design. 4. Verify that our proposed end-to-end model is superior to the idea of dual model prediction alone. We evaluated the experimental results using both MSE and PSNR criteria.

In order to verify the generality of our model in three scenarios and set the benchmark of VSRC dataset, we experimented separately in three scenes and merged three scenes. The experimental results are shown in Table 1. According to the visualization of the above data, it can be seen that the data distribution of the three scenes is different, so the results produced on the same model are also different. The two evaluation results of the three scenes' individual training and the fusion of the three scenes' training are consistent, which is determined by the data distribution itself.

In order to verify the validity of the loss function we designed, we made three different loss function comparison tests: the loss function is L_{cur}, $L_{cur} + L_{mot}$, and $L_{cur} + L_{mot} + L_{adv}$. The experimental results are shown in Table 2. The model test results optimized by the loss function L_{cur} are more inclined to optimize MSE, while the PSNR results are the worst of the three loss functions, and the loss function($L_{cur} + L_{mot} + L_{adv}$) designed by us has better results in PSNR.

In order to verify the validity of the model we designed, 1. Semantic Gap Destroy Path is effective, 2. We designed a method to predict the motion information of the current and surrounding frames by one frame, 3. Muti-scale Residual Module is valid. In terms of MSE evaluation results, the results of using the 3D convolution model and our model are the best. In the PSNR evaluation results, our model results are the best, and the inference time of the model using the 3D convolution is 0.094. And our model's inference time is 0.077. As we all know, 3D convolution can learn motion information from video sequences, while the experimental results of comparison Table 2 with L_{cur} loss function and Table 1 model using 3D convolution show that too many parameters in 3D convolution make the network over-fitting, and its performance is worse than that of model without considering motion information.

In order to validate our end-to-end model's idea of super-resolution and colorization, which is superior to the two-model prediction alone, we designed a comparative experiment. Our model uses LGN in VSRC dataset, that is to say, low-resolution gray images can directly predict high-resolution color images, HC as ground truth. The dual model needs to use LGN and HGN, HC and HG, LGN input to model 1, HG as ground truth, so that the network can learn high-resolution gray images from low-resolution gray images. HGN is input to model 2, HC is used as ground truth, Model 2 colors the gray image into a color image. We use the images predicted by model 1 to predict the high-resolution color images by model 2. Table 4 is the experimental evaluation results. We can see that the results of single model prediction are better than two-stages model both in MSE and PSNR. LCN is the image after color degradation, and LCN is used as input to train our model, the experimental results are all better than those of LGN is used as input, which also shows that it is more difficult to recover high-resolution color images from low-resolution gray images than from color-degraded images.

Qualitative Analysis. The following Fig. 6 shows the selected examples in the test pictures of each model. By observing, as indicated by the red frame in the figure, some models have problems such as color discontinuity. Relatively speaking, the color of the image predicted by our model is more realistic. The low-resolution color noise map retains part of the color information. With LCN as the input, the predicted image has the best appearance.

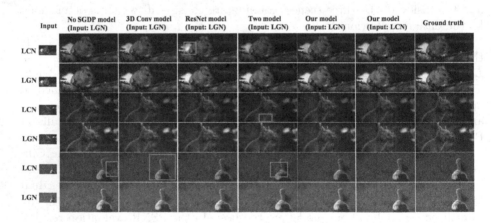

Fig. 6. The first six column shows the predicted images(lines 1, 3 and 5) of the model and the gray images(lines 2, 4 and 6) of the predicted images. The last column shows the ground truth (lines 1, 3 and 5) of the model and the gray images(lines 2, 4 and 6) of the ground truth.

5   Conclusion

In this paper, our main contributions include the production of VSRC dataset for super-resolution and colorization tasks, A new network model is designed to map the low-resolution gray image directly to the high-resolution color image, and a new method for acquiring motion information of objects in video. In future research, we will expand the VSRC dataset to add more scenes and further improve our model to restore more realistic images.

Acknowledgement. This work is supported by Shenzhen research grant (KQJSCX 20180330170311901, JCYJ20180305180840138 and GGFW2017073114031767).

References

1. Srivastava, N., Mansimov, E., Salakhudinov, R.: Unsupervised learning of video representations using lstms. In: International Conference on Machine Learning, pp. 843–852 (2015)
2. Villegas, R., Yang, J., Hong, S., Lin, X., Lee, H.: Decomposing motion and content for natural video sequence prediction. arXiv preprint arXiv:1706.08033 (2017)
3. Dong, C., Loy, C.C., He, K., Tang, X.: Image super-resolution using deep convolutional networks. IEEE Trans. Pattern Anal. Mach. Intell. **38**(2), 295–307 (2016)
4. Tran, D., Bourdev, L., Fergus, R., Torresani, L., Paluri, M.: Learning spatiotemporal features with 3d convolutional networks. In: Proceedings of the IEEE International Conference on Computer Vision, pp. 4489–4497 (2015)
5. Pérez-Pellitero, E., Sajjadi, M.S.M., Hirsch, M., Schölkopf, B.: Photorealistic video super resolution. arXiv preprint arXiv:1807.07930 (2018)
6. Goodfellow, I.: Generative adversarial nets. In: Advances in Neural Information Processing Systems, pp. 2672–2680 (2014)
7. Shi, W.: Real-time single image and video super-resolution using an efficient subpixel convolutional neural network. In: Proceedings of the IEEE Conference On Computer Vision and Pattern Recognition, pp. 1874–1883 (2016)
8. Tao, X., Gao, H., Liao, R., Wang, J., Jia, J.: Detail-revealing deep video super-resolution. In: Proceedings of the IEEE International Conference on Computer Vision, pp. 4472–4480 (2017)
9. Liu, D.: Robust video super-resolution with learned temporal dynamics. In: Proceedings of the IEEE International Conference on Computer Vision, pp. 2507–2515 (2017)
10. Kim, J., Kwon Lee, J., Mu Lee, K.: Deeply-recursive convolutional network for image super-resolution. In: Proceedings of the IEEE Conference on Computer Vision and Pattern Recognition, pp. 1637–1645 (2016)
11. Dong, C., Loy, C.C., Tang, X.: Accelerating the super-resolution convolutional neural network. In: European Conference On Computer Vision, pp. 391–407 (2016)
12. Jo, Y., Wug, O.S., Kang, J., Joo Kim, S.: Deep video super-resolution network using dynamic upsampling filters without explicit motion compensation. In: Proceedings of the IEEE Conference on Computer Vision and Pattern Recognition, pp. 3224–3232 (2018)
13. Sajjadi, M.S.M., Vemulapalli, R., Brown, M.: Frame-recurrent video super-resolution. In: Proceedings of the IEEE Conference on Computer Vision and Pattern Recognition, pp. 6626–6634 (2018)
14. Huang, Y., Wang, W., Wang, L.: Bidirectional recurrent convolutional networks for multi-frame super-resolution. In: Advances in Neural Information Processing Systems, pp. 235–243 (2015)
15. Sajjadi, M.S.M., Scholkopf, B., Hirsch, M.: Enhancenet: Single image super-resolution through automated texture synthesis. In: Proceedings of the IEEE International Conference on Computer Vision, pp. 4491–4500 (2017)

16. Cheng, Z., Yang, Q., Sheng, B.: Deep colorization. In: The IEEE International Conference on Computer Vision (ICCV) (2015)
17. Larsson, G., Maire, M., Shakhnarovich, G.: Learning representations for automatic colorization. In: European Conference on Computer Vision, pp. 577–593 (2016)
18. Iizuka, S., Simo-Serra, E., Ishikawa, H.: Let there be color!: joint end-to-end learning of global and local image priors for automatic image colorization with simultaneous classification. ACM Trans. Graph. (TOG) **35**(4), 110 (2016)
19. Zhang, R., Isola, P., Efros, A.A.: Colorful image colorization. In: European Conference on Computer Vision, pp. 649–666 (2016)
20. van der Laurens, M., Hinton, G.: Visualizing data using t-SNE. J. Mach. Learn. Res. **9**, 2579–2605 (2008)
21. Radford, A., Metz, L., Chintala, S.: Unsupervised representation learning with deep convolutional generative adversarial networks. arXiv preprint arXiv:1511.06434 (2015)

Data Aggregation Aware Routing for Distributed Training

Zhaohong Chen[1], Xin Long[1], Yalan Wu[1], Long Chen[1], Jigang Wu[1(✉)], and Shuangyin Liu[2]

[1] School of Computer Science and Technology, Guangdong University of Technology, Guangzhou, China
asjgwucn@outlook.com
[2] Guangzhou Key Laboratory of Agricultural Products Quality and Safety Traceability Information Technology, Zhongkai University of Agriculture and Engineering, Guangzhou, China

Abstract. For distributed training, the communication overhead for parameter synchronization is heavy in the network. Data aggregation can efficiently alleviate network overheads. However, existing works on data aggregation are based on the streaming message data, which can not well adapt to the discrete communication for parameter synchronization. This paper formulates a data aggregation aware routing problem, with the objective of minimizing training finishing time for global model under the constraint of cache capacity. The problem is formulated as a mixed-integer non-linear programming problem, and it is proved to be NP-Hard. Then we propose a data aggregation aware routing algorithm to solve the formulated problem, by transmitting the data to the closest aggregation node in greedy to reduce the network overhead. Simulation results show that, the proposed algorithm can reduce average training finishing time by 74%, and it can reduce the network overhead by 33% on average, compared with the shortest path algorithm.

Keywords: Distributed training · Data aggregation · Parameter synchronization · Routing · Network overhead

1 Introduction

In the last few years, distributed training of machine learning has been widely used in diverse applications, such as recommendation system, disease surveillance and traffic control. A critical issue of distributed training is the massive network overhead produced by parameter synchronization.

This work was supported in part by project of Guangdong Science and Technology Plan under Grant 2019B010121001, Guangzhou Innovation Platform Construction Plan under Grant 201905010006, National Natural Science Foundation of China under Grant 61871475, 61702115 and 62072118 and Jieyang R&D Foundation of Guangdong, China (2017xm037).

© Springer Nature Switzerland AG 2021
Y. Zhang et al. (Eds.): PDCAT 2020, LNCS 12606, pp. 241–250, 2021.
https://doi.org/10.1007/978-3-030-69244-5_21

There are many existing works using data aggregation to reduce network overhead. For example, [1] proposed a service-aware data aggregation scheme to reduce the transmission latency. [2] designed a routing protocol of data aggregations to keep the network overhead low. [3] dynamically selected an appropriate function of data aggregation to fit the accuracy constraint while reducing the network traffic. However, most of those existing works assume that the data packets are produced in continuous, while the communication data of distributed training is discrete due to the extensive computation time of model updates.

There are some challenges to use the data aggregation in the communication of distributed training. Firstly, the computing capabilities are various for different workers, leading to slowdowns of distributed training. Specifically, aggregating the data generated from the worker with powerful computing capabilities and the data generated from the worker with weak computing capabilities will result in a long waiting delay, which increases the training time. Secondly, data aggregation needs to store the data packets into the cache, while the cache capacity of network nodes are limited. Therefore, the size of aggregation data packets needs to satisfy the cache capacity constraint. Based on above challenges, this paper investigates the routing problem for the communication of distributed training.

The main contributions of this paper can be summarized as follows.

1) We formulate a data aggregation aware routing problem to minimize training finishing time with the cache capacity constraint. The problem is formulated as a mixed-integer non-linear programming problem, which is of NP-Hard.
2) We propose a data aggregation aware routing algorithm for distributed training, where data packets are aggregated at the closest node to reduce the network overhead.
3) The training finishing time can be reduced by 74%, and the network overhead can be reduced by 30%, compared with the shortest path routing algorithm.

The rest of paper is organized as follows. System model and problem formulation are presented in Sect. 2. Section 3 presents the proposed algorithms. Section 4 describes the performance evaluation. Section 5 concludes this paper with future remarks.

2   System Model

We consider a network, as shown in Fig. 1. Let graph $G(V, E)$ be the network, where V indicates the set of network nodes and E represents the set of network link between nodes. There are m nodes in the network, including n workers, the routers and the parameter server. In mobile edge computing, the workers are edge servers and the parameter server is a cloud server. This paper focuses on synchronous distributed training. The training process of an iteration includes pulling parameters, computing model updates and pushing parameters.

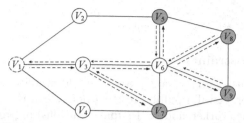

Worker ⬤   Router ◯   Parameter server ◌   ← - Gradient update   → - Parameter synchronization   -- Link

Fig. 1. System architecture.

2.1 Pulling Parameters

The distributed training starts when the parameter server commits the global model parameter to the worker nodes for each iteration. The training start time for each iteration is 0. Let $x_{i,j}^d \in \{0,1\}$. It represents the routing policy between node i and node j when the parameter server multi-casts the global model parameter to node j. Particularly, $x_{i,j}^d = 1$ means that the link between node i and node j is selected on the path of multi-casting global model parameters. Let d denote the size of global model parameters and let $w_{i,j}^d$ be the bandwidth between node i and node j when parameter server multi-casts the global model parameter. t_j^d is when the global model parameters are received by node j from the parameter server. For parameter server, t_j^d is 0. t_j^d can be expressed as,

$$t_j^d = \max_{0 \le i \le m} \left\{ x_{i,j}^d \cdot \left(\frac{d}{w_{i,j}^d} + t_i^d \right) \right\}. \tag{1}$$

2.2 Pushing Parameters

For worker node j, t_j^c is the computation time for the model updates. Let t_j^u be the time when node j begins to pushing the model updates to the parameter server. t_j^u is,

$$t_j^u = \max \left\{ \max_{0 \le i \le m} \left\{ x_{i,j}^u \cdot \left(\frac{d_g}{w_{i,j}^u} + t_i^u \right) \right\}, t_j^d + t_j^c \right\}, \tag{2}$$

where $x_{i,j}^u$ represents the pushing routing policy between node i and node j when node j pushes model updates to parameter server, d_g denotes the size of model updates and $w_{i,j}^u$ represents the upload bandwidth between node i and node j. When model updates from different worker nodes are routed to the same routing node, the routing node will aggregation those updates. The routing node can receive the model updates and aggregate updates in parallel. Therefore, the time cost on data aggregation can be ignored.

When the parameter server has received all model updates from worker nodes, one iteration of training ends. The finishing time T of one iteration of training can be,

$$T = \max_{0 \leq i \leq m} t_i^u. \tag{3}$$

2.3   Routing Constraints

A valid pushing routing policy can route all model updates of worker nodes to the parameter server. A valid pulling routing policy can route the global model parameter to all the worker nodes. To make the pushing and pulling routing policy valid, the routing constraints are proposed. We can get the reachable status of two nodes by using the breadth-first traversal [4] to search the routing policy's graph. Let $c_{i,j}^u$ represent the reachable states between node i and node j when worker nodes push the model updates. If two nodes are reachable from each other, $c_{i,j}^u = 1$. Otherwise, $c_{i,j}^u = 0$. Let $c_{i,j}^d$ represent the reachable states between node i and node j when the parameter server multi-casts the global model parameters to worker nodes. The routing constraint can be,

$$x_{i,j}^u \cdot c_{j,i}^u = 0, \quad x_{i,j}^d \cdot c_{j,i}^d = 0. \tag{4}$$

2.4   Cache Constraints

When the routing node aggregates the model updates from different nodes, it needs to store them in cache. However, the cache size of routing nodes is limited. The routing policy must satisfy the cache capacity constraint. Let g_j denote the cache size of node j, the cache capacity constraint can be expressed as,

$$\sum_{i=0}^{m} x_{i,j}^u \cdot d_g \leq g_j. \tag{5}$$

2.5   Problem Formulation

In this work, the goal is to minimize the finishing time of training with routing policy constraint and cache capacity constraint. The problem formulation can be defined as,

$$
\begin{aligned}
\textbf{(OPT)} \quad obj : \quad & \min T \\
C_1 : \quad & x_{i,j}^u \cdot c_{j,i}^u = 0, \quad x_{i,j}^d \cdot c_{j,i}^d = 0, \\
& 0 \leq i \leq m, \quad 0 \leq j \leq m, \\
C_2 : \quad & \sum_{i=0}^{m} x_{i,\varepsilon}^u \neq 0, \quad \sum_{j=0}^{n} x_{\varepsilon,j}^d \neq 0, \\
C_3 : \quad & \sum_{i=0}^{m} x_{i,j}^u \cdot w_{i,j}^u \leq w_j^u, \quad 0 \leq j \leq m, \\
& \sum_{j=0}^{m} x_{i,j}^d \cdot w_{i,j}^d \leq w_i^d, \quad 0 \leq i \leq m,
\end{aligned}
$$

$$C_4: \sum_{i=0}^{m} x_{i,j}^u \cdot d_g \leq g_j, \quad 0 \leq j \leq m,$$

where C_1 is routing policy constraint to make the pushing and pulling routing policy valid, C_2 is the constraint that worker nodes need to pushing the model updates to parameter server and parameter server needs to send the global model parameters to worker node, C_3 is the bandwidth constraint and C_4 is the cache capability constraint of network nodes.

Algorithm 1. Pushing routing

Input:
    $G(V, E)$: the network graph;
Output:
    MT^u: the routing path for gradient updating from worker nodes to the parameter server;
1: $MT \leftarrow \text{Dijkstra}(G, \varepsilon, N)$
2: **for** i in MT **do**
3:    computer the t_i^u by applying equation (2)
4: **end for**
5: $L^* \leftarrow \text{FindCriticalPath}(MT)$
6: $MT^u \leftarrow$ empty tree
7: add L^* to MT^u
8: $P \leftarrow MT^u.V$
9: $R \leftarrow N - P \cap N$
10: **for** j in R **do**
11:   **if** j not in MT^u **then**
12:     sort the nodes in P with the order in the hop count of shortest path between the node and j.
13:     **for** p in P **do**
14:       **if** $t_j^u + t_{j,p} \leq t_p^u$ and C_3, C_4 are satified **then**
15:         add $L(j, \varepsilon)$ to MT^u
16:         break
17:       **end if**
18:     **end for**
19:     **while** j not in MT^u **do**
20:       $L(j, \varepsilon) \leftarrow \text{SearchShortestPath}(G, j, \varepsilon)$
21:       **if** C_3 and C_4 are fitted **then**
22:         add $L(j, \varepsilon)$ to MT^u
23:         break
24:       **else**
25:         ban $L(j, \varepsilon)$ in G
26:       **end if**
27:     **end while**
28:   **end if**
29: **end for**

Theorem 1. *OPT is NP-Hard.*

Proof. We now show that a special case of the proposed problem OPT can be reduced to be the Steiner Tree problem. The Steiner Tree problem can be described as follows. Given an undirected graph with non-negative edge weights and a subset of vertices (i.e., terminals), the problem aims to generate a tree in the graph with minimum weight that contains all terminals (but may include additional vertices).

Let us consider a special case of the problem OPT, i.e., all constraints are removed. In this special case, an undirected graph G, consisting of workers, parameter server and routers, is given. Meanwhile, the terminals are workers and parameter server, and the edge weight between two nodes are communication delay between two nodes for transmitting parameters. Then, the objective of the problem OPT aims to generate a routing tree for minimizing the total communication delay for transmitting parameters, where all workers should be contained in the routing tree for completing the training. The objective is corresponding to generate a tree in the graph with minimum weight that contains all terminals. Therefore, the special case of the problem OPT is equivalent to the Steiner Tree problem. Noting that the Steiner Tree problem is NP-hard [5]. Thus, we conclude that the proposed problem OPT is NP-hard.

3   Algorithm

3.1   Pulling Routing Method

When the parameter server multi-casts the global model parameters to the worker nodes, the pulling routing method will generate a routing policy. With that routing policy, the parameter server sends the global model parameters to the worker nodes. The formula description of the algorithm is omitted due to the limit of paper length.

We design the pulling routing method based on Dijkstra Shortest Path algorithm (SPF) [6]. Firstly, for worker node i, we use SPF to find the shortest path $L_{\varepsilon,i}$ from parameter server to node i. Then, we denote C_l as the resident bandwidth of $L_{\varepsilon,i}$, which is the minimum link bandwidth in $L_{\varepsilon,i}$. Thirdly, to fit the constraint C_3, we traverse the links in path $L_{\varepsilon,i}$. If the bandwidth budget of nodes in $L_{\varepsilon,i}$ is less than C_l for each link, the algorithm will ban the link by setting the weight of link $e_{i,j}$ to infinity and the algorithm breaks to find a new shortest path. Finally, we add the $L_{\varepsilon,i}$ to the multi-cast tree MT. The time complexity of pulling routing method is $O\left(N \cdot E \cdot V^2\right)$.

3.2   Pushing Routing Method

The proposed pushing routing algorithm is formally described by Algorithm 1. Firstly, for each worker node i in graph G, the algorithm finds the shortest path from node i to the parameter server node with the Dijkstra algorithm. Then the algorithm constructs a multi-cast tree MT with those paths. Secondly, the algorithm employs pulling routing method to obtain the time when the latest model parameter data arrives at each node in MT, and the algorithm computes t_i^u by applying (2). Then the algorithm selects the path with the maximum t_i^u as the critical path and adds it to MT^u (MT^u is empty at the beginning). Thirdly, the algorithm starts the aggregation procedure according to MT^u, as shown from line 10 to line 29. For each worker node i, the algorithm searches the shortest paths between i and other nodes in MT^u. It should be noted that the path must satisfy $t_j^u + t_{j,p} \leq t_q^u$ and constraint C_3, C_4. Therefore, we can minimize the network overhead without increasing the transmission time. The algorithm may not find a valid path from i to the parameter server. In that case, the algorithm will break $t_j^u + t_{j,p}$ constraint and searches the shortest route to the parameter server under C_3, C_4 constraint. The time complexity of Algorithm 1 is $O\left(N \cdot V^3\right)$.

3.3   Routing Method for Distributed Training

The proposed routing method for distributed training is named RDT. RDT integrates the pulling routing method and the pushing routing method in the distributed training. In the pulling process, RDT uses the pulling routing method. In the pushing process, RDT uses the pushing routing method.

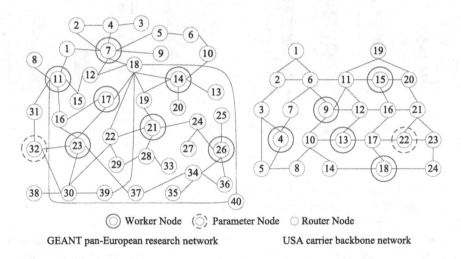

Fig. 2. Network graphs in the evaluation.

4   Numerical Results

To investigate the performance of proposed algorithms, we implement the algorithms on a high-performance work station with an Intel I7 processor at frequency 3.9 GHZ and 8G RAM. As shown in Fig. 2, we conduct the simulations on two networks [7]: USANet, which is a realistic carrier backbone network, and GEANT, which is a pan-European communications infrastructure. For each link of the network, the bandwidth constraint is randomly picked in [10, 15] MB/s. For each network node, the cache size constraint is set as three times size of model updates [8].

We use the two deep learning models to evaluate algorithm performance. The first one is GoogleNet [9]. The model size of GoogleNet is 218 MB, and its workload is 1550 million times of mult-adds operations [10]. The second one is MobileNet [11]. The model size of MobileNet is 128 MB, and its workload is 569 million times of mult-adds operations [12].

We implement our proposed algorithms in the pulling and pushing stage of distributed training. We compare RDT with the shortest path algorithm (SP) [13]. We evaluate the proposed algorithm in four cases. Case 1 is training MobileNet in USANet. Case 2 is training MobileNet in GEANTNet. Case 3 is training GoogleNet in USANet. Case 4 is training GoogleNet in GEANTNet.

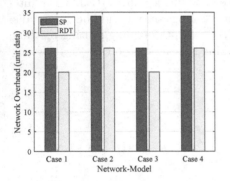
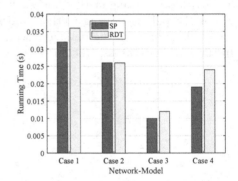

Fig. 3. Network overhead under different routing schemes in an iteration.

Fig. 4. Algorithms running time of different routing schemes in an iteration.

Figure 3 shows the network overhead of four cases in an iteration processed by SP and RDT. The network overhead processed by RDT is much lighter than that by SP. RDT outperforms than SP by 30% on average in terms of network overhead. The reason is that RDT uses data aggregation technology while SP does not. Therefore, we can conclude that data aggregation technology can reduce network overhead.

Figure 4 shows the running time of SP and RDT. The running time of RDT ranges from 0.016 s to 0.035 s. Compared with SP, RDT is slower. The reason is that RDT is designed based on SP. The running time of RDT is larger than SP by 30%. Compared with the training time in an iteration, the distributed training can ignore the running time of routing algorithms. The differences in running time of SP and RDT in four cases are small. The reason is that the difference in the number of nodes in the four cases is small.

Figure 5 shows the training time of four cases in an iteration processed by SP and RDT. The training time of SP is always longer than RDT in four cases as shown in Fig. 5. RDT outperforms SP by 74% on average in terms of training finishing time. The reason is that SP ignores data aggregation, which leads to massive network overhead, as shown in Fig. 3. The massive network overhead will increase the transition delay. Therefore, the training time of SP is longer.

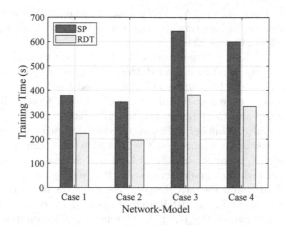

Fig. 5. Training time under different routing schemes in an iteration.

5   Conclusion

We have investigated the data aggregation aware routing problem for distributed training. The objective of the problem is to minimize the finishing time of training with cache capability constraints. The NP-hardness of the proposed problem has been proved. To solve the problem, we have proposed the algorithm RDT, which transmits the data to closest aggregation nodes in greedy to reduce the network overhead. Simulations have well demonstrated the promising potential of proposed algorithm in terms of reducing the network overhead and the finishing time of distributed training. In the future, we try to use machine learning techniques to solve the data aggregation aware routing problem.

References

1. Rahman, H., Ahmed, N., Hussain, M.I.: A qos-aware hybrid data aggregation scheme for internet of things. Ann. Telecommun. **73**(7), 475–486 (2018)
2. Redondi, A.E., Cesana, M., Fratta, L., Capone, A., Borgonovo, F.: A prediction-based approach for features aggregation in visual sensor networks. Ad Hoc Netw. **83**(1), 55–67 (2019)
3. Cui, J., Boussetta, K., Valois, F.: Classification of data aggregation functions in wireless sensor networks. Comput. Netw. **178**(1), 1–46 (2020)

4. Chen, C.C.Y., Das, S.K.: Breadth-first traversal of trees and integer sorting in parallel. Inf. Process. Lett. **41**(1), 39–49 (1992)
5. Segev, A.: The node-weighted steiner tree problem. Networks **17**(1), 1–17 (1987)
6. Johnson, D.B.: A note on dijkstra's shortest path algorithm. J. ACM **20**(3), 385–388 (1973)
7. Yang, S., Li, F., Trajanovski, S., Chen, X., Wang, Y., Fu, X.: Delay-aware virtual network function placement and routing in edge clouds. IEEE Trans. Mob. Comput. 1–14 (2019)
8. Li, C., Tang, J., Tang, H., Luo, Y.: Collaborative cache allocation and task scheduling for data-intensive applications in edge computing environment. Future Gener. Comput. Syst. **95**, 249–264 (2019)
9. Szegedy, C., et al.: Going deeper with convolutions. In: Proceedings of the IEEE Conference on Computer Vision and Pattern Recognition, pp. 1–9. IEEE (2015)
10. Qassim, H., Verma, A., Feinzimer, D.: Compressed residual-VGG16 CNN model for big data places image recognition. In: IEEE Annual Computing and Communication Workshop and Conference, pp. 169–175. IEEE (2018)
11. Howard, A.G., et al.: Mobilenets: efficient convolutional neural networks for mobile vision applications. arXiv preprint arXiv:1704.04861, 1–9 (2017)
12. Zhang, X., Zhou, X., Lin, M., Sun, J.: Shufflenet: an extremely efficient convolutional neural network for mobile devices. In: Proceedings of the IEEE Conference on Computer Vision and Pattern Recognition, pp. 6848–6856. IEEE (2018)
13. Garcia-Luna-Aceves, J.J.: A distributed, loop-free, shortest-path routing algorithm. In: Proceedings of the IEEE Conference on Computer Communications, pp. 1125–1137. IEEE (1988)

A New Integer Programming Model for the File Transfer Scheduling Problem

Jingwen Yang$^{(\boxtimes)}$, Jiaxin Li, and Xin Han

School of Software Technology, Dalian University of Technology, Dalian, China
jingwen_YANG@mail.dlut.edu.cn, lijiaxin_dlut@foxmail.com,
hanxin@dlut.edu.cn

Abstract. Scheduling the transfer of files in distributed computer systems is an increasing concern. The purpose of this paper is to propose a new time-index integer programming model for the file transfer scheduling problem with integer file length and port restrictions, which is to design a schedule to transfer series of files with different file length while minimizing overall completion time. By different formulations of variables and constraints, our new model contains fewer constraints and variables. And we also propose an enhanced technique to reduce the number of constraints further by utilizing the elementary lower bound and upper bound. Experimental results show that compared to existing best model, our model with the enhanced technique can solve the same instance with only 14.3% of the time. Moreover, our model can solve larger instances (up to 100 vertices and 600 files) to optimality that can not be solved by existing models.

Keywords: Integer programming · File transfer scheduling · Makespan

1 Introduction

The purpose of this paper is to present a new integer programming model for the *File Transfer Scheduling Problem* (FTSP). The FTSP can be illustrated as an undirected multigraph $G = (V, E)$, named *file transfer graph*. Vertices $v \in V$ represent computers or communication centers in a network, and every two vertices can communicate with each other directly. Each vertex $v \in V$ is associated with a positive integer $p(v)$, called port constraint, which denotes the maximum number of files transferring in v simultaneously. Edges $e \in E$ represent files that should be transferred from its one endpoint to another endpoint in G. Each edge $e \in E$ is associated with a positive integer $l(e)$, which denotes the transfer time of file. We assume that once a file starts to transfer, it will not stop until finishing the transfer of the whole file. The FTSP consists of designing a schedule of minimal overall completion time while satisfying the port constraints of all vertices.

Figure 1 is an example about a file transfer graph and a schedule in a time diagram. For example, vertex v_3 has a port constraint $p(v_3) = 2$ and three files

© Springer Nature Switzerland AG 2021
Y. Zhang et al. (Eds.): PDCAT 2020, LNCS 12606, pp. 251–263, 2021.
https://doi.org/10.1007/978-3-030-69244-5_22

(e_3, e_4 and e_5) are associated with it, which means that at most two files of these three files can be transferred simultaneously. Moreover, the time diagram represents such a schedule that file e_1, e_4 and e_5 start to be transferred at time 0, file e_3 starts at time 2 and file e_2 starts at time 3.

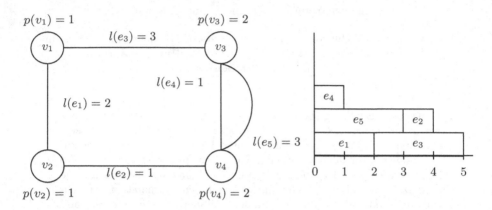

Fig. 1. A file transfer graph and a schedule

FTSP was first introduced by Coffman et al. [2]. They focused on the FTSP with undirected file transfer graph and didn't allow to forward files, that is, each file must be transferred directly between its endpoints without intermediate vertex forwarding. Such FTSP has been proved to be NP-complete. Besides, Whitehead [13] and Mao [10] extended the FTSP to two FTSP variants in which one allowed forwarding when no direct link existed between vertices and another considered directed file transfer graph. Both FTSP variants were proved to be NP-complete. Moreover, Jia et al. [9] considered the online version of FTSP and presented approximation algorithms and theoretical analysis.

Due to the complexity of FTSP, some studies focused on designing heuristic algorithms. Akbari et al. [1] proposed a model to solve FTSP based on a deterministic modified Hopfield model, which can solve instances of 39 vertices and 100 files. However, for larger instances, the neural network model did not converge. Moreover, *Variable Neighborhood Search* (VNS) can be used to solve continuous optimization problems and discrete optimization problems, and has achieved good results in practical applications [7,11]. So Dražić tried to solve the FTSP by the VNS algorithm and the *Gaussian Variable Neighborhood Search* (Gauss-VNS) algorithm successively [3,5]. Although both VNS algorithms obtained pretty good solutions, there is no guarantee to the gap with optimal solution.

To solve the FTSP to optimality, different time-index problem formulations were proposed. The time-index formulation has many applications in scheduling problem, like single machine schedule problems (see [12]). For FTSP, Higuero et al. [8] proposed a time-index formulation in multi-server and multi-user environments. In their paper, an approximate problem was proposed to reduce the complexity of the optimization. Furthermore, Dražić [6] reformulated the FTSP with integer file length and proposed a time-index integer programming model for the first time. This model can only solve small instances of 40 vertices and 50 files to optimality because of massive variables and constraints. Then Dražić [4] proposed an improved integer programming model which reduces the number of constraints by introducing extra variables. New model can solve larger instances with 50 vertices and 200 files. However, there is still room to design better integer programming model by reducing the number of constraints without increasing the number of variables.

We focus on the FTSP with integer file length and port constraints in this paper. Although there are some integer programming models, the scale of instances solved by existing models is still small and they take too much time. So we present a new integer programming model with fewer variables and constraints, which can solve larger scale of instances with less time. We formulate the decision variables and the port constraints in a different way. In this case, port constraints imply continuity constraints of file transfer, which greatly reduces the number of constraints while introducing no additional variables. We also propose an enhanced technique to reduce the number of constraints further by utilizing the elementary lower bound and upper bound. Furthermore, extensive experiments confirm the superiority of our models, which can solve larger instances and spend less time than the existing best model.

The rest of this paper is organized as follows. We first introduce the existing integer programming models in Sect. 2. Then our new time-index integer programming model is described in Sect. 3. In Sect. 4, we illustrate the experimental results. Finally, the conclusion is drawn in Sect. 5.

2   Existing Integer Programming Models

In this section, we will give a brief review of the existing integer programming models. First, we will introduce the problem formulation proposed by Dražić et al. [6]. Then two existing integer programming models based on the problem formulation are illustrated successively [4,6].

2.1   Problem Formulation

In this paper, we focus on the problem formulation proposed by Dražić et al. [6]. Given a file transfer graph $G = (V, E)$, each vertex $v \in V$ is associated with a positive integer port constraint $p(v)$ and each edge $e \in E$ is associated with a positive integer file length $l(e)$. The goal is to find a feasible schedule $s^\star$ with the minimal overall completion time, referred to as *makespan*.

Let $s(e)$ denote the start time of file $e \in E$. Then the *makespan* of a schedule s can be written as:

$$\max_{e \in E} s(e) + l(e).$$

Next we will introduce how to represent a feasible schedule. A schedule can be viewed as a function $s : E \rightarrow [0, \infty]$ that assigns to each file $e \in E$ a start time $s(e)$. Furthermore, the transfer time for each file $e \in E$ is an integer, in which case considering discrete time interval instead of continuous time is reasonable and easier. So two definitions about integer time intervals are presented as follows.

Definition 1. *Let $\tau \in Z^+$ represent the index of the time interval from $\tau - 1$ to τ.*

Definition 2. *We say that in schedule $s: E \rightarrow Z^+$, $e \in E$ is active in time interval with index τ if $s(e) \leq \tau$ and $s(e) + l(e) \geq \tau - 1$.*

Moreover, considering that the number of files is limited, a limited number of time intervals is sufficient to execute them. So Dražić et al. [6] proposed to limit the number of time intervals while guaranteeing that the best solution will not be missed. Let T_{MAX} be an upper bound of the optimal solution of the FTSP, which can be computed by approximation algorithms or heuristic algorithms. Therefore, a feasible schedule is viewed a function $s : E \rightarrow \{1, 2, ..., T_{MAX}\}$ which satisfies that, for all $\tau \in \{1, 2, ..., T_{MAX}\}$ and $v \in V$, $\{e : v$ is an end point of e and e is active in the time interval with index $\tau\} \leq p(v)$.

2.2   An Existing Integer Programming Model

Next, we will introduce the integer programming model proposed by Dražić et al. [6]. They defined a binary decision variable

$$x_{e\tau} = \begin{cases} 1, & \text{if transfer } e \text{ is active in time interval with index } \tau, \\ 0, & \text{otherwise.} \end{cases}$$

Let T_{MAX} be an upper bound of the optimal solution of the FTSP and z be a variable which represents the makespan of a feasible schedule. Then the integer programming model is as follows:

$$(M1) \ \min z \tag{1}$$

$$\text{s.t.} \sum_{\tau=1}^{T_{MAX}} x_{e\tau} = l(e), \ \forall e \in E; \tag{2}$$

$$\sum_{e \in E, \ e \ni v} x_{e\tau} \leq p(v), \ \forall v \in V, \ 1 \leq \tau \leq T_{MAX}; \tag{3}$$

$$z \geq \tau x_{e\tau}, \ \forall e \in E, \ 1 \leq \tau \leq T_{MAX}; \tag{4}$$

$$x_{ej} \geq x_{ei} + x_{ek} - 1, \ \forall e \in E, \ 1 \leq i < j < k \leq T_{MAX}; \tag{5}$$

$$x_{e\tau} \in \{0, 1\}, \ \forall e \in E, \ 1 \leq \tau \leq T_{MAX}; \tag{6}$$

$$z \in Z^+. \tag{7}$$

This model is denoted by $M1$ in the rest of this paper. Constraint (2) guarantees that each edge $e \in E$ is entirety transferred, and the transfer must be completed before T_{MAX}. Constraint (3) guarantees that the number of files transferred at one time does not exceed the port constraint. Constraint (4) ensures that no files are active after the time moment z. Constraint (5) guarantees that files are active in some continuous time intervals, i.e., if a file is active in time moment i and k, it has to be active for all moments j such that $i < j < k$. Constraint (6) guarantees that the variable $x_{e\tau}$ is a binary, and Constraint (7) guarantees that the variable z is an integer. The number of constraints and variables of this model is pseudo-polynomial, which is related to the upper bound T_{MAX}. Dražić et al. [6] generated the T_{MAX} by a VNS algorithm proposed by Dražić [3].

2.3   An Extension Integer Programming Model of $M1$

Dražić [4] improved the model $M1$ by adding some extra variables to reduce the number of constraints. The additional variables introduced were written as:

$$
y_{e\tau} = \begin{cases} 1, & \text{if transfer } e \text{ starts in the time interval with index } \tau, \\ 0, & \text{otherwise.} \end{cases}
$$

Then, the continuity of transmission can be expressed by the following constraints:

$$
\sum_{\tau=1}^{T_{MAX}} y_{e\tau} = 1, \ \forall e \in E; \tag{8}
$$

$$
x_{e\tau} - x_{e(\tau-1)} \leq y_{e\tau}, \forall e \in E, \ 2 \leq \tau \leq T_{MAX}; \tag{9}
$$

$$
x_{e1} = y_{e1}, \ \forall e \in E, \tag{10}
$$

which replaced the Constraint (5) in $M1$. This model is denoted by $M2$ in the rest of this paper.

3   New Time-Index Integer Programming Model

In this section, we proposed two new integer programming models to solve the FTSP. The first model we proposed is a time-index formulation and has a pseudo-polynomial number of variables and constraints. The second model we proposed is an enhanced time-index formulation, which utilizes the elementary lower bound to reduce the number of constraints further.

3.1 A New Time-Index Formulation for FTSP

Though Dražić et al. [6] has proposed an integer programming model $M1$, $M1$ can only solve small instances with only 40 vertices and 50 files because of a great deal of variables and constraints. Then Dražić [4] obtained another model $M2$ by improving $M1$, in which the author reduced the number of constraints of M1 by introducing extra variables. $M2$ can solve larger instances, however, there is still room to design better integer programming model by reducing the number of constraints without increasing the number of variables. That is what our new model has achieved.

Our new integer programming model is also a time-index formulation. The main difference is that we avoid the additional continuity constraints of file transfer by subtle formulations of the FTSP. Next, we will introduce our new integer programming model in detail. We first define the decision variable

$$y_{e\tau} = \begin{cases} 1, & \text{if transfer } e \text{ starts in the time interval with index } \tau, \\ 0, & \text{otherwise.} \end{cases}$$

Let T_{MAX} be an upper bound of the optimal solution of the FTSP and z be a variable which represents the makespan of a feasible schedule like before. Our new time-index integer programming model for FTSP is as follows:

$$(M3) \ \min z \tag{11}$$

$$\text{s.t.} \quad \sum_{\tau=1}^{T_{MAX}-l(e)+1} y_{e\tau} = 1, \ \forall e \in E; \tag{12}$$

$$\sum_{e \in E, \ e \ni v} \ \sum_{s=\max\{1, \ \tau-l(e)+1\}}^{\tau} y_{es} \leq p(v), \ \forall v \in V, \ 1 \leq \tau \leq T_{MAX}; \tag{13}$$

$$z \geq (l(e) + \tau - 1)y_{e\tau}, \ \forall e \in E, \ 1 \leq \tau \leq T_{MAX}; \tag{14}$$

$$y_{e\tau} \in \{0,1\}, \ \forall e \in E, \ 1 \leq \tau \leq T_{MAX}; \tag{15}$$

$$z \in Z^+. \tag{16}$$

This model is denoted by $M3$ in the rest of this paper. Constraint (12) ensures that each file $e \in E$ must have exactly one start transfer time. In other words, constraint (12) guarantees that each file must have been transferred exactly once. Constraint (13) guarantees that the number of files in some vertex transferred at one time does not exceed its port constraint. Besides, constraint (13) also implies that, once each file start to be transferred, it will occupy one port of its two endpoints for its whole transfer time $l(e)$. This guarantees that each file will finish its transfer without interruption. With constraint (13), port constraints and continuity constraints are both guaranteed. Constraint (14) ensures that the maximum completion time for all files $e \in E$ is at most z. Constraint (15) enforces that the variable $y_{e\tau}$ is a binary and constraint (16) guarantees that variable z is a positive integer.

3.2   An Enhanced New Time-Index Formulation for FTSP

Coffman et al. [2] proposed a lower bound for FTSP, called the *elementary lower bound*. The elementary lower bound can be used in our new integer programming model to reduce the number of constraints further. Let E_u represent the set of edges that has one endpoint at vertex u, and the degree of vertex u is $|E_u|$. Let $\sum_u$ represent the sum of the transfer time of all files corresponding to vertex v, i.e., $\sum_u = \sum_{e \in E_u} l(e)$. The time to transfer all files related to vertex u is at least $\lceil \sum_u /p(u) \rceil$. Therefore, the optimal solution for any file transfer graph $G = (V, E)$ must satisfy the following inequality:

$$OPT(G) \geq \max_u \lceil \textstyle\sum_u /p(u) \rceil. \tag{17}$$

We have introduced the elementary lower bound for FTSP, and now we can use the elementary lower bound to enhance the model $M3$. The elementary lower bound is denoted by ELB, and the enhanced model is denoted by $M4$ in the rest of this paper.

The Constraint (16) can be replaced by the following inequality:

$$ELB \leq z \leq T_{MAX}. \tag{18}$$

In this case, Constraint (14) can be replaced by:

$$z \geq (l(e) + \tau - 1)y_{e\tau}, \; \forall e \in E, \; ELB - l(e) + 1 \leq \tau \leq T_{MAX}. \tag{19}$$

3.3   Generation of the Upper Bound

We use an easy approximate algorithm to generate the upper bound T_{MAX} proposed by Coffman et al. [2], called List Scheduling (LS). In the LS algorithm, all files are arranged by ascending file length in a list. At time 0 and thereafter each time t a file is finished, the list is scanned to see if there is a file that has not started, and there are ports available on its both endpoints. When such a file e is found in the list, the start time of e is set to t, and the ports it uses become unavailable. Repeat this process until all files have been transferred. According to the above LS algorithm, the completion time of the last file will be used as T_{MAX} in our integer programming models.

4   Experimental Results

We will show the experimental results in this section. All tests are run on a desktop with Windows 10 system using Intel core i5-8250U and 8 GB RAM. Gurobi 7.5.2 is applied to solve the integer programming models and we set the maximum running time of each instance to 600 s.

4.1 Instances Generation

The instances used in this paper are generated randomly, and Table 1 summarizes the scale of the randomly generated instances. According to the number of files, the instances are partitioned into three parts, small-scale instances, medium-scale instances and large-scale instances. The instances generated include different number of vertices ($|V|$) and number of edges ($|E|$). The length of each file is uniformly distributed between $[1, e_{max}]$, where e_{max} is equal to $\lfloor |E|/2 \rfloor$. And the number of ports of each vertex is uniformly distributed between $[1, p_{max}]$, where p_{max} is equal to $\lfloor |E|/4 \rfloor$ or $\lfloor |E|/5 \rfloor$. Each randomly generated file transfer graph is a connected graph without self-looping.

Table 1. Instance scale

Instance type	Vertex number ($|V|$)	Edge number ($|E|$)
Small scale	5	10
	5	40
	10	40
	10	50
	20	50
	40	50
	30	100
	40	100
	50	100
Medium scale	30	200
	40	200
	50	200
	100	200
	30	300
	40	300
	50	300
	100	300
Large scale	50	400
	100	400
	150	400
	200	400
	100	500
	150	500
	200	500
	250	500
	300	500
	150	600
	200	600
	250	600
	300	600
	100	600

4.2 Performance Comparison Between Different Models

Comparisons are made between existing two integer programming models ($M1$ and $M2$) and our two integer programming models ($M3$ and $M4$). The detailed results about small-scale, medium-scale and large-scale instances are showed in Table 2, Table 3 and Table 4 respectively. The column "Instance" records the name of instance, which includes two numbers that the first number represents the number of vertices and the second number represents the number of files. For example, ftsp_5_40 represents an instance with 5 vertices and 40 files. The columns "p_{max}" and "e_{max}" contain the maximum of the number of port and file length respectively, which are used to generate the file transfer graph of instances. Besides, there are four blocks which records the experimental results of each model. Each block consists of two columns that one column "#Opt" contains the number of instances solved to optimality and another column "t(s)" contains the average time of solving instances to optimality by solver within 600 s. For each file transfer graph with the same number of vertices and edges, we will generate 10 instances. Each value in the corresponding position is the average value of all 10 instances. In addition, if Gurobi did not solve any instance to optimality (out of memory or out of time limit), we will denote the symbol "-" in the corresponding column.

Table 2. Comparison of different models on small-scale instances

Instance	p_{max}	e_{max}	$M1$		$M2$		$M3$		$M4$	
			#Opt	t(s)	#Opt	t(s)	#Opt	t(s)	#Opt	t(s)
fstp_5_40	10	20	0	-	10	59.06	10	2.12	10	0.12
fstp_5_40	8	20	0	-	10	50.56	10	0.76	10	0.09
fstp_10_40	10	20	4	420.93	10	25.37	10	0.22	10	0.06
fstp_10_40	8	20	0	-	10	42.21	10	1.41	10	0.13
fstp_10_50	12	25	0	-	10	189.99	10	6.91	10	0.39
fstp_10_50	10	25	0	-	10	54.29	10	1.36	10	0.13
fstp_20_50	12	25	2	312.08	10	4.77	10	0.11	10	0.07
fstp_20_50	10	25	0	-	10	63.93	10	3.01	10	0.21
fstp_40_50	12	25	10	30.04	10	0.08	10	0.02	10	0.01
fstp_40_50	10	25	9	123.11	10	3.32	10	0.06	10	0.02
fstp_30_100	25	50	0	-	8	137.23	10	5.76	10	0.70
fstp_30_100	20	50	2	10.59	10	2.08	10	0.22	10	0.18
fstp_40_100	25	50	0	-	10	10.26	10	0.68	10	0.28
fstp_40_100	20	50	0	-	10	51.73	10	2.52	10	0.36
fstp_50_100	25	50	0	-	10	5.00	10	0.24	10	0.15
fstp_50_100	20	50	5	8.14	10	5.91	10	0.43	10	0.14

Table 3. Comparison of different models on medium-scale instances

Instance	p_{max}	e_{max}	M1		M2		M3		M4	
			#Opt	t(s)	#Opt	t(s)	#Opt	t(s)	#Opt	t(s)
fstp_30_200	50	100	0	-	0	-	5	359.37	10	9.07
fstp_30_200	40	100	0	-	3	289.76	10	98.30	10	4.70
fstp_40_200	50	100	0	-	6	245.42	10	29.15	10	3.50
fstp_40_200	40	100	0	-	0	-	7	177.24	10	6.94
fstp_50_200	50	100	0	-	4	285.87	8	97.74	10	8.67
fstp_50_200	40	100	0	-	9	103.65	10	9.23	10	2.52
fstp_100_200	50	100	0	-	10	7.60	10	0.94	10	0.77
fstp_100_200	40	100	0	-	9	206.74	10	13.14	10	3.89
fstp_30_300	75	150	0	-	0	-	0	-	10	74.27
fstp_30_300	60	150	0	-	0	-	0	-	10	26.15
fstp_40_300	75	150	0	-	0	-	1	484.54	10	27.64
fstp_40_300	60	150	0	-	0	-	3	326.28	10	21.64
fstp_50_300	75	150	0	-	1	349.65	9	134.13	10	12.56
fstp_50_300	60	150	0	-	0	-	8	311.56	10	18.79
fstp_100_300	75	150	0	-	8	61.09	10	21.28	10	4.27
fstp_100_300	60	150	0	-	4	431.14	8	111.65	10	13.57

According to Table 2, we can see that, even for small-scale instances with at most 50 vertices and 100 files, there are only a small number of instances (32 out of 160) can be solved by $M1$. $M2$, $M3$ and $M4$ performed better than $M1$ and they mostly solved all small-scale instances to optimality. But our model $M3$ and $M4$ can solve faster than $M2$. For example, compared to $M2$, $M3$ took 3.6% of the time and $M4$ took 0.4% of the time in average. Moreover, as Table 3 and Table 4 showed, for medium-scale and large-scale instances, the number of instances solved by $M2$ are fewer than our models $M3$ and especially $M4$. For example, 54 out of 160 medium-scale instances and 129 out of 280 large-scale instances are solved by $M2$. $M3$ solved 109 out of 160 medium-scale instances and 202 out of 280 large-scale instances. And $M4$ has best performance in all instances which can solve all medium-scale instances and almost all large-scale instances (up to 100 vertices and 600 files) to optimality. And compared to $M2$, our model $M4$ took only 4.1% of the time for medium-scale instances and 19.0% of the time for large-scale instances in average. Furthermore, for all instances with different scale, compared to $M2$, our model $M4$ took only 14.3% of the time.

Table 4. Comparison of different models on large-scale instances

Instance	p_{max}	e_{max}	M1 #Opt	M1 t(s)	M2 #Opt	M2 t(s)	M3 #Opt	M3 t(s)	M4 #Opt	M4 t(s)
fstp_50_400	100	200	0	-	0	-	1	466.77	10	69.67
fstp_50_400	80	200	0	-	0	-	1	297.41	10	36.48
fstp_100_400	100	200	0	-	4	44.87	10	61.65	10	15.71
fstp_100_400	80	200	0	-	0	-	6	249.81	10	38.03
fstp_150_400	100	200	0	-	8	166.35	10	21.41	10	10.31
fstp_150_400	80	200	0	-	9	60.83	10	18.40	10	9.24
fstp_200_400	100	200	0	-	6	65.77	10	130.51	10	18.92
fstp_200_400	80	200	0	-	7	120.11	10	75.07	10	18.28
fstp_100_500	125	250	0	-	0	-	3	378.52	10	48.65
fstp_100_500	100	250	0	-	0	-	6	240.37	10	39.45
fstp_150_500	125	250	0	-	1	560.75	2	35.19	10	56.51
fstp_150_500	100	250	0	-	5	20.95	10	63.29	10	19.10
fstp_200_500	125	250	0	-	10	9.25	10	10.52	10	10.07
fstp_200_500	100	250	0	-	1	92.90	9	326.46	10	45.06
fstp_250_500	125	250	0	-	9	144.80	10	65.74	10	24.30
fstp_250_500	100	250	0	-	3	151.20	8	239.06	10	40.82
fstp_300_500	125	250	0	-	10	17.59	10	10.84	10	10.52
fstp_300_500	100	250	0	-	10	84.50	10	16.65	10	16.34
fstp_150_600	150	300	0	-	4	71.17	8	26.09	10	36.04
fstp_150_600	120	300	0	-	1	8.76	5	98.81	10	68.91
fstp_200_600	150	300	0	-	2	569.64	8	40.20	10	56.24
fstp_200_600	120	300	0	-	0	-	0	-	10	108.20
fstp_250_600	150	300	0	-	8	127.54	8	24.92	10	34.04
fstp_250_600	120	300	0	-	10	97.42	10	22.90	10	23.30
fstp_300_600	150	300	0	-	10	139.52	10	24.07	10	24.90
fstp_300_600	120	300	0	-	10	144.97	10	24.95	10	27.31
fstp_100_600	150	300	0	-	1	8.72	7	211.12	10	54.83
fstp_100_600	120	300	0	-	0	-	0	-	8	145.35

4.3 Comparison of Variables and Constraints Between Different Models

Table 5. Comparison of the number of variables and constraints of different models

Instance	p_{max}	e_{max}	M1 #Var	M1 #Cons	M2 #Var	M2 #Cons	M3 #Var	M3 #Cons	M4 #Var	M4 #Cons
fstp_5_10	2	5	145	4544	289	380	145	226	145	141
fstp_10_10	2	5	93	1488	185	296	93	194	93	147
fstp_10_15	3	7	252	11441	502	698	252	433	252	289
fstp_10_20	5	10	475	43623	949	1225	475	731	475	435
fstp_15_15	3	7	157	2636	313	498	157	327	157	257
fstp_15_20	5	10	253	7379	505	733	253	461	253	352
fstp_20_15	5	10	295	10946	589	922	295	608	295	459

We also compare the number of variables and constraints of different models. $M1$ can only solve small-scale instances to optimality, so to be able to compare, we focus on the number of variables and constraints of small-scale instances showed in Table 5. Columns "#Var" contain the number of variables and columns "#Cons" contain the number of constraints in the corresponding model. We can see that compared to $M1$, $M2$ reduced the number of constraints but increased the number of variables. Our models($M3$ and $M4$) reduced the number of constraints without introducing extra variables. More specifically, $M3$ reduced the number of constraints in $M1$ by about 96.37% and reduced the number of constraints in $M2$ by about 37.29% in average. $M4$ performs better with reducing the number of constraints in M1 by about 97.47% and reduced the number of constraints in $M2$ by about 56.23% in average. The number of variables and constraints greatly affect the difficulty of solving the model. The fewer constraints and variables the model contains, the faster the solver solves the model. That is why our models can solve the same instance faster and solve larger instances.

5   Conclusion

In this paper, we studied the FTSP with integer file length. We first proposed a new time-index integer programming model with fewer variables and constraints. Then an enhanced technique by utilizing the elementary lower bound and upper bound to reduce the number of constraints further was proposed.

To test the performance of our models, we randomly generated extensive instances with different vertices and files. We used Gurobi 7.5.2 solver to solve existing integer programming models($M1$ and $M2$) and our new models($M3$ and $M4$) with a time limit of 600 s. Experimental results showed that our models performed better than existing models, extremely our model with the enhanced technique, $M4$. Compared the existing best model $M2$, $M4$ can solve the same instance with only 14.3% of the time. Moreover, our model can solve larger instances(up to 100 vertices and 600 files) to optimality that can not be solved by existing models.

We will focus on whether this way can be applied to modeling other problems in the future. And we will also consider to solve such FTSP using branch and bound method.

References

1. Akbari, M.K., Nezhad, M., Kalantari, M.: A neural network realization of file transfer scheduling. CSI J. Comput. Sci. Eng. **2**, 19–29 (2004)
2. Coffman Jr., E.G., Garey, M.R., Johnson, D.S., LaPaugh, A.S.: Scheduling file transfers. SIAM J. Comput. **14**(3), 744–780 (1985)
3. Dražić, Z.: Variable neighborhood search for the file transfer scheduling problem. Serdica J. Comput. **6**(3), 333–348 (2012)
4. Dražić, Z.: Modifications of the variable neighborhood search method and their applications to solving the file transfer scheduling problem. Ph.D. thesis, University of Belgrade, Faculty of Mathematics (2014)

5. Dražić, Z.: Gaussian variable neighborhood search for the file transfer scheduling problem. Yugoslav J. Oper. Res. **26**(2), 173–188 (2016)
6. Dražić, Z., Savić, A., Filipović, V.: An integer linear formulation for the file transfer scheduling problem. TOP **22**(3), 1062–1073 (2014)
7. Hansen, P., Mladenović, N.: Variable neighborhood search: principles and applications. Eur. J. Oper. Res. **130**(3), 449–467 (2001)
8. Higuero, D., Tirado, J.M., Isaila, F., Carretero, J.: Enhancing file transfer scheduling and server utilization in data distribution infrastructures. In: 2012 IEEE 20th International Symposium on Modeling, Analysis and Simulation of Computer and Telecommunication Systems, pp. 431–438. IEEE (2012)
9. Jia, S., Jin, X., Ghasemiesfeh, G., Ding, J., Gao, J.: Competitive analysis for online scheduling in software-defined optical wan. In: IEEE INFOCOM 2017 - IEEE Conference on Computer Communications, pp. 1–9 (2017)
10. Mao, W.: Directed file transfer scheduling. In: ACM 31st Annual Southeast Conference (ACM-SE 1993), pp. 199–203 (1993)
11. Mladenović, N., Hansen, P.: Variable neighborhood search. Comput. Oper. Res. **24**(11), 1097–1100 (1997)
12. Sousa, J.P., Wolsey, L.A.: A time indexed formulation of non-preemptive single machine scheduling problems. Math. Program. **54**, 353–367 (1992)
13. Whitehead, J.: The complexity of file transfer scheduling with forwarding. SIAM J. Comput. **19**(2), 222–245 (1990)

Approximation Algorithms
for the General Cluster Routing Problem

Longkun Guo[1,2], Bin Xing[1], Peihuang Huang[3(✉)], and Xiaoyan Zhang[4]

[1] College of Mathematics and Computer Science, Fuzhou University, Fuzhou 350116,
People's Republic of China
longkun.guo@gmail.com
[2] School of Computer Science, Qilu University of Technology
(Shandong Academy of Sciences), Jinan 250353, People's Republic of China
[3] College of Data Science and Mathematics, Minjiang University, Fuzhou 350108,
People's Republic of China
peihuang.huang@foxmail.com
[4] School of Mathematical Science, Nanjing Normal University, Nanjing 210046,
People's Republic of China
zhangxiaoyan@njnu.edu.cn

Abstract. Graph routing problem (GRP) and its generalizations have
been extensively studied because of their broad applications in the real
world. In this paper, we study a variant of GRP called the general cluster routing problem (GCRP). Let $G = (V, E)$ be a complete undirected
graph with edge weight $c(e)$ satisfying the triangle inequality. The vertex set is partitioned into a family of clusters $C_1, ..., C_m$. We are given
a required vertex subset $\overline{V} \subseteq V$ and a required edge subset $\overline{E} \subseteq E$.
The aim is to compute a minimum cost tour that visits each vertex in
$\overline{V}$ exactly once and traverses each edge in $\overline{E}$ at least once, while the
vertices and edges of each cluster are visited consecutively. When the
starting and ending vertices of each cluster are specified, we devise an
approximation algorithm via incorporating Christofides' algorithm with
minimum weight bipartite matching, achieving an approximation ratio
that equals the best approximation ratio of path TSP. Then for the case
there are no specified starting and ending vertices for each cluster, we
propose a more complicated algorithm with a compromised ratio of 2.67
by employing directed spanning tree against the designated auxiliary
graph. Both approximation ratios improve the state-of-art approximation ratio that are respectively 2.4 and 3.25.

Keywords: Routing problem · Cluster routing problem ·
Approximation algorithm

1 Introduction

The graph routing problem [1,5] is a hot topic in the field of operations research,
while the traveling salesman problem (TSP) has played a crucial role in combinatorial optimization [6,11,12]. Since they are known NP-hard, there exist

© Springer Nature Switzerland AG 2021
Y. Zhang et al. (Eds.): PDCAT 2020, LNCS 12606, pp. 264–273, 2021.
https://doi.org/10.1007/978-3-030-69244-5_23

no exact algorithms with polynomial time unless $P = NP$, and hence many researchers consequently focus on their approximation algorithms. The mathematical description of the TSP problem is as follows. Let $G = (V, E)$ be a directed or undirected graph, V be a set of vertices which represents n cities, and E be a set of edges and for each edge $e \in E$, cost $c(e)$ indicates the distance between two cities. The purpose is to find a Hamilton cycle such that the total cost of the tour is minimized. TSP is NP-hard such that there is no exact algorithm in polynomial time. Fortunately, one can consider an approximation algorithm for TSP and other NP-hard problems, whose performance is measured by the approximate ratio which is defined as the ratio of the solution of the approximate algorithm to the optimal value. The long-standing best approximation ratio for TSP is $\frac{3}{2}$, which is achieved by Christofides' algorithm [4]. Besides TSP, researchers have also studied its variants, such as *the traveling salesman path problem (Path TSP)* [8] and *the clustered traveling salesman problem (CTSP)* [7].

Definition 1. *The general routing problem (GRP) [9]: Given a connected, undirected edge-weighted graph $G = (V, E)$, a cost $c(e)$ on each edge e satisfying triangle inequality, a set of required vertices $\overline{V} \subseteq V$, and a set of required edges $\overline{E} \subseteq E$. The aim is to find a minimum cost tour that passes through each vertex in $\overline{V} \subseteq V$ exactly once and visits each edge in $\overline{E} \subseteq E$ at least once.*

In 1992, Klaus Jansen [9] proposed a $\frac{3}{2}$-approximation algorithm for GRP by generalizing Christofides' algorithm. When $\overline{V} = \emptyset$, the GRP was reduced to *the rural postman problem (RPP)*, when $\overline{E} = E$, the GRP was reduced to *the Chinese postman problem (CPP)*.

Definition 2. *The clustered traveling salesman problem (CTSP): Given a complete undirected graph $G = (V, E)$ with edge weights $c(e)$ satisfying triangle inequality. The vertex set V is partitioned into m clusters $C_1, ..., C_m$. The problem is to obtain a minimum-cost Hamiltonian cycle that visits all vertices, while the vertices of each cluster are visited consecutively.*

A 3.5-approximation algorithm for the problem with given starting vertices was proposed by Arkin et al. [3]. A 1.9091-approximation algorithm for the problem in which the starting and ending vertices of each cluster are specified was proposed by Guttmann-Beck et al. [7]. Besides, for the problem, if two ending vertices are given for each cluster, one can be chosen as the starting vertex and the other one as the ending vertex, they developed a 1.8-approximation algorithm. Furthermore, when clusters are given without specific starting and ending vertices, they gave a 2.75-approximation algorithm.

In this paper, we focus on a more general problem, which is called *the general cluster routing problem (GCRP)*. In GCRP, given a complete undirected graph $G = (V, E)$ with edge cost $c(e)$ satisfying the triangle inequality. The vertex set V is partitioned into m clusters $C_1, ..., C_m$. $\overline{V} \subseteq V$ and $\overline{E} \subseteq E$ are required vertices subset and required edges subset, respectively. The aim is to find a minimum cost tour that visits each vertex in $\overline{V}$ exactly once and passes through each edge in $\overline{E}$ at least once. Besides, vertices of each cluster C_i $(i \in m)$ must be visited consecutively. For the problem, we consider two cases depending on whether there are specified starting and ending vertices or not. For the former, we give a $\frac{5}{3}$-approximation algorithm, while a $\frac{8}{3}$-approximation algorithm for the latter problem.

2   Preliminary

In this section, we will first describe Christofides' algorithm briefly, and then introduce some variants of TSP and the related algorithms for later use.

2.1   Christofides' Algorithm

It's well known that Christofides' algorithm achieves a ratio of $\frac{3}{2}$ for TSP on general metric spaces, and it's the best performance guarantee. It works as follows:

Firstly, find a minimum spanning tree T in G, which includes all vertices and has no circuits. Then find the set of vertices with odd degree V_{odd} in T and compute a minimum weight matching M against them. Next, we can find an Euler tour T_{Euler} in polynomial time, and finally obtain a solution S via releasing from the tour T_{Euler} by deleting the appearance of the repeated vertices except its first occurrence.

2.2   The Traveling Salesman Path Problem (Path TSP)

The traveling salesman path problem (Path TSP) is a variant of TSP and has been extensively studied [2,8,10,13–15,17,18]. The difference between them is whether the starting vertex s and ending vertex t are distinct or not. When $s = t$, the path TSP problem reduces to TSP. It can be formulated as follows: Given a complete undirected graph $G = (V, E)$ with starting vertex s and destination vertex t, and the edge weight $c(e)$ satisfying the triangle inequality. The purpose of the problem is to find a Hamiltonian path from s to t such that the total cost of the tour attains minimum. In the 90s, Hoogeveen [8] showed that the approximation ratio of the variant of Christofides' algorithm is $\frac{5}{3}$ for this problem. There was no improvement until An, Kleinberg, and Shmoys [2] proposed a $\frac{1+\sqrt{5}}{2}$-approximation algorithm. Then Sebő [13] further improved the ratio to $\frac{8}{5}$ for the metric s-t path TSP by simplifying and improving the approach of An, Kleinberg and Shmoys. Hereafter, through a recursive dynamic program, Traub and Vygen [15] obtained a $\left(\frac{3}{2} + \varepsilon\right)$-approximation guarantee for the s-t-path TSP for any fixed $\varepsilon > 0$. Inspiringly, soon after Zenklusen [18] improved the ratio to $\frac{3}{2}$ via novelly leveraging Traub and Vygen's dynamic programming, retaining the currently best performance guarantee for path TSP. Traub and Vygen [16] designed a polynomial-time algorithm for the $s - t$-path graph TSP with ratio 1.497. Table 1 demonstrates the existing results on path TSP.

Let $MST(G)$ denote a minimal spanning tree of G. Let OPT be the optimal solution for the problem. We have the following observations.

Table 1. Previous approximation ratios for path TSP.

Reference	Ratio
Hoogeveen [8]	$\frac{5}{3}$
An, Kleinberg and Shmoys [2]	$\frac{1+\sqrt{5}}{2}$
Sebő [13]	$\frac{8}{5}$
Traub and Vygen [15]	$\left(\frac{3}{2}+\varepsilon\right)$
Zenklusen [18]	$\frac{3}{2}$

Theorem 1. *There exists a polynomial-time algorithm for path TSP with two end vertices s and t, and two solutions S_1 and S_2 can be obtained satisfying the following inequalities [8]:*

$$c\left(S_1\right) \leq 2c\left(MST\left(G\right)\right) - c\left(s,t\right) \leq 2OPT - c\left(s,t\right),$$

$$c\left(S_2\right) \leq c\left(MST\left(G\right)\right) + \frac{1}{2}\left(c\left(s,t\right)+OPT\right) \leq \frac{3}{2}OPT + \frac{1}{2}c\left(s,t\right).$$

Corollary 1. *[8] The shorter of S_1 and S_2 is at most $\frac{5}{3}OPT$.*

Proof. Through Theorem 1, if $c\left(s,t\right) > \frac{1}{3}OPT$, then $c\left(S_1\right) \leq \frac{5}{3}OPT$. Otherwise, $c\left(s,t\right) \leq \frac{1}{3}OPT$, then $c\left(S_2\right) \leq \frac{5}{3}OPT$. □

Let's consider a general problem of path TSP: *The general traveling path problem (GTPP)*. It can be described as follows: given a weighted undirected graph $G = (V, E)$ with starting vertex s and ending vertex t. The two sets $\overline{V} \subseteq V$ and $\overline{E} \subseteq E$ are any given vertex subset and edge subset, respectively. The goal is to find a path from s to t which visits all vertices of $\overline{V}$ exactly once and traversals all edges of $\overline{E}$, such that the total cost of the path attains minimum. In this paper, we shall focus on the case when $s \neq t$.

First, we can add a dummy vertex to convert the required edge into the required vertex following the spirit of the method in [9]. Then, since this is a minimum cost problem, and the edge weight satisfies the triangle inequality, we can simplify the problem by deleting the vertices not in $\overline{V}$ and edges not in $\overline{E}$ of G for the sake of producing an approximation solution. Therefore, we consider the problem in a reduced graph:

$$\overline{G} = \left(\overline{V} \cup \{s\} \cup \{t\} \cup \{v|v \in e \in \overline{E}\}, \overline{E}\right).$$

First, we need to compute the connected components of $\overline{G}$, this can be done in polynomial time. Next, we can construct a new complete graph G', where edge cost is the longest distance between two components. Simply, we just only calculate the edge cost between vertices with 0-degree and 1-degree. Finally, by Hoogeveen's algorithm for path TSP, we can obtain a Hamiltonian path from s to t, consequently, a feasible solution can be found. The details are described in Algorithm 1.

Algorithm 1. GTPP algorithm with starting and ending vertices

---
Input:
1: A weighted undirected graph $G = (V, E)$ with cost $c(e)$;
2: A starting vertex s and an ending vertex t;
3: $\overline{V} \subseteq V$ and $\overline{E} \subseteq E$ are the required vertex subset and required edge subset, respectively.
Output: A GTPP path.
begin
1: Compute the connected components $N_1, N_2, ..., N_k$ of $\overline{G}$;
2: **For** $i = 1$ to k do
3:     **If** there exist required edges **then**
4:         Convert the edge by adding a dummy vertex;
5:     **EndIf**
6:   **EndFor**
7: Find the set of vertices with 0-degree and 1-degree, written as F;
8: Construct a new graph $G' = (k, E_k)$ with edge cost $c(e)$ of $e = (i, j)$ being the maximum weight between a vertex i in $N_i \cap F$ and a vertex j in $N_j \cap F$;
9: Find a Hamilton path T in G' by Hoogeveen's algorithm for path TSP;
10: **Return** a feasible solution S in G.
end

Let L be the sum of lengths of the paths of OPT through each connected component. Let $c(\overline{E})$ be the total cost of the required edges. Then $OPT = L + c(\overline{E})$. So we can obtain the following theorem:

Theorem 2. *Let S be the tour output by Algorithm 1, then*

$$c(S) \leq \min\left\{2OPT - c(s, t), \frac{3}{2}OPT + \frac{1}{2}c(s, t)\right\}.$$

Corollary 2. *The length of the tour output by Algorithm 1 is at most $\frac{5}{3}OPT$.*

Proof. Through Corollary 1, we can get $c(S) \leq \frac{5}{3}OPT$.   □

3   The General Cluster Routing Problem with Starting and Ending Vertices

The general cluster routing problem (GCRP) [19] can be described as follows: given a weighted undirected graph $G = (V, E)$ with edge cost $c(e)$ which satisfies the triangle inequality. The vertex set is partitioned into clusters $C_1, \ldots, C_m$. Inside each cluster C_i, $\overline{V}_i \subset \overline{V}$ and $\overline{E}_i \subset \overline{E}$ are required vertex subset and required edge subset, respectively. The goal is to find a minimum cost tour that visits the vertices in $\overline{V}$ exactly once and traverses edges in $\overline{E}$ at least once and each cluster is visited consecutively.

In this section, we consider a sub-problem that the starting and ending vertices are specified inside each cluster. We consider the two cases. In the first case,

---

Algorithm 2. GCRP algorithm with specified starting and ending vertices

---
Input:
1: A weighted undirected graph $G = (V, E)$ with cost $c(e)$.
2: A partition of V into clusters $C_1, ..., C_m$.
3: $\overline{V} \subseteq V$ and $\overline{E} \subseteq E$ are required vertices subset and required edges subset respectively.
4: A starting vertex s_i and an ending vertex t_i in each cluster C_i, $i = 1, ..., m$.
Output: A GCRP tour.
begin
1: **For** $i = 1$ to m **do**
2:     Apply Algorithm 1 to obtain a path p_i from s_i to t_i;
3:     Replace p_i by a directed edge (s_i, t_i), whose weight equals to the length of p_i;
   EndFor
4: Construct a bipartite graph with respect to two vertex sets $\{t_i | i = 1, \ldots, m\}$ and
   $\{s_i | i = 1, \ldots, m\}$, and then compute a minimum-weight perfect matching M therein;
5: **Return** the tour T_s composed by the set of path .$\{p_i | i = 1, \ldots, m\}$ and M.
end

---

all required edges are only distributed inside the clusters. In the other case, there exist some required edges connecting different clusters.

For each cluster C_i, we assume that starting vertex s_i and ending vertex t_i are pre-specified. Similar to the GTPP problem introduced before, we consider the problem in a new graph $G^* = \cup C_i^*$, in which

$$C_i^* = (V_i^*, E_i^*) = \left(\{v | v \in e \in \overline{E_i}\} \cup \{s_i\} \cup \{t_i\} \cup \overline{V_i}, \overline{E_i} \right).$$

We decompose the problem into two sub-problems based on the following lemma:

Lemma 1. *For any instance GCRP with specified starting and ending vertices, its optimal solution could be decomposed into two disjoint parts: a set of paths P where $p_i \in P$ connects s_i and t_i in C_i, and a perfect matching in the bipartite graph of $G[s_i | i = 1, \ldots, m; t_i | i = 1, \ldots, m]$.*

Therefore, for the first, we obtain a path p_i from s_i to t_i that visits all vertices of $\overline{V_i}$ and edges of C_i^* inside each cluster C_i^* through Algorithm 1; for the second, outside each cluster, we connect the paths, i.e. t_i to s_i by computing a minimum weight matching in the bipartite. The details are formally illustrated in Algorithm 2.

Let L be the sum of lengths of the paths of OPT through each cluster. Let A be the sum of lengths of these edges connecting different clusters. Let D be the total length of these directed edges (s_i, t_i) $(i = 1, ..., m)$. So $OPT = L + A$. Therefore, we can get the theorem:

Theorem 3. *Let T_s be the tour generated by Algorithm 2, then*

$$c(T_s) \leq \frac{5}{3} OPT.$$

Algorithm 3. GCRP algorithm without specified starting and ending vertices

---

Input:
1: A weighted undirected graph $G = (V, E)$ with cost $c(e)$;
2: A partition of V into clusters $C_1, ..., C_m$;
3: $\overline{V} \subseteq V$ and $\overline{E} \subseteq E$ are required vertices subset and required edges subset respectively.
Output: A GCRP tour.
begin
1: **For** $i = 1$ to m **do**
2:     Select every pair of vertices of $\overline{V}_i$, say s_i, t_i, and calculate a path p_i from s_i to t_i by Algorithm 1;
3:     **For** $j = 1$ to m and $j \neq i$ **do**
4:         Select one edge (t_i, v_j) whose one end v_j is in C_j^*;
5:         Shrink p_i and (t_i, v_j) as one new directed edge (s_i, v_j), the direction is $\overrightarrow{(s_i, v_j)}$ and the cost equals to the sum of the length of path p_i and the edge cost of (s_i, v_j);
6:     **EndFor**
7: **EndFor**
8: Compose a directed complete graph $\mathcal{G}' = \left(\mathcal{V}', \mathcal{E}'\right)$ by these directed edges and calculate a directed spanning tree $\mathcal{T}$ of $\mathcal{G}'$;
9: Add a matching $\mathcal{M}$ to $\mathcal{T}$ such that all vertices in $\mathcal{T}$ to compose a tour;
10: In $\mathcal{T} \cup \mathcal{M}$, compute a Hamilton tour T_H;
11: Replace the directed edge (s_i, v_j) by p_i and (t_i, v_j);
12: **Return** the resulting tour T_s.
end

From the above analysis, we can get that if we apply Zenklusen's algorithm [18] to design our algorithm, the ratio of Algorithm 2 is theoretically $\frac{3}{2}$.

Corollary 3. *The general cluster routing problem with specified starting and ending vertices admits an approximation algorithm with a ratio $\frac{3}{2}$ via employing Zenklusen's algorithm.*

4   The General Cluster Routing Problem Without Starting and Ending Vertices

In this section, we consider the problem in which the starting and ending vertices are not specified inside each cluster C_i. Similarly, we construct a reduced graph $G^* = \cup C_i^*$ which is defined as before:

$$C_i^* = (V_i^*, E_i^*) = \left(\{v | v \in e \in \overline{E_i}\} \cup \overline{V_i}, \overline{E_i}\right).$$

Our algorithm is based on the following idea. In the first step, we elaborate every pair of ending vertices s_i, t_i within each cluster C_i^*, then we obtain paths by Algorithm 1. Next, we consider a path p_i in C_i^* and with every edge (t_i, v_j) incident with cluster C_j^* ($i \neq j$) as one new directed edge (s_i, v_j). We use theses

Table 2. Computational results under the instance with 48 vertices.

Group	|V|	Cluster number	$\|\overline{V}\|$	$\|\overline{E}\|$	Algorithm 2		LP		Cost	Time
					Cost	Time(s)	Cost	Time(s)	Ratio	Ratio
1	48	5	9	11	97482.45	0.67	93027.89	4.53	1.048	0.148
2	48	5	12	13	98613.87	0.55	95564.59	0.96	1.032	0.573
3	48	5	10	12	98410.27	0.38	84727.92	1.12	1.161	0.339
4	48	5	8	13	94633.53	0.39	80233.08	1.23	1.179	0.317
5	48	5	6	12	90947.88	0.42	87196.5	1.1	1.043	0.382
6	48	5	9	13	88331.6	0.42	81290.63	0.85	1.087	0.494
7	48	5	11	12	94292.55	0.37	84798.5	1.06	1.112	0.349
8	48	5	11	11	81672.59	0.39	80759.96	1.31	1.011	0.298
9	48	5	9	9	75793.52	0.46	69975.44	1.02	1.083	0.451
10	48	5	9	7	67445.19	0.45	65741.32	0.85	1.026	0.529
11	48	5	8	15	113537.75	0.37	100019.07	0.87	1.135	0.425
12	48	5	12	13	82943.21	0.37	77554.56	1.33	1.069	0.278
13	48	5	10	12	94779.78	0.36	83709.98	0.75	1.132	0.480
14	48	5	6	10	83097.9	0.41	80344.86	0.8	1.034	0.513
15	48	5	10	13	117809.44	0.4	110574.88	1.25	1.065	0.320
16	48	5	9	7	64656.86	0.36	60178.3	0.9	1.074	0.400
17	48	5	9	11	94160.5	0.37	86855.93	0.83	1.084	0.446
18	48	5	12	8	85461.37	0.41	79519.73	1	1.075	0.410
19	48	5	11	13	106834.67	0.4	102115.09	1.04	1.046	0.385
20	48	5	7	8	87996.28	0.36	84689.21	0.87	1.039	0.414
Average						0.42		1.18	1.077	0.398

directed edges to compose an auxiliary graph. Then we calculate a directed spanning tree in the graph and a minimum weight matching. At last, we find a Hamilton tour from the graph that consists of the tree and the matching. The details are summarized in Algorithm 3.

We will analyze the approximation ratio of Algorithm 3. Similar to the previous analysis. Let L denote the sum of the lengths of path p_i within all clusters in OPT, and A denote the sum of the lengths of the remaining edges of OPT. We can get the following theorem:

Theorem 4. *Let T_s be the tour generated by Algorithm 3, then*

$$c(T_s) \leq \frac{8}{3}OPT.$$

We assume that the approximation ratio of path TSP is represented by ρ. Then, when $\rho = \frac{5}{3}$, the ratio of our algorithm is $\frac{8}{3}$, and when $\rho = \frac{3}{2}$, i.e. when we apply Zenklusen's algorithm [18] for path TSP, the ratio of Algorithm 3 is $\frac{5}{2}$. Because of the length limitation of the paper, we omit the detailed proof details.

5   Numerical Experiments

To evaluate our algorithms, we select some TSPLIB datasets, which are symmetric TSPs in the Euclidean distance. In the targeted problem, the vertex set is partitioned into clusters in each of which there exist some required vertices and required edges. In the experiments, we use the solution of integer linear programming as the optimal solution, and employ them to evaluate the running time and the solution quality of our algorithms.

In the experiment of Algorithm 2, we use the instance with 48 vertices and 5 clusters, we run our algorithm and integer programming in 20 different inputs for comparison. The results are shown in Table 2. Column $|\overline{V}|$ and Column $|\overline{E}|$ denote the number of required vertices and required edges, respectively. Table 2 demonstrates that Algorithm 2 is significantly faster than integer programming, while the quality of the solution is about 1.077 times than that of integer programming. Through the previous theoretical analysis, we know that the approximation ratio of Algorithm 2 is $\frac{5}{3}$, which means that the approximate solution is at most $\frac{5}{3}OPT$. We can see that the experimental results are much better in comparison to the theoretical performance guarantee. The reason is that, the theoretical performance guarantee is for the worst case while our experiments are for the average. In the end, the experiments also demonstrate that the solution to our algorithm is better than Zhang's algorithm [19].

6   Conclusion

In this paper, we mainly focused on the general cluster routing problem. We considered two sub-problems and presented an approximation algorithm for each of them. For the general cluster routing problem with starting and ending vertices, we present a $\frac{5}{3}$-approximation algorithm; while for the problem without ending vertices, we give a $\frac{8}{3}$-approximation algorithm. It is worth noting that our algorithms are combinatorial algorithms, and have performance gains compared to the state-of-art algorithms.

Acknowledgements. The first and the second author are supported by Natural Science Foundation of China (No. 61772005) and Outstanding Youth Innovation Team Project for Universities of Shandong Province (No. 2020KJN008). The third author is supported by Natural Science Foundation of Fujian Province (No. 2020J01845) and Educational Research Project for Young and Middle-aged Teachers of Fujian Provincial Department of Education (No. JAT190613).

References

1. Ahmed, H.: Graph routing problem using Euler's theorem and its applications (2019)
2. An, H.C., Kleinberg, R., Shmoys, D.B.: Improving Christofides' algorithm for the ST path TSP. J. ACM (JACM) **62**(5), 1–28 (2015)

3. Arkin, E.M., Hassin, R., Klein, L.: Restricted delivery problems on a network. Netw. Int. J. **29**(4), 205–216 (1997)
4. Christofides, N.: Worst-case analysis of a new heuristic for the travelling salesman problem. Carnegie-Mellon Univ Pittsburgh Pa Management Sciences Research Group, Technical report (1976)
5. Fischetti, M., Toth, P., Vigo, D.: A branch-and-bound algorithm for the capacitated vehicle routing problem on directed graphs. Oper. Res. **42**(5), 846–859 (1994)
6. Gutin, G., Punnen, A.P.: The Traveling Salesman Problem and its Variations. COOP, vol. 12. Springer, Boston (2006). https://doi.org/10.1007/b101971
7. Guttmann-Beck, N., Hassin, R., Khuller, S., Raghavachari, B.: Approximation algorithms with bounded performance guarantees for the clustered traveling salesman problem. Algorithmica **28**(4), 422–437 (2000)
8. Hoogeveen, J.: Analysis of Christofides' heuristic: Some paths are more difficult than cycles. Oper. Res. Lett. **10**(5), 291–295 (1991)
9. Jansen, K.: An approximation algorithm for the general routing problem. Inf. Process. Lett. **41**(6), 333–339 (1992)
10. Lam, F., Newman, A.: Traveling salesman path problems. Math. Program. **113**(1), 39–59 (2008)
11. Lenstra, J.K., Kan, A.R.: Some simple applications of the travelling salesman problem. J. Oper. Res. Soc. **26**(4), 717–733 (1975)
12. Neumann, F., Witt, C.: Adelaide research and scholarship: Bioinspired computation in combinatorial optimization: algorithms and their computational complexity (2010)
13. Sebő, A.: Eight-fifth approximation for the path TSP. In: Goemans, M., Correa, J. (eds.) IPCO 2013. LNCS, vol. 7801, pp. 362–374. Springer, Heidelberg (2013). https://doi.org/10.1007/978-3-642-36694-9_31
14. Sebő, A., Vygen, J.: Shorter tours by nicer ears: 7/5-approximation for the graph-tsp, 3/2 for the path version, and 4/3 for two-edge-connected subgraphs. Combinatorica **34**(5), 597–629 (2014)
15. Traub, V., Vygen, J.: Approaching 3/2 for the s-t-path TSP. J. ACM (JACM) **66**(2), 1–17 (2019)
16. Traub, V., Vygen, J.: Beating the integrality ratio for s-t-tours in graphs. SIAM J. Comput. (0), FOCS18-37 (2020)
17. Traub, V., Vygen, J., Zenklusen, R.: Reducing path TSP to TSP. In: Proceedings of the 52nd Annual ACM SIGACT Symposium on Theory of Computing, pp. 14–27 (2020)
18. Zenklusen, R.: A 1.5-approximation for path TSP. In: Proceedings of the Thirtieth Annual ACM-SIAM Symposium on Discrete Algorithms, pp. 1539–1549. SIAM (2019)
19. Zhang, X., Du, D., Gutin, G., Ming, Q., Sun, J.: Approximation algorithms for general cluster routing problem. In: Kim, D., Uma, R.N., Cai, Z., Lee, D.H. (eds.) COCOON 2020. LNCS, vol. 12273, pp. 472–483. Springer, Cham (2020). https://doi.org/10.1007/978-3-030-58150-3_38

Maximizing Group Coverage
in Social Networks

Yuting Zhong[1], Longkun Guo[1,2], and Peihuang Huang[3(✉)]

[1] College of Mathematics and Computer Science, Fuzhou University, Fuzhou 350116, People's Republic of China
yuting.zhong_zyt@foxmail.com, longkun.guo@gmail.com
[2] School of Computer Science, Qilu University of Technology
(Shandong Academy of Sciences), Jinan 250353, People's Republic of China
[3] College of Mathematics and Data Science, Minjiang University,
Fuzhou 350108, China
peihuang.huang@foxmail.com

Abstract. Groups play a crucial role in decision-making of social networks, since individual decision-making is often influenced by groups. This brings the Group Influence Maximization (GIM) problem which aims to maximize the expected number of activated groups by finding k seed nodes. The GIM problem has been proved NP-hard while computing the objective function is $\#P$-hard under Independent Cascade (IC) model. We propose an algorithm called Maximizing Group Coverage (MGC) which greedily selects the best node based on evaluating the contribution of nodes to the groups, ensuring the success of approximating the maximization of the number of activated groups. Finally, we compare the MGC algorithm with the baseline algorithm called Maximum Coverage (MC) through experiments, demonstrating that MGC outperforms MC under IC model regarding the average number of activated groups.

Keywords: Maximizing group coverage · Group influence maximization · Independent cascade

1 Introduction

With the rapid development of big data and communication technology, we have experienced 2G text age, 3G image age, 4G video age, and ushered in 5G age. The continuous development of Internet technology has brought convenience to communication between users in social networks. Social networks play an important part in the communication between people and the dissemination of ideas. Because social networks matter in the real world, they have many applications such as information dissemination, viral marketing, communicable disease control, etc.

People have sociality, thus forming groups according to their common interests or certain connections. For instance, a wechat user can be a member of the

© Springer Nature Switzerland AG 2021
Y. Zhang et al. (Eds.): PDCAT 2020, LNCS 12606, pp. 274–284, 2021.
https://doi.org/10.1007/978-3-030-69244-5_24

swimming group, family group, class group, work group and so on simultane-ously. As the saying goes, the minority is subordinate to the majority in the group, indicating that the decision-making of the individual will be often influ-enced by groups. Take a practical example, a class will purchase uniform exercise books, when most students therein reach an agreement about which kind of exer-cise book they will buy. There are many applications of group influence in real life, bringing out more research on the maximization of group influence.

1.1 Related Works

The Group Influence Maximization (GIM) problem and the Influence Maximiza-tion (IM) problem are closely related. Lots of researchers proposed heuristic algo-rithms and approximation algorithms to solve IM, which aims to seek k initial active nodes to maximize the expected number of activated nodes [1–6]. Kempe et al. [7] proved that the IM problem is NP-hard and proposed the greedy algo-rithm, which achieved $\left(1 - \frac{1}{e} - \varepsilon\right)$ approximation solution for any $\varepsilon > 0$. Never-theless, the greedy algorithm takes expensive computation time. Subsequently, taking advantage of the submodule of the objective function, plenty of scholars improved the greedy algorithms [8,9]. Although the greedy algorithms are better than heuristic algorithms regarding approximation ratio, the speed of heuristic algorithms is significantly faster than that of the greedy algorithms, so heuristic algorithms are more favored by many scholars. Recently, great breakthroughs in IM have been made by improving Reverse Influence Sampling algorithm [10], such as TIM, TIM$^+$, IMM algorithms proposed by Tang et al. [11,12], SSA and D-SSA raised by Nguyen et al. [13], etc.

Some scholars utilized the features of the community to maximize influence, due to that the nodes within communities have tight connection while the nodes among communities are not closely connected [14–18]. However, the task of GIM is to activate most groups not individuals, so the algorithms based on the features of the community are not entirely suitable for GIM.

Besides, the algorithms making use of the submodule of the objective function are not adapted to GIM of which the objective function is neither submodular nor supermodular [19]. For which the function is not submodular, researchers have put forward many solutions [20–22]. With regard to research on GIM, Zhu et al. presented that the GIM problem is NP-hard and computing the objective function is $\#P$-hard under IC model, then proposed a sandwich approxima-tion framework to obtain seed nodes. With more emphasis on groups in social networks, scholars are committed to studying efficient algorithms to solve GIM.

1.2 Contributions

In this paper, we propose the Maximizing Group Coverage (MGC) algorithm to solve GIM, focusing on the contribution of nodes to the groups. Then compared to the Maximum Coverage (MC) algorithm through experiments, our MGC algo-rithm is demonstrated to have better performance than MC regarding the aver-age number of activated groups under Independent Cascade (IC) model.

Table 1. Notations and Definitions

Notation	Definition
$G = (V, E, P, U)$	G is a graph,
	V is a set of nodes, E is a set of edges,
	P is a set of probabilities of edges,
	and U is a set of groups
p	The probability that a node
	activates its neighbor, $0 \leq p \leq 1$
$n = |V|$	The number of nodes in the graph
$m = |E|$	The number of edges in the graph
$l = |U|$	The number of groups in the graph
k	The number of initial seed nodes
β	The activation threshold of groups, $0\% < \beta\% \leq 100\%$
$\rho(S)$	The expected number of groups
	activated by seed nodes

1.3 Organizatioin

The context of the article is as follows: Sect. 2 introduces the MGC algorithm; Sect. 3 presents the application of MGC in social networks; Sect. 4 evaluates the MGC algorithm and the MC algorithm under IC model through experiments. Section 5 summarizes the paper. Table 1 is the notations and their definitions frequently used in this paper.

2 Maximizing Group Coverage Algorithm

We study the Group Influence Maximization (GIM) problem under Independent Cascade (IC) model. An instance of GIM is described as $G = (V, E, P, U)$. G is a graph that can be directed or undirected. V is the set of all nodes in G, and assume that there are n nodes in the graph, then $V = \{v_1, v_2, ..., v_n\}$. E is the set of all edges in G, and assume that there are m edges in the graph, then $E = \{e_1, e_2, ..., e_m\}$. P is the set of all probabilities of all edges in G, then $P = \{p_1, p_2, ..., p_m\}$, and $\forall p_i \in [0,1], 1 \leq i \leq m$, i is a positive integer. In the digraphs, for a directed edge (u, v), u is the source node and v is the target node. While in the undirected graphs, if there is an edge between two nodes, then the two nodes are neighbors to each other. U is the set of all groups in G, and assume that there are l groups in the graph, then $U = \{u_1, u_2, ..., u_l\}$, and u_j is a subset of V, $1 \leq j \leq l$, j is a positive integer.

The GIM problem aims to maximize the expected number of activated groups by seeking k nodes, and the group will be activated under the condition that $\beta\%$ members of the group are activated. The mathematical description of GIM is:

Algorithm 1. The Maximizing Group Coverage Algorithm

Input: An instance of GIM $G = (V, E, P, U)$, the number of seed nodes k, the activation threshold of groups β.

Output: the seed set S.

1: Set $S := \Phi$;

2: Calculate the f_c of each node following Equality 1.

3: Sort the nodes according to the computed f_c.

4: Add k nodes with maximum f_c into S.

5: **return** S.

$$\max \ \rho(S)$$
$$s.t. \ |S| \le k$$

where S is the initial seed set, and $\rho(S)$ is the expected number of activated groups. Observing that GIM is NP-hard, we propose a heuristic algorithm called Maximizing Group Coverage (MGC), which focuses on the contribution of nodes to groups.

The idea of MGC is to regard all nodes as seed nodes and sort the contribution of all seed nodes, then choose k seed nodes with the maximum contribution to groups. The contribution is determined by both group coverage and propagation influence within groups. The calculation of group coverage in digraphs and undirected graphs is the same, while the calculation of propagation influence within groups differs. The target nodes of the nodes influence the contribution in the digraphs, while the contribution is influenced by the neighbors of the nodes in the undirected graphs. Taking the digraph as an example, if a node and its target nodes belong to the same groups, it can make a great contribution to the groups due to that it helps to increase the chance of group activation. We use $f_c(v_i)$ to compute the contribution of v_i to groups covered by v_i. If the node does not cover any group, then $f_c = 0$. If the node covers at least one group, then f_c equals the sum of its contribution to groups. The value of f_c relates to both how many groups the node covers and whether or not the node and its target nodes belong to the same groups. We have $f_c(v_i)$

$$f_c(v_i) = \sum_j \frac{o_{v_i} + 1}{(1 - \beta\%) \times |u_j| + 1} \tag{1}$$

where j is the group number covered by v_i, u_j is the group covered by v_i, o_{v_i} is the number of target nodes or neighbors of v_i which also belong to u_j. In the numerator, we can not ignore the member v_i, so the numerator is $o_{v_i} + 1$. And o_{v_i} measures the diffusion influence of v_i in the group. $|u_j|$ is the total number of members of u_j, $(1 - \beta\%) \times |u_j|$ means that $(1 - \beta\%) \times |u_j|$ nodes in u_j can be deleted, and the smaller $(1 - \beta\%) \times |u_j|$ is, the more influence v_i has on u_j. Because the denominator can't be zero, we set the denominator to be $(1 - \beta\%) \times |u_j| + 1$. And the value of $(1 - \beta\%) \times |u_j| + 1$ must be positive integer, so we take the upper bound of the value.

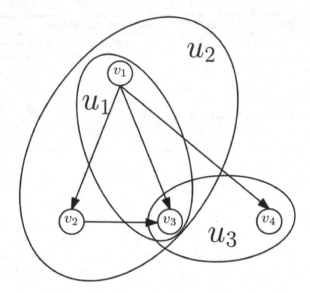

Fig. 1. An instance of MGC and MC

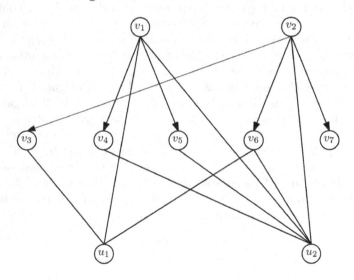

Fig. 2. A simple social network

The runtime of the MGC algorithm is $O(nl + n + m)$. MGC performs the following operations on the nodes numbered from 1 to n: traverse l groups and find out the groups covered by each node. The runtime of this step is $O(nl)$. Then compute the number of target nodes or neighbors belonging to the same groups with the current node, the runtime of the step is $O(n + m)$. Hence, the total runtime of MGC is $O(nl + n + m)$.

We take a simple example to explain MGC and compare it with Maximum Coverage (MC) algorithm which aims to select k nodes with maximum group coverage. As can be seen in Fig. 1, there are 4 nodes, 4 directed edges, 3 groups, and assume that the probability of each edge is 1. Then $V = \{v_1, v_2, v_3, v_4\}$, $E = \{(v_1, v_2), (v_1, v_3), (v_1, v_4), (v_2, v_3)\}$, $U = \{u_1, u_2, u_3\}$ and $u_1 = \{v_1, v_3\}$, $u_2 = \{v_1, v_2, v_3\}$, $u_3 = \{v_3, v_4\}$. Set the group activation threshold $\beta = 50$, meaning that the group will be activated when at least half of members in the group are active. If one seed node is needed, MC will choose $\{v_3\}$ on account of covering maximum groups, thus $\rho(S) = 2$. As for MGC, the contribution of each node is computed, and the result is $\{f_c(v_1) = 2, f_c(v_2) \approx 0.67, f_c(v_3) \approx 1.33, f_c(v_4) = 0.5\}$. According to the computed contribution, $\{v_1\}$ is chosen and $\rho(S) = 3$. Obviously, although $\{v_3\}$ covers the most groups, it does not spread influence, making little contribution to group activation.

3   Application of MGC in Social Networks

In this section, we will discuss the application of the Maximizing Group Coverage (MGC) algorithm in social networks. The node represents the user in the social network, and users can form groups based on common interests or some kind of relationship. The edge represents influence between two nodes, which can be directed or undirected. In Fig. 2, the simple social network with 7 nodes, 5 directed edges, 2 groups is given. Assume that the probability of each edge is 1. We can regard the simple social network as a two-layer model, where the upper layer is the nodes and their target nodes, and the lower layer is the groups and their members. Therefore, the upper layer in Fig. 2 is $layer_1 = \{v_1, v_2\}$ where v_1 influences $\{v_4, v_5\}$, v_2 influences $\{v_3, v_6, v_7\}$, the lower layer is $layer_2 = \{u_1, u_2\}$, where $u_1 = \{v_1, v_3, v_6\}$, $u_2 = \{v_1, v_2, v_4, v_5, v_6\}$. Obviously, $|u_1| = 3$ and $|u_2| = 5$.

Firstly, MGC needs to regard all nodes as seed nodes and compute the contribution of all nodes, assume that $\beta = 50$, then

$$f_c(v_1) = \left[\frac{0+1}{(1-50\%)\times 3+1} + \frac{2+1}{(1-50\%)\times 5+1}\right] \approx 1.08,$$

$$f_c(v_2) = \frac{1+1}{(1-50\%)\times 5+1} = 0.5,$$

$$f_c(v_3) = \frac{0+1}{(1-50\%)\times 3+1} \approx 0.33,$$

$$f_c(v_4) = \frac{0+1}{(1-50\%)\times 5+1} = 0.25, \tag{2}$$

$$f_c(v_5) = \frac{0+1}{(1-50\%)\times 5+1} = 0.25,$$

$$f_c(v_6) = \left[\frac{0+1}{(1-50\%)\times 3+1} + \frac{0+1}{(1-50\%)\times 5+1}\right] \approx 0.58,$$

$$f_c(v_7) = 0.$$

Table 2. Datasets information

	Type	Nodes	Edges	Groups	Average group size
Dataset1	Directed	6301	20777	234	33.84
Dataset2	Undirected	28281	92752	314	75.15

Secondly, sort the nodes according to the computed contribution, the result is $\{v_1, v_6, v_2, v_3, v_4, v_5, v_7\}$. If two seed nodes are needed, we will choose $\{v_1, v_6\}$ as seed nodes.

MGC obtains seed nodes according to the contribution depending on group coverage and the target nodes or neighbors of the nodes. In other words, in the two-layer model, the nodes not only cover maximum groups in the lower layer but also influence most target nodes or neighbors which also belong to the same groups in the upper layer may be potential seed nodes.

4   Numerical Experiments

Two data sets are used in our experiments, and Table 2 is about the information of them. Dataset1, 9 snapshots of the Gnutella peer-to-peer file sharing network, was collected in August 2002 from SNAP. Nodes represent hosts and edges represent connections between them. Dataset2 is a social network of European Deezer users, which was collected in March 2020. Nodes represent users and links are mutual follower relationships among users [23]. Besides, groups were randomly generated, each node can cover at least one group or be independent.

Because the value of β influences the number of activated groups, we performed different experiments by varying the value of β. In theory, the smaller the β is, the more groups can be activated, while the larger the β is, the fewer groups can be activated. Therefore, we set the value of the β from 10 to 25 at an interval of 5 for Dataset1, set the value of the β to 5, 8 for Dataset2 because it is more difficult to activate groups with bigger average group size than Dataset1. Meanwhile, in addition to the value of β, the size of the seed nodes k also affects the objective function. We also experimented by varying the size of k when β is fixed. Set the size of the k from 5 to 80 at an interval of 5. All programs were written in python 3.7.

Due to that the Maximizing Group Coverage (MGC) algorithm takes both static group coverage and dynamic propagation of nodes into consideration to choose seed nodes, MGC outperforms the Maximum Coverage (MC) in terms of the expected number of activated groups in theory. MGC and MC algorithms are tested under Independent Cascade (IC) model, experimental results are as follows, while the following conclusions can be drawn according to Fig. 3 and Fig. 4.

– With the value of k increases, the average number of activated groups increases. Owing to that with more seed nodes, more nodes are activated and the probability of group activation improves.

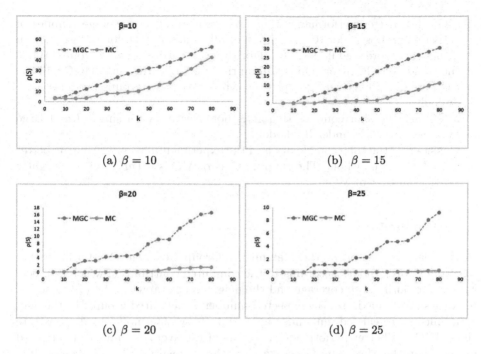

(a) $\beta = 10$ (b) $\beta = 15$

(c) $\beta = 20$ (d) $\beta = 25$

Fig. 3. Comparison of MGC and MC under IC model for Dataset1

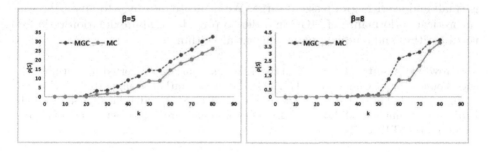

Fig. 4. Comparison of MGC and MC under IC model for Dataset2

- With the value of β grows, the average number of activated groups decreases. β represents the degree of difficulty to activate groups, if the value of β is greater, it is more difficult to activate groups successfully because more activated members of the groups are needed.
- The bigger average group size makes it more difficult to activate groups. That is, in a larger group, more members are required to be successfully influenced.

282 Y. Zhong et al.

- MGC has better performance than MC concerning the average number of activated groups under IC model. Especially when $\beta \geq 20$ for Dataset1, even if there are more seed nodes, MC algorithm can hardly activate groups. While the trend of MGC algorithm is still on the rise, indicating that MGC still has the potential to activate groups even when activating groups becomes more difficult.
- MGC has the shortcomings that it is not effective with small β and large average group size under IC model.
- Considering the time complexity and the experimental results, MGC is better than MC on the whole. The running time of MC is $O\,(nl)$, which is slightly lower than that of MGC.

5 Conclusion

In this paper, we propose the Maximizing Group Coverage (MGC) algorithm selecting k sorted nodes according to the computed contribution which is closely relevant to both group coverage and the neighbors or target nodes of the nodes, making sure to maximize the expected number of activated groups. In the end, experiments are carried out under independent cascade model to demonstrate that MGC has better performance in terms of the average number of activated groups than the Maximum Coverage algorithm known as a baseline algorithm. While the shortcomings of MGC are that it is not significantly effective when β is small and average group size is large. We dedicate ourselves to analyzing the theoretical performance of MGC in order to provide an approximation ratio for the algorithm and studying more efficient algorithms.

Acknowledgements. The first and the second author are supported by Natural Science Foundation of China (No. 61772005). The third author is supported by Natural Science Foundation of Fujian Province (No. 2020J01845) and Educational Research Project for Young and Middle-aged Teachers of Fujian Provincial Department of Education (No. JAT190613).

References

1. Pedro, D., Matt, R.: Mining the network value of customers. In: Proceedings of the Seventh ACM SIGKDD International Conference on Knowledge Discovery and Data Mining, pp. 57–66 (2001)
2. Pablo, A.E., Pablo, V., Kazumi, S.: Selecting the most influential nodes in social networks. In: 2007 International Joint Conference on Neural Networks, pp. 2397–2402. IEEE (2007)

3. Wei, C., Yajun, W., Siyu, Y.: Efficient influence maximization in social networks. In: Proceedings of the 15th ACM SIGKDD International Conference on Knowledge Discovery and Data Mining, pp. 199–208 (2009)
4. Wei, C., Yifei, Y., Li, Z.: Scalable influence maximization in social networks under the linear threshold model. In 2010 IEEE International Conference on Data Mining, pp. 88–97. IEEE (2010)
5. Amit, G., Wei, L., Laks, V.S.L.: Simpath: an efficient algorithm for influence maximization under the linear threshold model. In: 2011 IEEE 11th International Conference on Data Mining, pp. 211–220. IEEE (2011)
6. Guo-Liang, L., Ya-Ping, C., Jian-Hua, F., Yao-Qiang, X.: Influence maximization on multiple social networks. Chinese J. Comput. 39(4), 643–656 (2016)
7. David, K., Jon, K., Éva, T.: Maximizing the spread of influence through a social network. In: Proceedings of the Ninth ACM SIGKDD International Conference on Knowledge Discovery and Data Mining, pp. 137–146 (2003)
8. Jure, L., Andreas, K., Carlos, G., Christos, F., Jeanne, V., Natalie, G.: Cost-effective outbreak detection in networks. In Proceedings of the 13th ACM SIGKDD International Conference on Knowledge Discovery and Data Mining, pp. 420–429 (2007)
9. Amit, G., Wei, L., Laks, V.S.L.: Celf++ optimizing the greedy algorithm for influence maximization in social networks. In Proceedings of the 20th International Conference Companion on World Wide Web, pp. 47–48 (2011)
10. Christian, B., Michael, B., Jennifer, C., Brendan, L.: Maximizing social influence in nearly optimal time. In: Proceedings of the Twenty-Fifth Annual ACM-SIAM Symposium on Discrete Algorithms, pp. 946–957. SIAM (2014)
11. Youze, T., Xiaokui, X., Yanchen, S.: Influence maximization: near-optimal time complexity meets practical efficiency. In: Proceedings of the 2014 ACM SIGMOD International Conference on Management of Data, pp. 75–86 (2014)
12. Youze, T., Yanchen, S., Xiaokui, X.: Influence maximization in near-linear time: a martingale approach. In: Proceedings of the 2015 ACM SIGMOD International Conference on Management of Data, pp. 1539–1554 (2015)
13. Hung, T.N., My, T.T., Thang, N.D.: Stop-and-stare: optimal sampling algorithms for viral marketing in billion-scale networks. In: Proceedings of the 2016 International Conference on Management of Data, pp. 695–710 (2016)
14. Tianyu, C., Xindong, W., Song, W., Xiaohua, H.: Oasnet: an optimal allocation approach to influence maximization in modular social networks. In: Proceedings of the 2010 ACM Symposium on Applied Computing, pp. 1088–1094 (2010)
15. Yu, W., Gao, C., Guojie, S., Kunqing, X.: Community-based greedy algorithm for mining top-k influential nodes in mobile social networks. In: Proceedings of the 16th ACM SIGKDD International Conference on Knowledge Discovery and Data Mining, pp. 1039–1048 (2010)
16. Jin-chao, J., Lan, H., Zhe, W., Hong-ming, L., San-yi, L.: A new approach to maximizing the spread of influence based on community structure. J. Jilin University (Science Edition), 1, 23 (2011)
17. Huang, H., Shen, H., Meng, Z., Chang, H., He, H.: Community-based influence maximization for viral marketing. Appl. Intell. 49(6), 2137–2150 (2019). https://doi.org/10.1007/s10489-018-1387-8
18. Bozorgi, A., Samet, S., Kwisthout, J., Wareham, T.: Community-based influence maximization in social networks under a competitive linear threshold model. Knowledge-Based Syst. 134, 149–158 (2017)
19. Zhu, J., Ghosh, S., Weili, W.: Group influence maximization problem in social networks. IEEE Trans. Comput. Soc. Syst. 6(6), 1156–1164 (2019)

20. Schoenebeck, G., Tao, B.: Beyond worst-case (in) approximability of nonsubmodular influence maximization. ACM Trans. Comput. Theory (TOCT) **11**(3), 1–56 (2019)
21. Francis, B., et al.: Learning with submodular functions: a convex optimization perspective. Found. Trends® Machine Learn. **6**(2–3), 145–373 (2013)
22. Wei, L., Wei, C., Laks, V.S.L.: From competition to complementarity: comparative influence diffusion and maximization. Proc. VLDB Endowment, **9**(2), 60–71 (2015)
23. Benedek, R., Rik, S.: Characteristic functions on graphs: birds of a feather, from statistical descriptors to parametric models. In: The 29th ACM Conference on Information and Knowledge Management, abs/2005.07959 (2020)

LightLayers: Parameter Efficient Dense and Convolutional Layers for Image Classification

Debesh Jha[1,2(✉)], Anis Yazidi[3], Michael A. Riegler[1], Dag Johansen[2],
Håvard D. Johansen[2], and Pål Halvorsen[1,3]

[1] SimulaMet, Oslo, Norway
debesh@simula.no
[2] UIT The Arctic University of Norway, Tromsø, Norway
[3] Oslo Metropolitan University, Oslo, Norway

Abstract. Deep Neural Networks (DNNs) have become the de-facto standard in computer vision, as well as in many other pattern recognition tasks. A key drawback of DNNs is that the training phase can be very computationally expensive. Organizations or individuals that cannot afford purchasing state-of-the-art hardware or tapping into cloud hosted infrastructures may face a long waiting time before the training completes or might not be able to train a model at all. Investigating novel ways to reduce the training time could be a potential solution to alleviate this drawback, and thus enabling more rapid development of new algorithms and models. In this paper, we propose LightLayers, a method for reducing the number of trainable parameters in DNNs. The proposed LightLayers consists of LightDense and LightConv2D layers that are as efficient as regular Conv2D and Dense layers but uses less parameters. We resort to Matrix Factorization to reduce the complexity of the DNN models resulting in lightweight DNN models that require less computational power, without much loss in the accuracy. We have tested Light-Layers on MNIST, Fashion MNIST, CIFAR 10, and CIFAR 100 datasets. Promising results are obtained for MNIST, Fashion MNIST, and CIFAR-10 datasets whereas CIFAR 100 shows acceptable performance by using fewer parameters.

Keywords: Deep learning · Lightweight model · Convolutional neural network · MNIST · Fashion MNIST · CIFAR-10 · CIFAR 100 · Weight decomposition

1 Introduction

Deep learning (DL) techniques have revolutionized the field of Machine Learning (ML) and gained immense research attention during the last decade. Deep neural networks provide state-of-the-art solution in several domains such as image recognition, speech recognition, and text processing [20]. One of the most popular techniques within deep learning is Convolutional Neural Network (CNN), which possesses a structure that is well-suitable specially for image and video processing. A CNN [16] comprises a convolution layer and dense layer. CNN has

© Springer Nature Switzerland AG 2021
Y. Zhang et al. (Eds.): PDCAT 2020, LNCS 12606, pp. 285–296, 2021.
https://doi.org/10.1007/978-3-030-69244-5_25

emerged as powerful techniques for solving many classification [14] and regression [12] tasks. Additionally, CNN has produced promising results in various applications areas, including in the medical domain, with applicability in diabetic retinopathy prediction [3], endoscopic disease detection [23], and breast cancer detection [19].

Recently, developing deeper and larger architectures has been a common trend in the development of state-of-the-art methods [4]. Most of the time, we can observe that deeper networks especially with large and complex datasets lead to better performance. One of the major drawbacks of CNNs are that they often require an immense amount of training time compared to other classical ML algorithms. Hyperparameter optimization for fine-tuning the model is another challenging task that increases dramatically the overall training time to achieve optimum results from any model. CNN models often require powerful Graphical Processing Units (GPUs) for training, which can span over days, weeks, and even months, with no guarantee that the model will produce satisfactory results. A long training process also consumes a lot of energy and is not considered environmentally friendly. Furthermore, long training is demanding in terms of resources as a large amount of memory is required which renders it difficult to deploy on low-power devices [11]. The requirements for the expensive hardware and high training time complicate the use of models with large number of trainable parameters to be deployed on portable devices or conventional desktops [20].

A potential way to address these issues is the introduction of lightweight models. A lightweight model can potentially be built by reducing the number of trainable parameters within the layers. In an effort towards reducing the training time and complexity of CNN models, we propose LightLayers, which is a combination of LightDense and LightConv2D layers, that focuses on CNNs and more particularly on creating both a lightweight convolutional layer and a lightweight dense layer that are both easy to train. Lightweight CNN models are computationally cheap and can be used in various applications for carrying out online estimation. Therefore, the main goal of this paper is to present a general model to reduce the number of parameters in a CNN model so that it can be used in various image processing or other applicable tasks in the future.

The main contributions of the paper are:

- LightLayers, a combination of LightConv2D and LightDense layers, is proposed. Both layers are based on matrix decomposition for reducing the number of trainable parameters of the layers.
- We have investigated and tested the proposed model with four different publicly available datasets: MNIST [16], Fashion MNIST [26], and CIFAR10 [13], CIFAR100 [13], and we have showed that the proposed method is competitive in terms of both accuracy and efficiency when the number of training parameters used are taken into consideration.
- We experimentally show that good accuracy can be achieved by using a relatively small number of trainable parameters with MNIST, Fashion MNIST,

and CIFAR 10 datasets. Moreover, we found there was a significant reduction in the number of trainable parameters as compared to Conv2D.

2   Related Work

In the context of reducing the cost of network model training, several approaches have been presented. For example, Xue et al. [27] presented a Deep Neural Network (DNN) technique for reducing the model size while maintaining the accuracy. For achieving this goal, they used singular value decomposition (SVD) on the weight matrix in DNN, and reconstructed the model based on inherent sparseness of the original matrices. The application of DNNs for mobile applications has become increasingly popular. The computational and storage limitation should be taken into account while deploying DNN on such devices.

To address this need, Li et al. [17] proposed two techniques for effectively learning from DNNs with a smaller number of hidden nodes and smaller number of senones set. The details about these techniques can be found in the literature [17]. Similarly, Xue et al. [28] introduced two SVD based techniques to solve the issue related to DNN personalization and adaptation. Garipov et al. [8] developed a tensor factorization framework for compressing fully connected layers. The focus of their work was to compress convolutional layers which would potentially excel in image recognition tasks by reducing the memory complexity and high computational cost. Later, Kim et al. [10] proposed an energy-efficient kernel decomposition architecture for binary-weight CNNs.

Ding et al. [7] proposed CIRCNN, an approach for representing the weights and processing neural networks using block-circulant matrices. CIRCNN utilizes Fast Fourier Transform based fast multiplication operation which simultaneously reduces the computational and storage complexity causing negligible loss in accuracy. Chai et al. [25] proposed a model for reducing the parameters in DNNs via product-of-sums matrix decomposition. They obtained good accuracy on the MNIST and Fashion MNIST datasets with a smaller number of trainable parameters. Another similar work is by Agrawal et al. [2], where they designed a lightweight deep learning model for human activity recognition that is sufficiently computationally efficient to be deployed on edge devices. For more recent works on matrix and tensor decomposition, we refer the reader to [6,15].

Kim et al. [11] proposed a method for compressing CNNs to be deployed as a mobile application. Mariet et al. [18] proposed another efficient neural network architecture that reduces the size of neural networks without hurting the overall performance. Novikov et al. [20] converted dense weight matrices of fully connected layers to Tensor Train [21] format such that the number of parameters are reduced by a huge factor by preserving the expressive power of the layer.

Lightweighted networks have gained attention in computer vision (for instance, in the area of real-time image segmentation [9,22,24,29]). Real-time applications are growing because the lightweight models can be an efficient

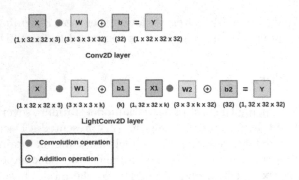

Fig. 1. Comparative diagram of Conv2D layer and LightConv2D layer.

solution for resource constraints and mobile devices. Only a lightweight model demands lower memory that leads to a lower computation and faster speed. Therefore, developing a lightweight model can be a good idea for achieving real-time solutions, and it can be used for other applications too.

The above studies show that there is great potential for lightweight networks for computer-vision tasks. With large amounts of training data, it is likely that a model with huge numbers of trainable parameters will outperform the smaller models—if one can afford the high training costs and resource demands at inference time. However, there is a need for models with low-cost computational power and small memory footprints [11], especially for mobile applications [11] and portable devices. In this respect, we propose LightLayers that is based on the concept of matrix decomposition. LightLayers uses fewer trainable parameters and shows the state-of-the-art tradeoff between parameter size and accuracy.

3   Methodology

In this section, we introduce the proposed layers. Figure 1 shows the comparison of a Conv2D and a LightConv2D layer. In the LightConv2D layer, we decompose the weight matrix W into $W1$ and $W2$ on the basis of hyperparameter k, which leads to a reduction of the total number of trainable parameters in the network. We follow the same strategy for the LightDense layer. The block diagram of the LightDense layer is shown in Fig. 2.

The main objective of building the model is to compare our LightLayers (i.e., the combination of LightConv2D and LightDense layers) with the conventional Conv2D and SeparableConv2D layers. For comparing the performance of the various layers, we have built a simple model from scratch. The block diagram of the proposed model is shown in Fig. 3. We used the same hyperparameters and setting for all the experiments. For the LightLayers experiments, we used LightConv2D and LightDense layers (see Fig. 3). For the other experiments, we replaced LightConv2D with Conv2D or SeparableConv2D and LightDense with a regular Dense layer.

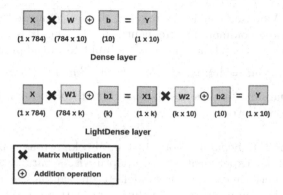

Fig. 2. Comparative diagram of Dense layer and LightDense layer.

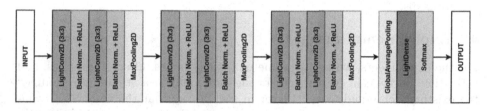

Fig. 3. Block diagram of the architecture used for comparison of the proposed Light-layers with regular convolution and dense layers. In the case of regular layers, we use regular convolution and dense layers instead of Lightlayers.

The model architecture used for experimentation (see Fig. 3) comprises two 3×3 convolution layers, each followed by a batch-normalization and ReLU non-linearity as the activation function. We have introduced 2×2 max-pooling, which reduces the spatial dimension of the feature map. We have used three similar blocks of layers in the model followed by the GlobalAveragePooling, LightDense layers with $k = 8$, and a softmax activation function for classifying the input image.

3.1   Description of Convolution Layers

Conv2D. A convolution layer is the most common layer used in any computer vision task and is applied extensively. This layer uses a multidimensional kernel as the weight, which is used to perform convolution operation on the input to produce an output. If the bias is used, then a $1D$ vector is added to the output. Finally, the activation is applied to introduce the non-linearity into the neural network. In this paper, we worked on a $2D$ convolution layer, which uses a $4D$ tensor as the weight.

$$Output = Activation((Input \otimes Weight) + Bias) \qquad (1)$$

In the above equation, $\otimes$ represents the convolution operation, and weight represents the kernel.

Dense Layer. A dense layer is the regular, deeply connected neural-network layer. It is the most common and frequently used layer. It is also known as a fully-connected layer as each neuron receives input from the previous layer.

$$Output = Activation((Input \oplus Weight) + Bias) \tag{2}$$

In the above equation, $\oplus$ represents the matrix multiplication instead of convolution operation as above.

Separable Conv2D. Separable convolution, also known as depth-wise convolution, is used in our experiment. We use depth-wise separable 2D convolution to compare the performance of our model. It first applies a depth-wise spatial convolution, i.e., performing a convolution operation on each input channel independently. After that, it is followed by a point-wise convolution, i.e., a 1×1 convolution. Pointwise, convolution controls the number of filters in the output feature maps.

4   Experimental Setup

For the experiments, we use the same number of layers, filters, filter sizes, and activation functions in every model for the individual dataset. We have modified the existing Dense and Conv2D layer in such a way that the number of trainable parameters decreases with some decrease in the accuracy of the model. In particular, we use three types of layers for this experiment, i.e., Conv2D, SeparableConv2D, and LightLayers. First, we run the model using Conv2D layers. The Conv2D layer is replaced by SeperableConv2D and run again. Again, we replace SeperableConv2D with the LightLayers and run the model.

In the modified layers, we introduced the hyperparameter k to control the number of trainable parameters in the LightDense and LightConv2D layer. In the LightDense layer, we set k to 8. In the LightConv2D layer, k varies between 1 to 6, and more could be set depending on the requirement. The values of the k are chosen empirically. We only replace the Conv2D layer with the LightConv2D layer and Dense layer with the LightDense layer of the proposed lightweight model. The rest of the network architecture remains the same.

4.1   Implementation Details

We have implemented the proposed layers using the Keras framework [5] and TensorFlow 2.2 [1] as backend. The implementation can be found at GitHub[1]. We performed all the experiments on an NVIDIA GEFORCE GTX 1080 system, which has 2560 NVIDIA CUDA Cores with 8 GB GDDR5X memory. The system was running on Ubuntu 18.04.3 LTS. We used a batch size of 64. All the experiments were run, keeping all the hyperparameters (i.e., learning rate, optimizer, batch size, number of filters, and filter size) the same. We have trained all the models for 20 epochs. After each convolution layer, batch normalization is used, which is activated by the Rectified linear unit (ReLU).

---

[1] https://github.com/DebeshJha/LightLayers.

4.2   Datasets

To evaluate LightConv2D layer and LightDense layer, we have performed experiments using various datasets.

MNIST Database. Modified National Institute of Standards and Technology (MNIST) [16] is the primary dataset for computer vision tasks introduced by LeCun et al. in 1998. MNIST comprises 10 classes of handwritten digits with 60,000 training and 10,000 testing images. The resolution of the images in the MNIST dataset is 28 × 28. There is a huge recent advancement in ML and DL algorithms. However, the MNIST remains a common choice for learners and beginners. The reason is that it is easy to deploy, test, and compare an algorithm on a publicly available dataset. The dataset can be downloaded from http:// yann.lecun.com/exdb/mnist/.

Fashion MNIST Database. Fashion MNIST [26] is a 10 class of 70,000 grayscale images of size 28 × 28. Xiao et al. released a novel image dataset that could be used for benchmarking ML algorithms. Their goal was to replace the MNIST database with a new database. The images of the Fashion MNIST database are more challenging as compared to the MNIST database. It contains natural images such as t-shirt/top, trouser, pullover, dress, coat, sandal, shirt, sneaker, bag, and ankle boot. The database can be downloaded from https:// github.com/zalandoresearch/fashion-mnist.

CIFAR-10 Database. CIFAR-10 [13] is a commonly established dataset for computer-vision tasks. It is especially used for object recognition tasks. CIFAR-10 contains 60,000 color images of size 32 × 32. It also has 10 classes of images. Each class contains 6,000 images per class. The classes contain datasets of cars, birds, cats, deer, dogs, horses, and trucks. The dataset can be downloaded from https://www.cs.toronto.edu/~kriz/cifar.html.

CIFAR-100 Database. CIFAR-100 [13] is also collected by the team of Alex Krizhevsky, Vinod Nair, and Geoffrey Hinton. This database is similar to the previous CIFAR-10 database. The 100 classes of the database consist of images such as beaver, dolphin, flatfish, roses, clock, computer keyboard, bee, forest, baby, pine, tank, etc. Each class of the database contains 600 images each. This dataset contains 500 training examples and 100 testing examples per class. The dataset can be found on the same webpage as CIFAR-10.

5   Results

In this section, we present and compare the experimental results of the Conv2D, SeperableConv2D, and LightLayers models on the MNIST, Fashion MNIST, CIFAR-10, and CIFAR 100 datasets. Table 1 shows the summary of result

Table 1. Results on **MNIST** test dataset (Number of epochs = 10, Batch size = 64, Learning rate = 1e-3, Number of filters = [8, 16, 32]).

Method	Parameters	Test Accuracy	Test Loss
Conv2D	18,818	0.9887	0.018
SeparableConv2D	3,611	0.9338	0.2433
LightLayers ($K = 1$)	2,649	0.9418	0.1327
LightLayers ($K = 2$)	4,392	0.9749	0.0554
LightLayers ($K = 3$)	**6,135**	**0.9775**	**0.0513**
LightLayers ($K = 4$)	7,878	0.9720	0.0704

Table 2. Results on **Fashion MNIST** test dataset (Number of epochs = 10, Batch size = 64, Learning rate = 1e-3, Number of filters = [8, 16, 32]).

Method	Parameters	Test accuracy	Test loss
Conv2D	18,818	0.9147	0.1468
SeparableConv2D	3,611	0.8725	0.3175
LightLayers ($K = 1$)	2,649	0.789	0.6752
LightLayers ($K = 2$)	4,392	0.8452	0.4247
LightLayers ($K = 3$)	6,135	0.8695	0.3708
LightLayers ($K = 4$)	7,878	0.8623	0.6184
LightLayers ($K = 5$)	**9,621**	**0.8820**	**0.2810**
LightLayers ($K = 6$)	11,364	0.8733	0.3986

comparison of Conv2D, SeperableConv2D, and LightLayers on MNIST dataset. Based on Conv2D and SeperableConv2D, we propose Layers and show improvement over both layers. The concept of LightLayers is based on weight matrix decomposition. This is the main motivation behind comparison of the proposed layers with Conv2D and SeperableConv2D.

The hyperparameters used are described in the caption of the Table 1. We can see that the result of the proposed LightLayers is comparable to that of Conv2D and SeperableConv2D in terms of test accuracy. When we compare the LightLayers with Conv2D, in terms of the number of parameters used, it uses only $\frac{1}{3}$ of parameters of Conv2D, which is more efficient with only 1% drop in terms of test accuracy. LightLayers with hyperparameter $k = 3$ achieves the highest test accuracy. However, for the other values of k as well there is only minimal variation in test accuracy.

Table 2 shows the results for different layers for the model trained on the Fashion MNIST dataset. From the table, we can observe that the proposed model (LightLayers) with hyperparameter $k = 5$ uses only half of the parameters with around 3% drop in terms of test accuracy with the Fashion MNIST dataset. However, when we compare the quantitative results with SeperableConv2D, our

Table 3. Evaluation results on test set of **CIFAR10** dataset (Number of epochs = 20, Batch size = 64, Learning rate = 1e-4, Number of filters = [8, 16, 32, 64]). The 'Params' in the bold represents total number of parameters.

Method	Parameters	Test accuracy	Test loss
Conv2D	76,794	0.6882	0.9701
SeparableConv2D	14,440	0.5953	1.3263
LightLayers ($K = 1$)	5,937	0.3686	1.6723
LightLayers ($K = 2$)	9,592	0.4596	1.5372
LightLayers ($K = 3$)	13,247	0.4937	1.5287
LightLayers ($K = 4$)	16,902	0.5319	1.3214
LightLayers ($K = 5$)	**20,557**	**0.5576**	**1.2122**

Table 4. Evaluation on **CIFAR100** test set (Number of epochs = 20, Batch size = 64, Learning rate = $1e - 4$, Number of filters = [8, 16, 32, 64]).

Method	Parameters	Test accuracy	Test loss
Conv2D	82,644	0.3262	2.6576
SeparableConv2D	20,290	0.2207	3.2108
LightLayers ($K = 1$)	6,747	0.0275	4.2391
LightLayers ($K - 2$)	10,402	0.0398	4.1836
LightLayers ($K = 3$)	14,057	0.0559	4.0304
LightLayers ($K = 4$)	17,712	0.0551	3.9978
LightLayers ($K = 5$)	**21,367**	**0.0589**	**4.0009**

proposed LightLayers achieves better test accuracy with the trade-off in number of trainable parameters.

Table 3 shows the results on the CIFAR 10 dataset. On this dataset as well, the proposed method is 3.75 times computationally efficient in terms of parameters it uses. However, there is a drop in accuracy of around 13%. Nevertheless, for some tasks the efficiency can be more important than the reduced accuracy.

Similarly, we have trained and tested the proposed model on the CIFAR 100 dataset, where the test accuracy of the proposed layers is much lower as compared to the Conv2D. This is obvious because CIFAR 100 consists of 100 classes of images that are difficult to generalize with such a small number of trainable parameters. However, the total number of parameters used is still around 4 times less than that of Conv2D. The total number of trainable parameters for Conv2D is 82,644, and for LightLayers, it is only 21,367. We refer to Table 4 for more details on the test accuracy and test loss.

From the experimental results, we are convinced that LightLayers has the following advantages:

- It requires less trainable parameters than Conv2D, which is an important factor to implement in different applications where heavy trainable parameters could not be beneficial.
- Due to less parameters, the space taken by the weight file is smaller, which makes it more suitable to devices where storage space is limited.

6   Ablation Study

Let us consider that the input size is 784, and the number of output features is 10. Therefore, the weight matrix W is 784×10 resulting in $7,840$ trainable parameters. Now, in the LightDense layer, we decompose the weight matrix W into two smaller matrix $W1$ and $W2$ of lower dimension using the hyperparameter k.

Here, $W1 = [784, k]$ and $W2 = [k, 10]$ values from the above example, the total number of trainable parameters in the LightDense layer becomes $786 \times k$ $+ k \times 10$. Now, if $k = 1$, then trainable parameters are 796, and if $k = 2$ the number of trainable parameters becomes $1,588$, and so on.

Next, consider the weight decomposition in the LightConv2D layer. If the input is $32 \times 32 \times 3$, the number of filters is 32, and the kernel size is 3×3, then the filters size becomes $3 \times 3 \times 3 \times 32$. This means that the total number of trainable parameters is 864. Now, we will decompose the kernel W into $W1$ and $W2$ using hyperparameter k. Here, $W1$ is $3 \times 3 \times 3 \times k$ and $W2$ is $3 \times 3 \times k \times 32$. If k is 1, then the total number of trainable parameters becomes $27 + 288$, which is equal to 315.

From the ablation study, we deduce that the number of trainable parameters used is less in LightLayers compared to the Conv2D and Dense layers. Overall, we can argue that the proposed LightLayers approach has the potential to be a powerful solution to solve the problem of excessive parameter used by traditional DL approaches. However, our LightLayers model needs further improvement for successfully implementing it on a larger dataset with high resolution images. We can conclude that further investigating matrix weight decomposition is important and other similar studies are necessary to reach the goal of lightweight models in the near future.

7   Conclusion

In this paper, we propose the LightLayers model, which uses matrix decomposition to help to reduce the complexity of the DLN. With the extensive experiments, we observed that changing the value of hyperparameter k yields a tradeoff between model complexity in terms of the number of trainable parameters and performance. We compare the accuracy of the LightLayers model with Conv2D. An extensive evaluation shows the tradeoffs in terms of parameter uses, accuracy, and computation. In the future, we want to train LightLayers on other publicly available datasets. We also aim to develop efficient techniques for finding the optimal value of k automatically. Further research will be required to find suitable algorithms and implementations that will scale this approach to a biomedical dataset.

References

1. Abadi, M., Barham, P., et al.: Tensorflow: a system for large-scale machine learning. In: Proceedings of the Symposium on Operating Systems Design and Implementation (OSDI), pp. 265–283 (2016)
2. Agarwal, P., Alam, M.: A lightweight deep learning model for human activity recognition on edge devices. arXiv preprint arXiv:1909.12917 (2019)
3. Arcadu, F., Benmansour, F., Maunz, A., Willis, J., Haskova, Z., Prunotto, M.: Deep learning algorithm predicts diabetic retinopathy progression in individual patients. NPJ Digit. Med. **2**(1), 1–9 (2019)
4. Brown, T.B., Mann, B., Ryder, N., Subbiah, M., Kaplan, J., Dhariwal, P., Neelakantan, A., Shyam, P., Sastry, G., Askell, A., et al.: Language models are few-shot learners. arXiv preprint arXiv:2005.14165 (2020)
5. Chollet, F., et al.: Keras (2015). https://keras.io
6. Denton, E.L., Zaremba, W., Bruna, J., LeCun, Y., Fergus, R.: Exploiting linear structure within convolutional networks for efficient evaluation. In: Advances in Neural Information Processing Systems, pp. 1269–1277 (2014)
7. Ding, C., et al.: Circnn: accelerating and compressing deep neural networks using block-circulant weight matrices. In: Proceedings of the IEEE/ACM International Symposium on Microarchitecture, pp. 395–408 (2017)
8. Garipov, T., Podoprikhin, D., Novikov, A., Vetrov, D.: Ultimate tensorization: compressing convolutional and fc layers alike. arXiv preprint arXiv:1611.03214 (2016)
9. Jiang, W., Xie, Z., Li, Y., Liu, C., Lu, H.: Lrnnet: A light-weighted network with efficient reduced non-local operation for real-time semantic segmentation. In: Proceedings of International Conference on Multimedia & Expo Workshops (ICMEW), pp. 1–6 (2020)
10. Kim, H., Sim, J., Choi, Y., Kim, L.S.: A kernel decomposition architecture for binary-weight convolutional neural networks. In: Proceedings of the Annual Design Automation Conference, pp. 1–6 (2017)
11. Kim, Y.D., Park, E., Yoo, S., Choi, T., Yang, L., Shin, D.: Compression of deep convolutional neural networks for fast and low power mobile applications. arXiv preprint arXiv:1511.06530 (2015)
12. Kleinbaum, D.G., Dietz, K., Gail, M., Klein, M., Klein, M.: Logistic regression (2002)
13. Krizhevsky, A., Hinton, G., et al.: Learning multiple layers of features from tiny images (2009)
14. Krizhevsky, A., Sutskever, I., Hinton, G.E.: Imagenet classification with deep convolutional neural networks. In: Proceedings of Advances in Neural Information Processing Systems, pp. 1097–1105 (2012)
15. Lebedev, V., Ganin, Y., Rakhuba, M., Oseledets, I., Lempitsky, V.: Speeding-up convolutional neural networks using fine-tuned cp-decomposition. arXiv preprint arXiv:1412.6553 (2014)
16. LeCun, Y., Bottou, L., Bengio, Y., Haffner, P.: Gradient-based learning applied to document recognition. Proc. IEEE **86**(11), 2278–2324 (1998)
17. Li, J., Zhao, R., Huang, J.T., Gong, Y.: Learning small-size dnn with output-distribution-based criteria. In: Proceedings of the Conference of the International Speech Communication Association (2014)
18. Mariet, Z., Sra, S.: Diversity networks: Neural network compression using determinantal point processes. arXiv preprint arXiv:1511.05077 (2015)

19. McKinney, S.M., et al.: International evaluation of an AI system for breast cancer screening. Nature **577**(7788), 89–94 (2020)
20. Novikov, A., Podoprikhin, D., Osokin, A., Vetrov, D.P.: Tensorizing neural networks. In: Advances in Neural Information Processing Systems, pp. 442–450 (2015)
21. Oseledets, I.V.: Tensor-train decomposition. SIAM J. Sci. Comput. **33**(5), 2295–2317 (2011)
22. Paszke, A., Chaurasia, A., Kim, S., Culurciello, E.: Enet: A deep neural network architecture for real-time semantic segmentation. arXiv preprint arXiv:1606.02147 (2016)
23. Thambawita, V., et al.: An extensive study on cross-dataset bias and evaluation metrics interpretation for machine learning applied to gastrointestinal tract abnormality classification. ACM Trans. Comput. Healthcare 1(3), 1–29 (2020)
24. Wang, Y., et al.: Lednet: a lightweight encoder-decoder network for real-time semantic segmentation. In: Proceedings of International Conference on Image Processing (ICIP), pp. 1860–1864 (2019)
25. Wu, C.W.: Prodsumnet: reducing model parameters in deep neural networks via product-of-sums matrix decompositions. arXiv preprint arXiv:1809.02209 (2018)
26. Xiao, H., Rasul, K., Vollgraf, R.: Fashion-mnist: a novel image dataset for benchmarking machine learning algorithms. arXiv preprint arXiv:1708.07747 (2017)
27. Xue, J., Li, J., Gong, Y.: Restructuring of deep neural network acoustic models with singular value decomposition. In: Interspeech, pp. 2365–2369 (2013)
28. Xue, J., Li, J., Yu, D., Seltzer, M., Gong, Y.: Singular value decomposition based low-footprint speaker adaptation and personalization for deep neural network. In: Proceedings of International Conference on Acoustics, Speech and Signal Processing (ICASSP), pp. 6359–6363 (2014)
29. Yu, C., Wang, J., Peng, C., Gao, C., Yu, G., Sang, N.: Bisenet: bilateral segmentation network for real-time semantic segmentation. In: Proceedings of the European Conference on Computer Vision (ECCV), pp. 325–341 (2018)

The Hybrid Navigation Method in Face of Dynamic Obstacles

Kaidong Zhao[1,2] and Li Ning[1,2(✉)]

[1] Shenzhen Institutes of Advanced Technology,
Chinese Academy of Science, Beijing, China
li.ning@siat.ac.cn
[2] University of Chinese Academy of Sciences, Beijing, China

Abstract. With the development of society and lifestyle changes. Robotic navigation based on traditional navigation and positioning techniques can hardly cope with navigation in complex dynamic scenarios. Traditional navigation and positioning algorithms, such as Dijkstra and A*, cannot perform path planning in dynamic scenarios. In this paper, we present an improved reinforcement learning-based algorithm for local path planning that allows it to still perform well when there are more dynamic obstacles. The algorithm has the advantage of low cost to transform into others situation and fast training speed over the navigation algorithm based entirely on reinforcement learning.

Keywords: SLAM · Reinforcement learning · Dynamic obstacles · Agent navigation

1 Introduction

As mobile robotics has evolved, SLAM technology [1] has come up with greater challenges. In a new environment, robots need to accurately identify their location in the scene and plan a correct and safe route to reach a given location. There are several difficulties in this problem: first, although the robot can use sensors to correct its actual position, it still needs human intervention when the robot is turned on. second, the robot can't avoid obstacles in a dynamic scene. In this paper, the combination of the reinforcement learning algorithm, A* [2], and AMCL algorithm [3] is used to realize a fast and efficient navigation method for robots in dynamic scenarios.

AMCL Algorithm. AMCL (Adaptive Monte Carlo Localization) is a probabilistic localization system for robots moving in dimensions. It uses particle filters to track the robot's position in an already known map and works well

This work is supported by National Natural Science Foundation of China 12071460, Shenzhen research grant (KQJSCX20180330170311901, JCYJ20180305180840138 and GGFW2017073114031767).

© Springer Nature Switzerland AG 2021
Y. Zhang et al. (Eds.): PDCAT 2020, LNCS 12606, pp. 297–303, 2021.
https://doi.org/10.1007/978-3-030-69244-5_26

for a wide range of localization problems. The localization of the robot is very important because if the current position of the robot cannot be located correctly, the path to the destination planned later based on the wrong starting point must also be wrong.

Reinforcement-Learning Based Navigation. Many navigation systems are based entirely on reinforcement learning, with their superior navigation performance and their ability to cope with complex situations in the scene. This reinforcement learning-based navigation approach allows the robot to rely on its own sensor data (e.g., laser, odometer, GPS, etc.) to perform a series of navigation operations in combination with the given orientation of the target point and the scene information from the network memory.

The key point of reinforcement-based learning navigation is how to get neural networks to remember and understand scene information. Most current approaches to reinforcement learning navigation are based on an extensive exploration of specific scenarios to compute the best action strategy. However, this strategy is scene-specific, and replacing it with a different one is often not as effective as expected.

Traditional global navigation-based methods such as (Dijkstra and A*, etc.) are highly practical for arbitrary scenarios, but these static map-based navigation methods are often not well suited for dynamic scenarios. Some local path planning systems based on traditional algorithms (e.g., DWA) are applicable to dynamic scenarios. However, these methods are often quite effective in avoiding obstacles that are relatively stationary and unmarked in static maps while in motion. However, when faced with complex situations, they are especially helpless when faced with actively impacting obstacles.

Therefore, in this paper, we propose a method that combines global navigation and reinforcement learning methods for local navigation. This method has high applicability to different scenarios and does not require an additional neural network training process when changing scenarios.

2   Relate Work

Navigation Stack [4]. In the ROS [5] system, traditional navigation is based on a global path planned by A*'s algorithm based on the global map information. Based on the global path, the DWA [6] algorithm calculates the execution path for the processing of the real-time scene information. For DWA algorithm's ability to correct in the face of differences between the real scene and map information. However, with today's diversity of scenes, there may be a large number of dynamic obstacles (e.g., cats, dogs, people who do not notice the robot, disabled people, etc.) and the robot needs to avoid possible collisions and plan a reasonable and safe path to its destination based on the given global path.

With the development of reinforcement learning [7] and digital simulation (twin) technology [8], there are a large number of researchers engaged in the use

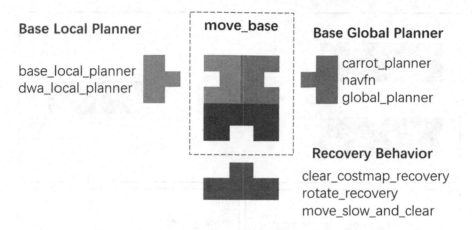

Fig. 1. Move base has three parts, include base local planner, base global planner and recovery behavior. Users only need to choose the part they want, then the move base can plan a path to the destination.

of reinforcement learning and other methods through digital simulation technology to enable robots to learn a large amount of information about specific real-life scenarios before starting up. There are a lot of researchers who use reinforcement learning and other methods to make robots learn a large amount of information about specific real scenes through digital simulation technology before starting so that robots can take quick actions in real scenes. However, there are some drawbacks to the approach of using reinforcement learning for navigation and path control: it is labor-intensive to build different simulated scenes similar to the real scenes, it requires a lot of computing power and time to train, and the models between different scenes may be difficult to migrate and have low re-usability.

In this paper, by combining the features of the above two algorithms, an augmentation algorithm based on reinforcement learning for navigation stack that can be adapted to contain more dynamic obstacles is proposed.

3 Method

The DDPG algorithm is base on the Actor-Critic [9] algorithm. In this model, we design two critics, seen as Fig. 2. One will estimate the probability of the collision and teach the actor how to avoid a collision. The other critic will minimize the difference between input velocity and output velocity, while will teach the actor how to get the destination.

At the begging of the training, we use a bigger learning rate in the critic which is teaching how to get the goal, and a smaller learning rate in teaching avoid a collision. In this way, the model can quickly figure out how to get to the destination. Then using a similar learning rate lets the actor learning how to avoid obstacles on the way to the target point.

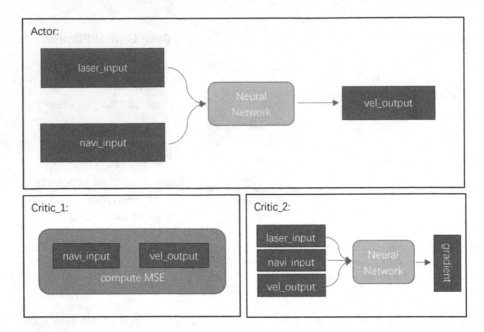

Fig. 2. The architecture the method in this article. The laser-input is from the robot sensor. The Navi-input is from global navigation. The vel-output is used for the mobile base to control the robot. The compute MSE and the gradient is used for the actor's neural network update.

The algorithm in this paper is a combination of the reinforcement learning approach and the A* algorithm. The A* algorithm plans the optimal route without any obstacles based on the existing scene map and generates the motion control instruction V1. the network model NET1 extracts feature by line processing the data from the laser sensor and generate the control instruction V2 by the DDPG [10] reinforcement learning algorithm local obstacle avoidance is achieved during the robot's operation. The speed instructions are then merged through a second network. The network outputs the speed control command V3 to the robot chassis. Then, the odometer and encoder are used to send the error of the robot's operation back to the algorithm for error adjustment. This enables closed-loop control of navigation.

In the actor-network, as shown in Fig. 3, we design two 1D convolutional layers to process the data obtained from the LIDAR data, the convolution layers use 32 one-dimensional filters with kernel size = 3, stride = 2 over the three input scans and applies ReLU activation. Then we use a fully-connected layer with 128 rectifier units and apply ReLU activation. The output fully-connected layer is concatenated with the other input and then fed into the last hidden layer, the fully-connected layer with 64 units, and applies ReLU activation. the output layer with two units which is the linear and rotation speed with the tanh

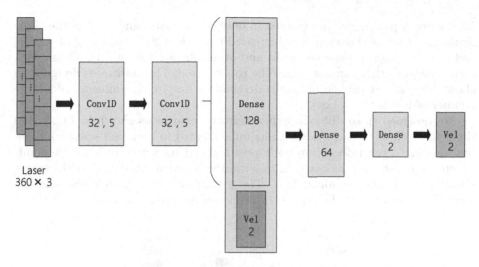

Fig. 3. The architecture of the collision avoidance neural network. The network input laser data with length 360 and have 3 channel and velocity with length 2 generated from global navigation. The output velocity with the same shape from velocity input is used for the mobile base

activation [11]. We limit the linear speed in $[-0.5 , 0.8]$ and the rotation speed to $[-1 , 1]$, which is a little more scale than the global navigation speed.

The critic network has similarities to the actor-network. We first use two layers 1D convolutional as same as the actor's to process the laser data. Secondly, use 128 units fully-connect layer with ReLu activation. Then the outputs concatenated the Navi-input and the vel-output into another two dense layers with 64 units and ReLu activation. Finley, use 1 unit dense layer to the critic network's output used to actor's network update.

4 Experiments

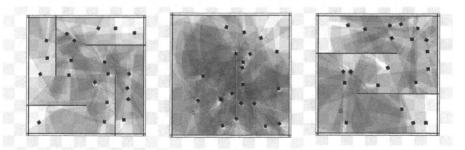

Fig. 4. Test environment for three experiments. part 3, part 1, part 2 (from left to right).

The network parameters are trained in only one scenario and used in the three situations. Each of these scenes is increased more than 20 free-walking obstacles, and obstacles can be observed walls and other obstacles to avoid collision with each other, but they can not notice the robot in red. The network is designed to allow safe and fast navigation to the destination despite the influence of a large number of interfering robots.

We designed three different experimental scenarios as shown in Fig. 4. 100 times for each of the three algorithms implemented in each test scenario. Each test randomly generates the robot's position and navigates to a target point, which the robot needs to navigate to within a radius of 0.5 m around the target point through different algorithms. The average number of collisions and the navigation success rate during the 100 times of navigation were recorded.

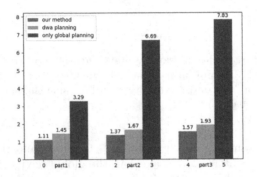

Fig. 5. This is the average number of collisions for the two algorithms and a blank control group (global navigation only) for 100 navigation in three different experimental scenarios.

From the experimental results Fig. 5 and Fig. 6, we can find that the A* algorithm based on reinforcement learning enhancement can perform well for the navigation problem in dynamic environments. However, we also found that

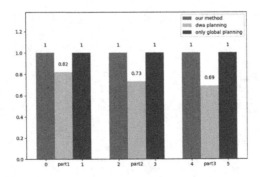

Fig. 6. This is the success rate of two algorithms and a blank control group (using only global navigation) for 100 navigation in three different experimental scenarios.

when the only path in a static scene changes, the robot will not be able to find a second possible path by changing the path.

5 Conclusion

This article presents a navigation method that mixes traditional and reinforcement learning algorithms. We use A* as the path planning for global navigation, which is highly adaptable to different scenarios. We choose a local path planning method using reinforcement learning, which allows the robot to make the right decisions on a local scale through extensive experiments.

Then, we hope to further improve the efficiency of robot navigation by combining a multi-robot framework to achieve navigation and control of robot clusters. The integration with the physical environment allows the algorithm to be deployed on a physical robot cluster on a large scale.

References

1. Hess, W., Kohler, D., Rapp, H., Andor, D.: Real-time loop closure in 2d lidar slam. In: 2016 IEEE International Conference on Robotics and Automation (ICRA), pp. 1271–1278. IEEE (2016)
2. Duchoň, F., et al.: Path planning with modified a star algorithm for a mobile robot. Procedia Eng. **96**, 59–69 (2014). Modelling of Mechanical and Mechatronic Systems. http://www.sciencedirect.com/science/article/pii/S187770581403149X
3. Hennes, D., Claes, D., Meeussen, W., Tuyls, K.: Multi-robot collision avoidance with localization uncertainty. In: AAMAS, pp. 147–154 (2012)
4. Konolige, K., Marder-Eppstein, E., Marthi, B.: Navigation in hybrid metric-topological maps. In: IEEE International Conference on Robotics and Automation, vol.2011, pp. 3041–3047 (2011)
5. Quigley, M., et al.: ROS: an open-source robot operating system. In: ICRA Workshop on Open Source Software, vol. 3, no. 3.2, p. 5, Kobe, Japan (2009)
6. Lopes, A., Rodrigues, J., Perdigao, J., Pires, G., Nunes, U.: A new hybrid motion planner: applied in a brain-actuated robotic wheelchair. IEEE Robot. Autom. Mag. **23**(4), 82–93 (2016)
7. Sutton, R.S., Barto, A.G.: Reinforcement Learning: An Introduction. MIT press, Cambridge (2018)
8. Tao, F., Cheng, J., Qi, Q., Zhang, M., Zhang, H., Sui, F.: Digital twin-driven product design, manufacturing and service with big data. Int. J. Adv. Manuf. Technol. **94**(9), 3563–3576 (2017). https://doi.org/10.1007/s00170-017-0233-1
9. Peters, J., Schaal, S.: Natural actor-critic. Neurocomputing **71**(7), 1180–1190 (2008). progress in Modeling, Theory, and Application of Computational Intelligenc. http://www.sciencedirect.com/science/article/pii/S0925231208000532
10. Lillicrap, T.P., et al.: Continuous control with deep reinforcement learning (2019)
11. Long, P., Fan, T., Liao, X., Liu, W., Zhang, H., Pan, J.: Towards optimally decentralized multi-robot collision avoidance via deep reinforcement learning. In: IEEE International Conference on Robotics and Automation (ICRA), vol. 2018, pp. 6252–6259 (2018)

A Relaxed Balanced Lock-Free Binary Search Tree

Manish Singh[1]([✉]), Lindsay Groves[2], and Alex Potanin[2]

[1] Wellington Institute of Technology, Wellington, New Zealand
manish.singh@weltec.ac.nz
[2] Victoria University of Wellington, Wellington, New Zealand
{lindsay.groves,alex.potanin}@ecs.vuw.ac.nz

Abstract. This paper presents a new relaxed balanced concurrent binary search tree using a single word compare and swap primitive, in which all operations are lock-free. Our design separates balancing actions from update operations and includes a lock-free balancing mechanism in addition to the insert, search, and relaxed delete operations. Search in our design is not affected by ongoing concurrent update operations or by the movement of nodes by tree restructuring operations. Our experiments show that our algorithm performs better than other state-of-the-art concurrent BSTs.

Keywords: Concurrent data structure · Lock-free · Binary search tree

1 Introduction

Recently, the chip manufacturing company, AMD, has released the Threadripper CPU with 64 cores and 128 threads for desktops. As CPU manufacturers embed more and more cores in their multi-core processors, the need to design scalable and efficient concurrent data structures has also increased. Considerable research has been done towards making both blocking and non-blocking concurrent versions of sequential data structures. Unlike blocking a concurrent data structure design is non-blocking, or lock-free, if it ensures at any time at least one thread can complete its operation. Performance has always been an important factor and will drive the design of these data structures.

The Binary Search Tree (BST) is a fundamental data structure. Many concurrent BST, both blocking and non-blocking, have been proposed [1,4–6,14,15,17]. However, only a few designs include self-balancing operations. Most of the published work tries to emulate a sequential implementation. This results in performance compromise as strict sequential invariants must be maintained. In a concurrent environment the effect of some operations might cancel out the effects of other operations. In the case of a self-balancing AVL tree each insert or delete operation requires a balancing operation to be performed immediately to preserve the height-balanced property. A balancing operation might cancel out the effects of other balancing operations on the same set of nodes.

© Springer Nature Switzerland AG 2021
Y. Zhang et al. (Eds.): PDCAT 2020, LNCS 12606, pp. 304–317, 2021.
https://doi.org/10.1007/978-3-030-69244-5_27

We have designed a lock-free relaxed balanced BST, hereafter referred as RBLFBST, in which balancing operations are decoupled from the regular insert and delete operations[1]. The idea of relaxing some structural properties of a sequential data structure in their concurrent version has been tried in several previous works. Lazy deletion [9,10] is an example of relaxing the requirement that nodes are immediately removed from the data structure. In a relaxed self-balancing such as AVL [11], chromatic trees [16], rotation operations, are performed separately from the insertion and deletion operation.

2   Related Work

The first lock-free BST dates back to the 90's [19] where the author suggested a design based on a threaded binary tree representation, but did not discuss the implementation of his design. Fraser et al. [7] described a lock-free implementation of an internal BST using multiple CAS (MCAS) to update multiple parts of the tree structure atomically. This algorithm uses a threaded representation of a BST. A delete operation required up to 8 memory locations to be updated atomically which added an appreciable overhead to the design. Based on the cooperative technique of [2], Ellen et al. [6] developed the first specific lock-free external BST algorithm, where internal nodes only act as routing nodes, which simplifies the deletion operation.

Using a helping technique similar to [6], Howley and Jones [17] presented a non-blocking algorithm for internal BSTs. They added an extra field *operation* to each node of the BST, in which all the information related to a particular update operation can be stored. A delete operation in this work could use as many as 9 CAS instructions. However, the paper gives experimental results in which this design outperforms [6,15]. Aravind et el. [14] presented a lock-free external BST similar to [6], but using edge-marking. Being a leaf-oriented tree this design also has a smaller contention window for insert and delete operations and uses fewer auxiliary nodes to coordinate among conflicting operations. Another contemporary paper [18] describes a general template for non-blocking trees. This template uses multi-word versions of LL/SC and VL primitives, making it easier to update multiple parts of the tree. The paper also presented a fine-grained synchronized version of a leaf-oriented chromatic tree based on their template.

The concurrent AVL tree of Bronson et al. [15] uses lock coupling which uses per-node locks to remove conflicts between concurrent updates, and a relaxed balancing property which does not enforce the strict AVL property. This design uses a partially external tree such that by default nodes behave as per an internal tree but in order to simplify the removal of nodes with two children, the removed node objects are allowed to remain in the structure and act as routing nodes. This avoids the problem of locking large parts of the tree if a delete operation was implemented exactly as in a sequential BST. These deleted (logically) nodes can then be either removed during a later re-balancing operation or, if the node's key

---

[1] A poster describing the design of RBLFBST was presented in ICPP 2019, Kyoto, Japan.

is reinserted, the key can be made part of the set again. This partially external tree design was experimentally shown to have a small increase (0–20%) on the total number of nodes required to contain the key set. Crain et al. [4] present a lock-based tree in which balancing operations are decoupled from the insert and delete, and are done by a separate dedicated thread. Keys that are deleted are not immediately removed from the tree but are only marked as deleted as in [15]. Later, a dedicated thread can remove nodes with deleted keys that have single or no child. The balancing mechanism used closely mirrors that in [3]. Despite claiming performance improvement by more than double to that of [15] another lock-based design this tree still is lock-based. More recently, Drachsler et al. [5] proposed a lock-based internal BST supporting wait-free search operations and lock-based update operations via logical ordering. Their design uses a similar threaded binary tree representation as in [19].

All the designs of concurrent BST with balancing operations described above use locks to synchronize concurrent updates and therefore are not immune to problems that are associated with locking in general. Based on techniques used in previous research we present a concurrent BST in which all the update operations are lock-free. To our knowledge, our design is the first AVL tree based lock-free partially external BST which includes balancing operation for all the update operations.

3   Algorithm Description

3.1   Overview

We implement a set using a BST. Our implementation supports concurrent executions of *search(k)*: to determine whether the key k is in the set, *insert(k)*: to add the key k to the set, *delete(k)*: to remove the key k from the set. To ensure that the tree does not become unbalanced, causing the operations to have linear cost, our design supports a relaxed tree balancing mechanism.

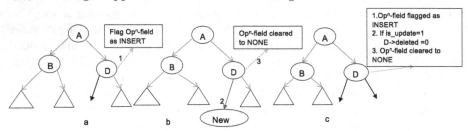

Fig. 1. Left child pointer of Node D is identified as an insertion point for a 'new' key. (a). An *operation-descriptor* flagged as INSERT is inserted in to the *operation* field of node D (1). (b). After successful insertion of *operation-descriptor*, the node having key 'new' as a child of node D is physically inserted tiny (2) and then the *operation* field of node D is cleared by setting its flag to NONE (3). (c) Shows the case when the key to be inserted is found in the tree but is deleted. In this case, the *deleted* field is set to false.

Any thread attempting to insert a *key*, upon finding the insertion point first announces its intention. The thread first collects the information required in an *operation-descriptor* and tries to insert it in the *operation* field of the node identified as the insertion point (the parent of the new node). Once the *operation-descriptor* is inserted into the node the *key* is considered to be part of the set. Then the new node is inserted using a CAS to the appropriate child pointer of the parent node as shown in Fig. 1(a) and (b).

To remove a key, first the node having the key to be removed is flagged as deleted (*deleted* bit is set to true) and then physically removed later. Once the *deleted* bit is set the key is considered to be removed (logically) from the tree. The physical removal is started by a separate maintenance thread by marking the deleted nodes having at most one child. Once marked any other thread can physically remove the node from the tree. Deleted nodes with two children are not physically removed from the tree until they have less than two children. This relaxed deletion is done to reduce contention in the tree as the delete algorithm does not need to locate and update the node to be deleted and the replacement node, which might involve several restarts. This relaxed deletion approach also means that a thread doing an insert operation may find the intended key already in the tree but the node is flagged as deleted. In this case, the key can be made part of the set again by setting the *deleted* bit off.

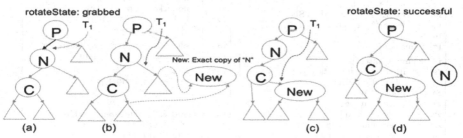

Fig. 2. Right rotation example (similarly for the left). (a) Shows the case where the maintenance thread has found a balance condition violation at node N and has successfully inserted a rotation *operation-descriptor* in the *operation* field of the parent of N, P, N, and child of N, C. After that a *new* node having key and other fields exactly as the node N is created. (b) Then the right and the left pointer of the *new node* are allocated to those children which the node N would have got after the rotation. (c) The next step is to insert the *new* node to its position after *rotation*. In this case, the *new* node would be the right-child of node C. (d) The third and last step is to connect node the parent P to child C effectively removing node N from the tree. A thread T1 carrying out search is oblivious to the movement of nodes by rotations

Our balancing mechanism is based on heights of nodes and rotation operations, as in sequential AVL trees. However, balancing is carried out separately to the update operations, similar to that of [4,12]. The balancing adjustments are performed by the maintenance thread. If the balancing condition at any node is found to be violated then the maintenance thread collects all the information required to do rotation operation in to an *operation-descriptor* and tries to insert it in to the *operation* field of the parent of the node. After successfully inserting the operation descriptor, the maintenance thread (or any other thread) tries to do the same for the node and the appropriate child of the node involved in the rotation. Figure 2 shows the remaining steps for the rotation operation (right rotation) once the *operation-descriptor* is inserted in all the three nodes involved. Due to the ongoing concurrent updates, it is difficult to determine the exact height of a node at any point in the execution. Thus, the balancing condition in RBLFBST is based on apparent local heights of the node and its children. It should also be noted that the rotation technique shown in Fig. 2 makes ongoing concurrent search operations oblivious to any node moving up or down the tree.

```
 1  struct node_t {
 2      int key
 3      node_ t* left, right
 4      operation_t* op
 5      int local_height, lh, rh
 6      bool deleted,removed
 7  }

 8  union operation_t {
 9      insert_op_t insert_op
10      rotate_op_t rotate_op
11  }

12  struct insert_op_t {
13      bool_t is_left
14      bool_t is_update=FALSE
15      node_t* expected
16      node_t* new
17  } insert_op_t

18  struct rotate_op_t {
19      int /*volatile*/ state =
                UNDECIDED
20      node_t* parent
21      node_t* node
22      node_t* child
23      operation_t* pOp
24      operation_t* nOp
25      operation_t* cOp
26      bool rightR
27      bool dir
28  } rotate_op_t

29  //operation marking status
30  NONE 0, MARK 1, ROTATE 2,
        INSERT 3
```

Algorithm 1: Basic structures used

```
Input: int k, node_t* root
Output: true, if the key is found in
        the set otherwise false

 1  node_t* node, next
 2  operation_t* node_op
 3  bool result = FALSE
 4  node = root
 5  node_op = node→op
 6  int node_key
 7  next = node→right
 8  while !ISNULL(next) do
 9  │   node = next
10  │   node_op = node→op
11  │   node_key = node→key
12  │   if k < node_key then
13  │   └   next = node→left
14  │   else if  k > node_key then
15  │   └   next = node→right
16  │   else
17  │   │   result = TRUE
18  │   └   break

19  if result && node→deleted then
20  │   if GETFLAG(node→op) ==
    │       INSERT then
21  │   │   if
    │   │       node_op→insert_op.new→key
    │   │       == k then
22  │   │   └   return TRUE
23  │   return FALSE
24  return result
```

Algorithm 2: Search

All the operations in RBLFBST are lock-free and linearisable. Once an operation is flagged with the corresponding FLAG and the *operation-descriptor* is inserted, it is considered to be logically completed. Any other thread can complete the announced operation using the information available in the *operation-descriptor*. Updating multiple locations using a single word CAS while preserving atomicity is a challenging task. Especially, in case of a rotation operation, three locations are needed to be updated atomically. We achieve this by careful design of our operation, described in detail in later sections. Another notable feature of this design is that we managed to keep the most frequently used operation in BSTs, *search*, free of additional coordination.

3.2   Detailed Algorithm

Structures: Algorithm 1 shows various structures used in the implementation. As in a sequential BST, a node has key, right, and left pointers to its corresponding child which are set to NULL for every new node. To synchronize various concurrent operations, each node has an operation pointer field, *op*, that is used to announce the intended operation and contains a pointer to that particular *operation-descriptor*. A node in RBLFBST has 3 heights, *local-height*: updated by adding one to the maximum *local-height* of its children, *right-height*: *local-height* of right child, and *left-height*: *local-height* of left child, which are stored in fields *local-heights*, *lh* and *rh* respectively. In addition to that it has two boolean fields *deleted* and *removed* which are initialized to false. If a node has its *deleted* field set that means it is deleted from the set but could still be in the tree, while the *removed* field is set when the node is removed from the tree.

```
int seek (int key, node_t** parent,
    operation_t** parent_op,
    node_t** node, operation_t**
    node_op, node_t* aux_root,
    node_t* root)
1   int result, nodekey
2   node_t* next
3   retry:
4   result = NOT_FOUND_L
5   *node = aux_root
6   *node_op = *node→op
7   if GETFLAG(*node_op) ≠ NONE
    then
8       if auxroot == root then
9           bst_help_insert((operationt*)
            UNFLAG(*node_op), *node)
10          goto retry

11  next = (nodet*) (*node)→right
12  while !ISNULL(next) do
13      *parent = *node
14      *parent_op = *node_op
15      *node = next
16      *node_op = (*node)→op
17      nodekey = (*node)→key
18      if key < nodekey then
19          result = NOT_FOUND_L
20          next = (*node)→left
21      else if key > node_key then
22          result = NOT_FOUND_R
23          next = (*node)→right
24      else
25          result = FOUND
26          break

27  if GETFLAG(*node_op) ≠ NONE
    then
28      help(*parent, *parent_op, *node,
        *node_op)
29      goto retry
30  return result
```

Algorithm 3: Seek

```
Input: int key, node_t* root
Output: true: if the key is added to
        the set, false otherwise
1   node_t* parent
2   node_t* curr
3   node_t* new_node = NULL
4   operation_t* parent_op
5   operation_t* node_op
6   operation_t* cas_op
7   int result
8   while TRUE do
9       result = seek(key, &parent,
            &parent_op, &node, &node_op,
            root, root)
10      if result == FOUND && !
            node→deleted then
11          return FALSE

12      if new_node == NULL then
13          new_node = alloc_node(k,v,0)

14      bool_t is_left = (result ==
            NOT_FOUND_L)
15      node_t* old
16      if is_left then
17          old = node→left
18      else old = node→right

19
20      cas_op = alloc_op()
21      if result == FOUND &&
            node→deleted then
22          cas_op→insert_op.is_update =
            TRUE

23      cas_op→insert_op.is_left = is_left
24      cas_op→insert_op.expected = old
25      cas_op→insert_op.new = new_node
26      if CAS(&node→op, node_op,
            FLAG(cas_op, INSERT)) ==
            node_op then
27          help_insert(cas_op, node)
28          return TRUE
```

Algorithm 4: Insert

The *operation-descriptor* is either an, *insert_op* containing information for an intended insert, or a *rotate_op*: containing information for a rotation. An *insert_op* contains, *new*: pointer to the new node to be inserted, *expected*: previous node, *isLeft*: indicating whether the new node would be inserted as a left-child or right-child, and *is_update*: indicating whether it is a new node to be inserted or an existing deleted node to be made part of the set again. A *rotate_op* descriptor contains in addition to the nodes involved in the corresponding operation, a *state* field which can have three values, UNDECIDED: all nodes involved in the operation have not been grabbed yet, GRABBED: all the nodes that are needed have been grabbed, ROTATED: indicating that the rotation has been completed.

As in [9,17], we use two least significant bits of a pointer address to store the operation status. The operation pointer, *op*, of a node could be in one of following statuses, NONE: meaning no operation is currently intended, MARK: this node can be physically removed, ROTATE: a rotate operation is intended, and INSERT: an insert operation is intended. The following macros are used to modify *op*'s status, *FLAG(op, status)*: sets the operation pointer to one of the above status, *GET-FLAG(op)*: returns the status of *op*, *UNFLAG(op)*: resets the status to NONE.

Search: The *search* algorithm, outlined in Algorithm 2, is mostly similar to the sequential search with additional checks due to other concurrent inserts. It traverses from the *root*[2] to a leaf node looking for the *key*. If found, it breaks out of the loop as shown at lines 17–18. A further check is needed to see if the *key* is deleted from the set. In this case, if the corresponding node's *deleted* field is set and its *operation* field is flagged as INSERT then the new *key* which is to be inserted, is compared with the *key* that the *search* is looking for at lines 19–21. If a match is found then the algorithm returns *true* line 22. If *deleted* bit is not set and the node is found then the algorithm simply returns the true at line 24. If the *key* is not found the *search* algorithm returns false line 24. Synchronization of any form is not needed in this *search* algorithm. Also, the *search* algorithm never restarts.

Seek: Algorithm 3 outlines the *seek* method which is used by both insert and delete algorithms to locate the position or potential position of a *key*, starting from a specified point *aux_root* in the tree. The position is returned in the variables pointed to by arguments *parent* and *node* and the values of their *operation* fields are returned in the variables pointed to by arguments *parent_op* and *node_op*. The result of the seek can be one of three values, FOUND: if the key was found, NOT_FOUND_L: if the *key* was not found but would be positioned at the left child of the *node* if it were inserted, NOT_FOUND_R: similar to NOT_FOUND_L but for the right child. The variable *next* is used to point to the next node along the seek path, the *parent* and *parent_op* are used to record previous node. The check at lines 7–10 handles the case when the root has an ongoing operation to add to the empty tree. The seek loop traverses nodes one at a time until either the *key* is found or a null link is reached. Line 27 checks whether the node that is found does has an ongoing operation. If the node has an ongoing operation, the appropriate helping is done and seek restarts at lines 28–29.

---

[2] For implementation purpose, the first *node* to be inserted in the empty tree is kept as the right child of a fixed node root which has *key* assumed to be infinity.

Insert: The *Insert* algorithm, Algorithm 4, begins by calling the *seek* method at line 9. Depending on the result of the *seek* method : case 1: the *key* is found and the corresponding *node* is not *deleted*; case 2: the *key* is found and the corresponding *node* is *deleted*; case 3: the key is not found in the tree. For case 1, the *insert* method returns false at line 11. For both cases 2 and 3, a *new-node* and an *insert_op* *operation-descriptor* are allocated with all the necessary information at lines 12–13 and 20–25 respectively. Furthermore, for case 2 the *is_update* field of *insert_op* descriptor is set to true at lines 21–22. Then a CAS is used to flag the *operation* field of the node to INSERT. If successful, the *help_Insert* method is called to complete the physical insertion, otherwise it retries. If the *help_insert* method finds *is_update* field set it simply sets off the *deleted* bit of the node to make it part of the set again otherwise it inserts the *new-node*.

Delete: Similar to the *Insert*, the *Delete* method starts by calling the *seek* method at line 6. If a *node* with the desired *key* is not found, it simply returns false at lines 7–8. If the *key* is found and is *deleted*, it further checks whether the node is flagged as INSERT, since an ongoing *insert* operation might be trying to insert the same *key*. If the *node* is not flagged as INSERT, the *delete* method returns false as the *node* is already deleted at lines 9–11. If the key is found and not deleted, the *delete* method tries to set the *node*'s *deleted* field to *true* and returns *true* on successful CAS at line 12–15.

Tree-Maintenance: The tree maintenance actions involve checking for balance condition violations, rotation operations, and physical removal of deleted nodes having at most one child. The latter two operations are started by a maintenance-thread and can be completed by any other thread[3].

The maintenance-thread repeatedly performs a depth-first traversal in the background in which, at each node, it looks to see if the node can be removed and then adjusts heights. If it finds any node with *deleted* field set and at most one child, it then tries to flag the *operation* field of the node to MARK. If marking of the node is successful, the *help_marked* method outlined in algorithm 7 is called to physically remove the node from the tree. After adjusting heights, a check is performed to see if a balance violation has occurred. At any node, the balance condition is said to be violated if *right-height* and *left-height* of the node differ by 2 or more. If there is a balance violation at any node further checks are performed to determine whether a single rotation (left or right) or double rotations are required. Once the type (left or right) and number (single or double) of rotation operations are determined the rotation process is started as listed in the *left_rotate*, algorithm 6, for the left rotation (similarly for the right rotation).

In the *left_rotate* method checks are performed to ensure that nodes that are involved in rotation operation are still intact at lines 2–6. The check at line 7 is performed to see whether double rotations are needed. After this check, a *rotate_op* descriptor is allocated at line 10 using appropriate values.

---

[3] Detailed explanations and full algorithm for tree-maintenance can be found in [13].

Input: int k, node_t* root
Output: *true*: if the *node* is found (not already deleted) and deleted from the set; else *false*

```
1  node_t* parent
2  node_t* node
3  operation_t* parent_op
4  operation_t* node_op
5  while TRUE do
6      int res = seek(k, &parent,
            &parent_op, &node, &node_op,
            root, root)
7      if (res ≠ FOUND) then
8          return FALSE

9      if node→deleted then
10         if GETFLAG(node→op) ≠
               INSERT then
11             return FALSE

12     else
13         if GETFLAG(node→op) ==
               NONE then
14             if CAS(&node→deleted,
                   FALSE, TRUE) ==
                   FALSE then
15                 return TRUE
```

Algorithm 5: delete

---

int leftrotate(node_t* parent, int dir, bool_t rotate)

```
1  node_t* node
2  if parent→removed then
3      return 0

4  node = (dir == 1) ?
       (node_t*)parent→right :
       (node_t*)parent→left
5  if ISNULL(node) ||
       ISNULL(node→right) then
6      return 0

7  if (node→right→lh -node→right→rh)
       > 0 && !rotate then
8      return 3; //double rotation

9  if GETFLAG(parent→op) == NONE
       then
10     operation_t* rotateOp
11     rotateOp =
           alloc_rotateop(parent,node,
           node→right, parent→op,
           node→op,
           node→right→op,dir,FALSE)
12     if CAS(&(parent→op),
           rotateOp→rotate_op.pOp,
           FLAG(rotateOp,ROTATE)) ==
           rotateOp→rotate_op.pOp then
13         help_rotate(rotateOp, parent,
               node, node→right);

14 return 1;
```

Algorithm 6: left_rotate

---

void help(node_t* parent, operation_t* parent_op, node_t* node, operation_t* node_op)

```
1  if GETFLAG(node_op) == INSERT
       then
2      help_insert(UNFLAG(node_op),
           node)
3  else if GETFLAG(parent_op) ==
       ROTATE then
4      help_rotate((UNFLAG(parent_op),
           parent,node,
           parent_op→rotate_op.child)
5  else if GETFLAG(node_op) ==
       MARK then
6      help_marked(parent, parent_op,
           node)
```

void help_insert(operation_t* op, node_t* dest)

```
1  if op→insert_op.update then
2      CAS(&dest→deleted, TRUE,
           FALSE)
3  else
4      node_t** address = NULL
5      if op→insert_op.is_left then
6          address = (node_t**)
               &(dest→left)
7      else
8          address = (node_t**)
               &(dest→right)
9      CAS(address,
           op→insert_op.expected,
           op→insert_op.new)
10 CAS(&(dest→op), FLAG(op,
       INSERT), FLAG(op, NONE))
```

void help_marked(node_t* parent, operation_t* parent_op, node_t* node)

```
1  node_t* child, address
2  if ISNULL((node_t*) node→left)
       then
3      if ISNULL((node_t*)
           node→right) then
4          child =
               (node_t*)SETNULL(node)
5      else
6          child = (node_t*) node→right

7  else
8      child= (node_t*) node→left

9  node→removed = TRUE
10 operation_t* cas_op = alloc_op()
11 cas_op→insert_op.is_left = (node ==
       parent→left)
12 cas_op→insert_op.expected = node
13 cas_op→insert_op.new = child
14 if CAS(&(parent→op), parent_op,
       FLAG(cas_op, INSERT)) ==
       parent_op then
15     help_insert(cas_op, parent)
```

Algorithm 7: help, help_insert, help_marked

Then, an attempt is made to insert *rotate_op* to the *parent* of the *node* where the violation has occurred using CAS. If successful, a call is made to the *help_rotate* method.

Input: operation_t* op, node_t* parent, node_t* node, node_t* child

```
 1  int seen_state = op→rotate_op.state
 2  retry:
 3  if seen_state== UNDECIDED then
 4      if GETFLAG(node→op)== NONE ||
        GETFLAG(node→op) == ROTATE
        then
 5          if GETFLAG(node→op)==NONE
            then
 6              CAS(&(op→rotate_op.node→op),
                node→op,FLAG(op,ROTATE))
 7          if GETFLAG(node→op) ==
            ROTATE then
 8              nodegrabbed:
 9              if GETFLAG(child→op)
                ==NONE ||
                GETFLAG(child→op)
                ==ROTATE then
10                  if GETFLAG(child→op)
                    ==NONE then
11                      CAS(&(op→rotate_op.
                        child→op), child→op,
                        FLAG(op,ROTATE))
12                  if GETFLAG(child→op)
13                  == ROTATE then
14                      CAS(&
                        (op→rotate_op.state),
                        UNDECIDED,
                        GRABBED)
15                      seen_state =GRABBED
16                  else goto nodegrabbed
17
18              else
19                  help(node,node→op,child,
                    child→op)
20                  goto nodegrabbed
21          else
22              goto retry
23      else
24          help(parent,parent→op,node,
            node→op)
25          goto retry
```

```
26  if seen_state==GRABBED then
27      if op→rotate_op.rightR then
28          // right rotation
29          //allocate newnode with
30          appropriate children and heights
31          //carry out rotation steps
32          using CAS
33      else
34          // left rotation
35          ///allocate newnode with
36          appropriate children and heights
37          //carry out rotation steps
38          using CAS
39      //parent pointer swing
40      if op→rotate_op.rightR then
41          CAS(&op→rotate_op.parent→left,
            op→rotate_op.node,child)
42      else
43          CAS(&op→rotate_op.parent
44          →right, op→rotate_op.node,child)
45      //adjust child and parent heights
46      CAS(&(op→rotate_op.state),
47      GRABBED, ROTATED)
48      seen_state=ROTATED;
49  if seen_state==ROTATED then
50      //Clear parent,node,child operation
51      fields from ROTATE to NONE
52  bool_t result = (seen_state == ROTATED)
53  return result;
```

Algorithm 8: help_rotate

The *help_rotate* method, outlined in algorithm 8 can be called by any thread if it finds that the *operation* field of a node is flagged as ROTATE. Any thread executing *help_rotate* tries to the grab remaining nodes namely, the *node* and the *child* by flagging their *operation* fields to ROTATE at lines 2–24. Failing to flag any node means that there is an ongoing operation. In this case, the thread helps the ongoing operation first, lines 18 and 23. Once both *node* and *child* are flagged the *state* field of *rotate_op* descriptor is updated to a value GRABBED at line 13. The rest of the operation is carried out (lines 25–40) as shown in Fig. 2.

It should be noted that all of the methods involved in tree balancing except the *help_rotate* are not exposed to any thread other than the maintenance-thread.

Help_insert/Marked: Both these methods are called only after inserting appropriate flags to the *operation* field of the *node*. Once a *node* is flagged it will never be un-flagged until the operation is completed. The former adds a new node as a child of the node that was flagged as INSERT or simply sets off the *deleted* field depending on the *is_update* field of the node, and the later physically removes the marked node from the tree.

3.3   Correctness

Lock-Freedom: The non-blocking property of the algorithm is explained by describing the interactions that can occur between the threads that are executing read and write operations. A *search* will either locate the *key* it is searching for or terminate at a leaf node. The search operation never restarts. An *Insert* or *Delete* operation retries until it gets clean nodes through the *seek* method. Then, the *insert* tries to flag node as INSERT which cannot be undone until the physical insertion is completed. The *Delete* operation will only set the *deleted* bit when there is no other *insert* operation going on concurrently. The *seek* method restarts when it finds the leaf node has an ongoing operation when called from *Insert*, or if it finds the node having the *key* to be deleted has an ongoing operation when called from *Delete*. In both cases if the *operation* field of the node contains INSERT, ROTATE or MARK, this means an operation has been applied to the tree so system-wide progress was achieved. Assuming that there is no infinite addition of nodes on its search path, the *seek* method will eventually return clean nodes. An *Insert* operation could also restart if it fails to flag the node returned by *seek* to INSERT. In this case also it has encountered an ongoing operation. Any thread executing an operation will help the ongoing operation to its completion before restarting its own operation. Similarly, a rotation starts only when the maintenance-thread successfully flags the parent of the node where balance violation has occurred. If flagging of other nodes involved rotation fails in the *help_rotate* method, this means there is an ongoing operation and the process completes that first before coming back to flagging the node to ROTATE.

Linearisability: To prove the linearisability of RBLFBST, we first define the linearisation points of the *Search, Insert,* and *Delete* operations. The *search* operation can have two possible outcomes: a *key* is found in the set or not. The linearisation point for finding a *key* is the point at which an atomic read of the *key* has occurred at line 11. As our design allows a deleted *key* to be present in the tree it has to pass check at line 19. If the search does not find the *key*, it will linearise reading a NULL link at either line 13 or 15. If the tree is empty the search will linearise at reading the null link at line 7. A successful *Insert* operation will linearise when the *operation-descriptor* is successfully inserted to the *node* returned by the seek algorithm at line 26. Failure of insert operation will have linearisation point at line 17 of the *seek* algorithm where it reads the *key* to be inserted already present in the tree. However, it has to verify if the *key* is logically deleted by failing the test at line 11 of the *Insert* algorithm. Similarly, the successful *Delete* operation will linearise at the successful CAS at line 14 of the delete algorithm. Linearisation point of the failed *Delete* will occur at line 7 of the delete algorithm where it is verified that the *key* is not present in the tree or is logically deleted by passing the checks at lines 10–12. An elaborate correctness discussion can be found in [13].

4   Experimental Results

To evaluate performance of RBLFBST we compared it with following recently published implementations : (i) non-blocking internal BST denoted by bst-howley [17]. (ii) lock-free external binary BST denoted by, bst-Aravind [14]. (iii) lock-based partially internal BST with balancing operations denoted by bst-bronson [15]. (iv) lock-based internal BST with balancing operations denoted by bst-drachsler [5].

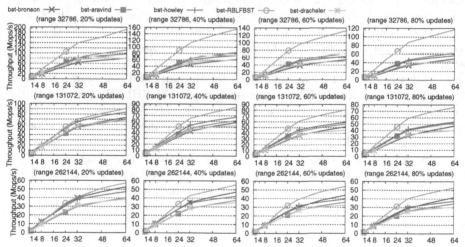

Fig. 3. Throughput in Mops (million of operation per second) for different algorithms for varying parameters. Each row shows results for variations in the distribution of operations. Each column shows results for different key ranges.

All experiments were conducted on a machine with 32 core AMD Ryzen Threadripper 2990WX processor, 64 hardware threads, and 32 GB RAM running x86_64 version of Ubuntu 18.04. All codes were implemented in C and compiled using gcc version 7.5.0 using optimization level O3. The source codes for other implementations were obtained from ASCYLIB [8].

The comparison of RBLFBST with other implementation was done varying three parameters: the key range ([0, 32786], [0, 131072] and [0, 262144]), the distribution of operations, and the number of threads concurrently executing on the data structures (1 to 64). Operations distribution considered were: (a) Write dominated workload: 80% updates, 20% search (b) mixed workload: 60% updates, 40% search (c) mixed workload: 40% updates, 60% search (d) read dominated workload: 20% updates, 80% search. Update operations had an equal number of insert and delete operations. All the operation types and the keys were generated using a pseudo-random number generator which were seeded with different values for each run. Each run was carried out for 5 s, and the results were collected averaging throughput over 20 runs. To mimic steady state the tree was filled with half the key range for each run.

The graphs in Fig. 3 show the comparisons of other implementations against RBLFBST. Overall, RBLFBST scales very well as the number of threads increases. For smaller key range 2^{15}, RBLFBST outperforms its nearest competitor bst-aravind by a maximum of 90% and 85% in read-dominated (80% and 60% search respectively), by 87% and 70% (60% and 80% updates respectively) in write-heavy workloads. Our tree beats the other two lock-based trees with balancing operations(bst-drachsler and bst-bronson) by 142% in 80% search workload and more than 85% in update heavy workload for the same key range. The performance of RBLFBST for 2^{17} key range again is better by 43% and 37% than its nearest lock-free competitor bst-howley in 60% and 80% update workload respectively. Similarly, it performs 56% and 61% better than the closest lock-based competitor bst-bronson in 60% and 80% update workload respectively. For the key range 2^{18}, RBLFBST beats bst-bronson by 24–38% and 20–30%, bst-howely by 10–31% and 20–30%, bst-aravind by 11–53% and 47–50% and bst-drachsler by 11–72% and 2–65% for the workload containing 80% and 60% updates respectively. The better performance of RBLFBST is due to the fact that it uses less number of expensive operations (CAS) than other lock-free implementations for insert and delete operations combined, thereby allowing more concurrency. The performance of our design goes down relatively for read-heavy workloads (80% read) particularly for the larger key ranges (2^{18}, 2^{17}). Larger key range will grow tree longer and more rotations will be performed. This is when the effect of threads helping rotation operation which can use up to 10 CAS instructions is clearly visible. However, the performance of RBLFBST in such cases is still better than all other implementations.

5   Conclusion and Future Work

In this work, we presented a relaxed balanced lock-free binary search tree. In our design, all the set operations are lock-free. The search operation in our algorithm is oblivious to any structural changes done by other operations and is also free of any additional synchronization. Our results show its concurrent performance to be very good compared with other concurrent BSTs. We have discussed the correctness of RBLFBST operations. We are working on formal verification of our algorithm. We are also planning to apply the separation of tree balancing operations as well as relaxing delete operation to other lock-free balanced binary trees designs as future work.

References

1. Afek, Y., Kaplan, H., Korenfeld, B., Morrison, A., Tarjan, R.E.: Cbtree: a practical concurrent self-adjusting search tree. In: Aguilera, M.K. (ed.) Distributed Computing, pp. 1–15. Springer, Berlin, Heidelberg (2012)
2. Barnes, G.: A method for implementing lock-free shared-data structures. In: Proceedings of the Fifth Annual ACM Symposium on Parallel Algorithms and Architectures, pp. 261–270. SPAA 1993 (1993)
3. Bougé, L., Vallés, J., Peypoch, X.M., Schabanel, N.: Height-relaxed avl rebalancing: a unified, fine-grained approach to concurrent dictionaries (1998)
4. Crain, T., Gramoli, V., Raynal, M.: A contention-friendly binary search tree. In: Wolf, F., Mohr, B., an Mey, D. (eds.) Euro-Par 2013. LNCS, vol. 8097, pp. 229–240. Springer, Heidelberg (2013). https://doi.org/10.1007/978-3-642-40047-6_25
5. Drachsler, D., Vechev, M., Yahav, E.: Practical concurrent binary search trees via logical ordering. In: Proceedings of the 19th ACM SIGPLAN Symposium on Principles and Practice of Parallel Programming, pp. 343–356, PPoPP 2014. Association for Computing Machinery, New York, NY, USA (2014)
6. Faith Ellen, P., Fatourou, E.R., van Breugel, F.: Non-blocking binary search trees. In: Proceedings of the 29th ACM SIGACT-SIGOPS Symposium on Principles of Distributed Computing, PODC 2010, pp. 131–140 (2010)
7. Fraser, K.: Practical Lock freedom. Ph.D. thesis, King's College, University of Cambridge September 2003
8. Guerraoui, R.: Ascylib, July 2020. https://dcl.epfl.ch/site/optik
9. Harris, T.L.: A pragmatic implementation of non-blocking linked-lists. In: Welch, J. (ed.) DISC 2001. LNCS, vol. 2180, pp. 300–314. Springer, Heidelberg (2001). https://doi.org/10.1007/3-540-45414-4_21
10. Herlihy, M., Shavit, N.: The Art of Multiprocessor Programming. Morgan Kaufmann Publishers Inc., San Francisco, CA, USA (2008)
11. Kessels, J.L.W.: On-the-fly optimization of data structures. Commun. ACM 26(11), 895–901 (1983)
12. Larsen, K.S.: Avl trees with relaxed balance. J. Comput. Syst. Sci. 61(3), 508–522 (2000)
13. Singh, M., Lindsay Groves, A.P.: A relaxed balanced lock-free binary search tree. Tech. rep. (2020). https://ecs.wgtn.ac.nz/Main/TechnicalReportSeries
14. Natarajan, A., Mittal, N.: Fast concurrent lock-free binary search trees. In: Proceedings of the 19th ACM SIGPLAN Symposium on Principles and Practice of Parallel Programming, pp. 317–328. PPoPP 2014, New York, NY, USA (2014)
15. Bronson, N.G., Casper, J., Chafi, H., Olukotun, K.: A practical concurrent binary search tree. In: ACM SIGPLAN Symposium on Principals and Practice of Parallel Programming (2010)
16. Nurmi, O., Soisalon-Soininen, E.: Uncoupling updating and rebalancing in chromatic binary search trees. In: Proceedings of the Tenth ACM SIGACT-SIGMOD-SIGART, pp. 192–198. PODS 1991, New York, NY, USA (1991)
17. Shane V. Howley, J.J.: A non blocking internal binary tree. SPAA, June 2012
18. Brown, T., Ellen, F., Ruppert, E.: A general technique for non-blocking trees. In: Proceedings of the 19th ACM SIGPLAN Symposium on Principles and Practice of Parallel Programming (PPoPP) (2014)
19. Valois, J.D.: Lock-Free Data Structures. Ph.D. thesis, Rensselaer Polytechnic Institute, Troy, NY, USA (1996)

A Dynamic Parameter Tuning Method for High Performance SpMM

Bin Qi[(✉)], Kazuhiko Komatsu, Masayuki Sato, and Hiroaki Kobayashi

Tohoku University, Sendai, Miyagi, Japan
{qi.bin.t,komatsu,masa,koba}@dc.tohoku.ac.jp

Abstract. Sparse matrix-matrix multiplication (SpMM) is a basic kernel that is used by many algorithms. Several researches focus on various optimizations for SpMM parallel execution. However, a division of a task for parallelization is not well considered yet. Generally, a matrix is equally divided into blocks for processes even though the sparsities of input matrices are different. The parameter that divides a task into multiple processes for parallelization is fixed. As a result, load imbalance among the processes occurs. To balance the loads among the processes, this paper proposes a dynamic parameter tuning method by analyzing the sparsities of input matrices. The experimental results show that the proposed method improves the performance of SpMM for examined matrices by up to 39.5% and 12.3% on average.

Keywords: Sparse matrix-matrix multiplication · Parallel execution · Parameter tuning.

1 Introduction

1.1 Background and Objective

Sparse matrix-matrix multiplication (SpMM) is one of fundamental operations in several areas of sciences such as machine learning [10], data analyzing [20], linear solver [23], and graph problems [8]. There have been many previous studies for SpMM on different platforms, such as CPU [9], GPU[16], VE (Vector Engine) [14], and the other multicore processors [3]. To make efficient use of resources provided by each platform, some researchers apply parallel programming techniques to improve performance [7,19], and utilize cache memories to reduce redundant data transfer [14].

However, a parallelization parameter that separates the matrix into blocks is fixed for any kind of input matrices and parallel processes. A fixed parameter will affect load balance among parallel processes, especially for imbalanced input matrices whose non-zero entries are different across rows. Therefore, parameter tuning must be taken into account for the sparsities of two input sparse matrices.

© Springer Nature Switzerland AG 2021
Y. Zhang et al. (Eds.): PDCAT 2020, LNCS 12606, pp. 318–329, 2021.
https://doi.org/10.1007/978-3-030-69244-5_28

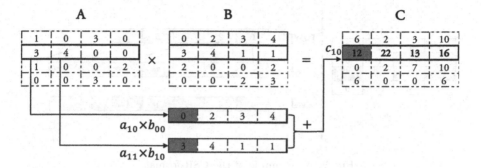

Fig. 1. The basic structure of the SpMM algorithm.

The contribution of this work is to balance loads among the parallel process for the acceleration of SpMM. To this end, this paper proposes a dynamic parameter tuning method. Each parallel process analyzes the sparsities of slices before the start of the calculations. Then, each process decides the block size to fit the capacity of the last-level cache (LLC) of a multi-core processor. Blocking with the dynamically-determined size can reduce the execution time of each parallel process.

The rest of the paper is organized as follows. Section 2 describes the related work on various optimizations for SpMM. Section 3 introduces the proposed dynamic parameter tuning method to improve the performance of SpMM. Section 4 evaluates the proposal. Finally, Sect. 5 concludes this paper.

2 Related Work

Figure 1 demonstrates a basic SpMM algorithm represented by operations on rows of the matrices. For each non-zero entry a_{ij} in matrix $\mathbf{A}$, b_{j*} in matrix $\mathbf{B}$ is loaded, and $a_{ij}b_{j*}$ is calculated. a_{ij} denotes the entry in the i-th row and the j-th column of matrix $\mathbf{A}$, and b_{j*} denotes a vector consisting of the j-th row of matrix $\mathbf{B}$. Intermediate results, namely $\hat{C}$, are then added up to c_{i*} in matrix $\mathbf{C}$. Most of the SpMM researches are based on this basic algorithm and improve the data structure for storing intermediate results [4,9].

CSR(Compressed Storage Row) [2] is the standard storage format for sparse matrices in SpMM algorithms. In this paper, input and output matrices are stored in the CSR format. The CSR format stores a sparse matrix by three 1-D arrays called as *rowptr*, *col*, and *val*. Figure 2 represents an example of a sparse matrix in the CSR format. The length of arrays *val* and *col* become $nnz(\mathbf{A})$, where $nnz(\mathbf{A})$ denotes the number of non-zero elements in matrix $\mathbf{A}$. Array *val* stores numerical values of each non-zero entry in the matrix while array *col* stores corresponding column index of each element in array *val*. Array *rowptr* stores pointers to the start and end positions of each row. Thus, the number of rows in a matrix can be known by getting the length of array *rowptr*.

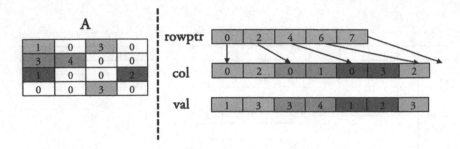

Fig. 2. An example of the CSR format.

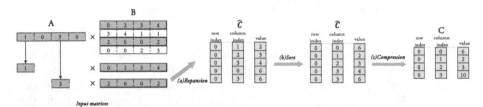

Fig. 3. An overview of the ESC algorithm.

There are many researches on accelerating SpMM on various computing architectures. Most of the researches focus on optimizations using CPUs and GPUs. These researches can be divided into five categories according to the data structure of intermediate results: Dense accumulation, Hash-based solutions, Merge methods, ESC (expansion, sorting, and compression) methods, and Hybrid methods.

Dense accumulations store and accumulate intermediate elements in dense arrays. Gilbert et al. [9] have proposed a classic CPU SpMM algorithm, also known as the Matlab algorithm. It uses a dense vector-based sparse accumulator (SPA) to temporarily store one current active row. However, if multiple rows of **A** are processed in parallel, the memory capacity required for the SPA becomes too large because of multiple dense vectors. Matam et al. [17] have developed a similar Matlab algorithm for a GPU implementation.

Hash-based solutions use hash tables to store and accumulate intermediate elements. The cuSPARSE [4] and nsparse [18] are classic hash-based libraries. Using a hash table to handle atomic operations in a scratchpad memory of GPU is very efficient. Therefore, the hash-based solutions perform well for rows that fit into a scratchpad memory to avoid storing hash tables in the global memory [6]. It is necessary to estimate memory requirements before the real calculation of SpMM.

The Merge methods use sorted lists to store intermediate results and combine lists using merge algorithms on GPU [11,22]. Elements with the same column index will be added up. Gremse et al. [12] have proposed an RMerge algorithm that iteratively merges submatrices in matrix **B** into the resulting matrix **C**. It is suitable for input matrices with evenly distributed short rows. However, since the space complexity is related to the size of lists, for irregular input matrices, load imbalance makes this method inefficient.

The ESC method is developed by Bell et al. [1]. Figure 3 shows an overview of the ESC algorithm. First, all candidates of non-zero entries generated by arithmetic operations are expanded into intermediate sparse vectors $\hat{C}$. Then, intermediate sparse vectors $\hat{C}$ are sorted according to the row and column indices. Finally, non-zeros at the same positions are compressed into the resulting matrix $\mathbf{C}$. Since the time complexity of the ESC algorithm is related to the size of $\hat{C}$, this method can achieve high performance when the number of non-zero entries to be processed in the rows is small.

The hybrid method is a mixture of the above algorithms. Liu et al. [16] have fully utilized GPU registers and the shared memory to achieve SpMM with a good load balance by combining the dense accumulation, the ESC algorithm, and their merge method. Xie et al. [24] have applied a deep learning model to predict the best algorithm for arbitrary sparse matrices. However, the cost of the preparation stage by a deep learning model is high. Mathias et al. [21] have proposed a policy to dynamically choose the best fitting algorithm for each row of the matrix. The policy is supported by a lightweight and multi-level analysis for matrices.

Besides, many researches utilize new multi-core architectures to achieve efficient SpMM. Chen et al. [3] have proposed a model that selects the most appropriate compressed storage formats for the sparse matrices to achieve the optimal performance of SpMM on the Sunway TaihuLight. Li et al. [14] have optimized the SpMM kernel on the vector engine of SX-Aurora TSUBASA [13]. They have proposed a novel hybrid method that combines SPA and ESC.

Most of the researches apply parallel programming techniques to make efficient use of resources provided by platforms. The first step of SpMM parallelization is to decide how to partition input sparse matrices. Yang et al. [25] have proposed a partitioning strategy of sparse matrices based on probabilistic modeling [15] of non-zero elements in a row for SpMV. However, this strategy is effective only for one input sparse matrix in the SpMV algorithm. Li et al. [14] take into account the characteristics of two input matrices. In the implementation of MPI parallelization, matrix $\mathbf{A}$ is evenly divided into *slices* according to the $nnz(\hat{C})$. By assigning each slice to an MPI process, the computations can be conducted in parallel on processing cores.

For the ESC algorithm, because the space complexity is related to the size of $\hat{C}$, Li et al. [14] further separate a slice into blocks that fit the size of the last-level cache (LLC) of a multicore processor. According to the evaluation results, they set the block size to 4096 as the best default parallelization parameter value for SX-Aurora TSUBASA.

However, the block size is fixed for any sparse matrix $\mathbf{A}$ and any parallel processes. The computation cost of a process by using an improper block size becomes high. As a result, load imbalance among parallel processes occurs.

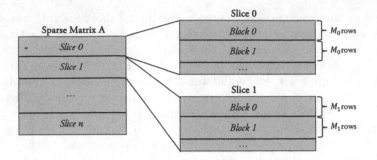

Fig. 4. SpMM parallel execution.

Algorithm 1. Parameter Tuning on i-th parallel process.

Require:

The number of intermediate non-zero entries that can be processed without cache misses: $nnz(cache)$;

The total intermediate nnz of $slice_i$: $nnz(slice_i)$;

The CSR arrays of $slice_i$: $rowptr$, col, val;

Ensure:

The number of rows in a block: M_i;

1: // Step1, get the number of rows in $slice_i$: $rows$

2:   $rows = rowptr.length - 1$;

3: // Step2, get the average number of intermediate non-zero entries in $slice_i$: $average$

4:   $average = nnz(slice_i)/rows)$;

5: // Get the nearest 2^n numver of $average$ by bit operations: Avg

6:   $Avg = theNearestPowerOfTwo(average)$;

7: // Step3, return the parameter that separate $slice_i$ into blocks: M_i

8:   $M_i = nnz(cache)/Avg$;

9: **return** M_i;

Therefore, it is necessary to dynamically adjust the block size for different MPI processes according to their assigned matrices.

3   Dynamic Parameter Tuning Method for High Performance SpMM

3.1   Concepts of the Proposed Method

To solve the load imbalance among parallel processes, this paper proposes a dynamic parameter tuning method of the block size for the ESC algorithm by analyzing the sparsities of input matrices. Figure 4 shows an overview of the proposed SpMM parallel execution. First, a sparse matrix **A** is evenly divided into n slices according to $nnz(\hat{C})$. A slice and matrix **B** are allocated to an

MPI process. $slice_i$ refers to the slice assigned to the i-th process. Then, each process separates the slice into blocks by dynamically defined parameter M_i to effectively utilize the LLC. M_i indicates the number of rows to be processed in the $slice_i$ at once. Finally, processes complete the calculation of slices, which means that non-zero entries of results are filled into matrix $\mathbf{C}$. The matrix $\mathbf{C}$ is a final result.

Even though each slice has the same number of intermediate elements, the distribution of these elements across rows in a slice may still be uneven. The imbalanced distribution of elements leads to the harmful effect of blocking. For a slice with very sparse element distribution, if parameter M_i is too small, the cost of the real execution stage becomes too high. Then, processes with inappropriate parameters lead to inefficiency in parallel processing. Blocking with proper parameter M_i can reduce the execution time of each process. Thus, instead of using a fixed default M_i, a dynamic parameter tuning method is required.

3.2 Dynamic Parameter Tuning Method

Algorithm 1 describes the dynamic parameter tuning method to decide M_i for each process. The dynamic parameter tuning method consists of three steps: First, each MPI process collects information for the slice. Second, the average nnz of rows is calculated. Third, the parameter M_i is decided.

As the first step, the number of rows in the slice is collected. Since matrix $\mathbf{A}$ is divided into slices according to the number of the intermediate non-zero entries in advance, each parallel process already knows the total intermediate nnz of each slice.

In the second step, each MPI process can obtain the average nnz of rows in a slice by calculation. This value represents the sparsity of the slice. To effectively use hardware resources, the nearest 2^n number to the average nnz of the rows are taken as Avg value. For example, if the average nnz of rows is calculated as 31, the proposed method employs 32 as Avg in the following step.

In the third step, each process dynamically sets a proper parameter M_i according to the diversity of sparse matrices and the cache of the computing architecture. First, this method assumes that $nnz(cache)$ to be the number of intermediate non-zero entries that can be processed without cache misses. Since $nnz(cache)$ depends on the cache size of the computer used, it can be set by empirical experiments in advance. Then, by considering the cache size, parameter M_i can be set as $\frac{nnz(cache)}{Avg}$.

Here, the cost of the proposed parameter tuning method is considered low. In Step 1, the number of rows can be obtained by only reading the array length in the CSR storage structure; in Step 2, bit operation can be used to obtain the nearest 2^n number; in Step 3, there are mainly division operations. However, these computational costs are not large compared with SpMM calculations.

Table 1. Evaluation matrices

Name	N	Intermed.*nnz*	Matrix	Name	N	Intermed.*nnz*	Matrix
Economics	207K	7.6M		Wind Tunnel	218K	626.1M	
FEM/Ship	141K	450.6M		FEM/Harbor	47K	156.5M	
Protein	36K	555.3M		FEM/Spheres	83K	463.8M	
2cubes_sphere	102K	27.5M		M133-b3	200K	3.2M	
Majorbasis	160K	19.2M		Offshore	260K	71.3M	
FEM/Accelerator	121K	79.9M		Cage12	130K	34.6M	
Hood	221K	562.M		Poisson3Da	14K	11.8M	
Circuit	171K	8.7M		eu-2005	86K	649M	
mario002	390K	12.8M		in-2004	138K	91M	
cit-Patents	377K	8.2M		Webbase	1M	69.5M	

Table 2. Hardware Specification

	Model	Frequency (GHz)	Physical cores	Performance per core	Library	Memory capacity(GB)	LLC cache(MB)
Vector engine	SX-Aurora TSUB-ASA	1.40	8	537.6 Gflop/s (SP) 268.8 Gflop/s (DP)	Frovedis	48	2 × 8

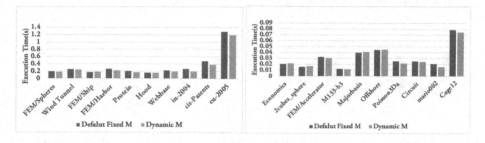

Fig. 5. Performance comparison.

4 Evaluation

4.1 Experimental Environment

The experiments are performed on the Vector Engine (VE) of SX-Aurora TSUB-ASA. Table 2 shows the specification of the hardware. The system architecture of SX-Aurora TSUBASA mainly consists of VEs and a vector host (VH) of an x86 Linux node. The VE is used as a primary processor for executing applications, whereas the VH is used as a secondary processor for executing a basic operating system (OS) function, i.e., Linux system calls. The VE has eight powerful vector cores, and each of them provides 537.6 GFlop/s of single-precision performance running at a 1.40 GHz frequency.

To evaluate the proposed method, 20 widely-used sparse matrices are selected from the University of Florida Sparse Matrix Collection [5]. Table 1 presents the basic information of the testing sparse matrices. $C = A^2$ is used to examine the performance, and all the experiments are carried out in single-precision. The proposed method is compared with the ESC that separates matrices by the default fixed parameter value [14].

4.2 Experimental Results and Discussion

Figure 5 shows the evaluation results of the SpMM kernel. From the comparison between the proposed method (ESC with dynamic M_i) and ESC with fixed M_i, the ESC with the proposed method improves almost all the matrices by up to 39.5% and 12.3% on average. The proposed method significantly improves performance of *in-2004*, *mario002*, and *cit-Patents*. These matrices have more imbalanced sparse structures between rows and MPI processes than the other matrices.

Table 3 shows *Avg* and M_i values on 8 MPI processes of *in-2004*. The difference in sparsity between slices is large, with $P7$ processing the sparsest slice of all. Thus, separating slices into the same size blocks on all processes is not proper. Figure 6 shows execution times on 8 vector cores of *in-2004*. When using default fixed M, $P7$ is the slowest process to calculate the slice, and much slower than the others. This leads to overall performance degradation. However,

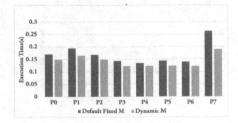

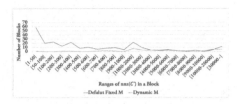

Fig. 6. Execution times on 8 vector cores.

Fig. 7. Frequency distribution of blocks with *in-2004*.

Table 3. Parameter values of parallel processes with *in-2004*

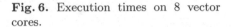

Process	P0	P1	P2	P3	P4	P5	P6	P7
Avg	2^9	2^7	2^8	2^{17}	2^{10}	2^{17}	2^{17}	2^3
M_i	2^9	2^{11}	2^{10}	2	2^8	2	2	2^{15}

the proposed method reduces the execution time of each process, especially *P7*. The slice allocated to *P7* is split into 243 blocks when using default fixed M, and 39 blocks when using the proposed method. Figure 7 shows the frequency distribution of these blocks on *P7*. The x-axis represents the ranges of $nnz(\hat{C})$ in a block. The y-axis represents the number of blocks. In the original ESC method, most of the blocks contain a very small number of non-zero entries with a default fixed parameter. After applying the proposed method, the ranges of $nnz(\hat{C})$ in blocks are more concentrated and then *P7* handles blocks more efficiently.

Figure 8 shows the average LLC cache hit ratios of 8 processes. The proposed method improves the LLC hit ratio by up to 4.1% and 1.9% on average for almost matrices. This increase in the LLC hit ratio contributes to performance improvement. As for *in-2004*, *mario002* and *cit-Patents*, even though the average LLC cache hit ratio of processes is decreased, the performance is improved by reducing the overhead due to too many blocks on the slowest parallel process.

In summary, the matrices that do not benefit from the proposed method are mainly regular matrices (belong to the first ten matrices in Table 1). For imbalanced matrices like *in-2004*, the proposed method can achieve performance improvements.

4.3   Cost of the Proposed Method

The execution of SpMM kernel has three stages: dynamic parameter tuning, real execution, and merge. For regular matrices, the dynamic parameter tuning stage is an extra cost. However, as mentioned in Sect. 3, the computational costs of the proposed parameter tuning method are low. Table 4 shows time costs of three stages in SpMM kernel in the case of *in-2004*. On average, the cost of dynamic parameter tuning on each parallel process is only 0.5% of the total execution time, which is not large compared with SpMM calculation.

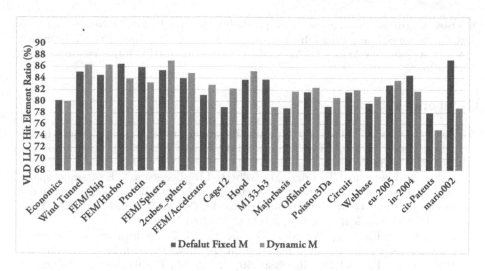

Fig. 8. LLC cache hit ratio.

Table 4. Time costs of three stages in SpMM kernel with *in-2004*

Process	Dynamic parameter tuning(ms)	Execution(ms)	Merge(ms)
P0	0.597	145.969	0.862
P1	0.409	161.216	1.099
P2	1.058	147.091	0.768
P3	0.721	122.303	0.236
P4	0.713	121.581	0.32
P5	0.716	122.125	0.354
P6	0.748	120.788	0.427
P7	1.086	178.564	1.199

5   Conclusions

In this paper, a dynamic parameter tuning method has been proposed for high performance SpMM by reducing the overhead of blocking and increasing the LLC hit ratio. As a result, the loads between MPI processes become balanced. The experimental results show that the ESC algorithm using our dynamic parameter tuning method improves the performance by up to 39.5% and 12.3% on average. Especially, the proposed method performs well for imbalanced input matrices. In the future, the scalability and portability of the proposed method will be studied, and more detailed performance analyses of the proposed method will be conducted. Then, the proposed method will be applied to a practical application.

Acknowledgment. This research was partially supported by MEXT Next Generation High Performance Computing Infrastructures and Applications R&D Program, entitled "R&D of A Quantum-Annealing-Assisted Next Generation HPC Infrastructure and its Applications."

References

1. Bell, N., Dalton, S., Olson, L.: Exposing fine-grained parallelism in algebraic multigrid methods. SIAM J. Sci. Comput. **34**(4), C123–C152 (2012). https://doi.org/10.1137/110838844
2. Buluç, A., Fineman, J.T., Frigo, M., Gilbert, J.R., Leiserson, C.E.: Parallel sparse matrix-vector and matrix-transpose-vector multiplication using compressed sparse blocks, pp. 233–244. SPAA 2009, Association for Computing Machinery, New York, NY, USA (2009). https://doi.org/10.1145/1583991.1584053
3. Chen, Y., Li, K., Yang, W., Xiao, G., Xie, X., Li, T.: Performance-aware model for sparse matrix-matrix multiplication on the sunway taihulight supercomputer. IEEE Trans. Parallel Distrib. Syst. **30**(4), 923–938 (2019). https://doi.org/10.1109/TPDS.2018.2871189
4. Dalton, S., Olson, L., Bell, N.: Optimizing sparse matrix–matrix multiplication for the gpu. ACM Trans. Math. Softw. **41**(4) (2015). https://doi.org/10.1145/2699470
5. Davis, T.A., Hu, Y.: The university of florida sparse matrix collection. ACM Trans. Math. Softw. textbf38(1) (2011). https://doi.org/10.1145/2049662.2049663
6. Deveci, M., Trott, C., Rajamanickam, S.: Performance-portable sparse matrix-matrix multiplication for many-core architectures. In: 2017 IEEE International Parallel and Distributed Processing Symposium Workshops (IPDPSW), pp. 693–702 (2017). https://doi.org/10.1109/IPDPSW.2017.8
7. Forum, M.P.: MPI: a message-passing interface standard. Technical report, USA (1994)
8. Gilbert, J., Reinhardt, S., Shah, V.: A unified framework for numerical and combinatorial computing. Comput. Sci. Eng. **10**, 20–25 (2008). https://doi.org/10.1109/MCSE.2008.45
9. Gilbert, J.R., Moler, C., Schreiber, R.: Sparse matrices in matlab: design and implementation. SIAM J. Matrix Anal. Appl. **13**(1), 333–356 (1992). https://doi.org/10.1137/0613024
10. Graf, D., Labib, K., Uznański, P.: Hamming distance completeness and sparse matrix multiplication (2018)
11. Green, O., Mccoll, R., Bader, D.: GPU merge path: a GPU merging algorithm (2014). https://doi.org/10.1145/2304576.2304621
12. Gremse, F., Höfter, A., Schwen, L.O., Kiessling, F., Naumann, U.:Gpu-accelerated sparse matrix-matrix multiplication by iterative row merging. SIAM J. Sci. Comput. **37**(1), C54–C71 (2015).https://doi.org/10.1137/130948811
13. Komatsu, K., et al.: Performance evaluation of a vector supercomputer sx-aurora tsubasa. In: Proceedings of the International Conference for High Performance Computing, Networking, Storage, and Analysis. SC 2018, IEEE Press (2018)
14. Li, J., Wang, F., Araki, T., Qiu, J.: Generalized sparse matrix-matrix multiplication for vector engines and graph applications. In: 2019 IEEE/ACM Workshop on Memory Centric High Performance Computing (MCHPC). pp. 33–42 (2019)
15. Li, K., Yang, W., Li, K.: Performance analysis and optimization for SPMV on GPU using probabilistic modeling. IEEE Trans. Parallel and Distrib. Syst. **26**(1), 196–205 (2015). https://doi.org/10.1109/TPDS.2014.2308221

16. Liu, W., Vinter, B.: A framework for general sparse matrix-matrix multiplication on GPUs and heterogeneous processors. J. Parallel Distrib. Comput. **85**, 47–61 (2015). https://doi.org/10.1016/j.jpdc.2015.06.010

17. Matam, K., Krishna Bharadwaj Indarapu, S.R., Kothapalli, K.: Sparse matrix-matrix multiplication on modern architectures. In: 2012 19th International Conference on High Performance Computing, pp. 1–10 (2012). https://doi.org/10.1109/HiPC.2012.6507483

18. Nagasaka, Y., Nukada, A., Matsuoka, S.: High-performance and memory-saving sparse general matrix-matrix multiplication for NVIDIA pascal GPU. In: 2017 46th International Conference on Parallel Processing (ICPP), pp. 101–110 (2017). https://doi.org/10.1109/ICPP.2017.19

19. Vingelmann, P., Fitzek, F.H.: NVIDIA Cuda, release: 10.2.89 (2020). https://developer.nvidia.com/cuda-toolkit

20. Ordonez, C., Zhang, Y., Cabrera, W.: The gamma matrix to summarize dense and sparse data sets for big data analytics. IEEE Trans. Knowl. Data Eng. **28**(7), 1905–1918 (2016). https://doi.org/10.1109/TKDE.2016.2545664

21. Parger, M., Winter, M., Mlakar, D., Steinberger, M.: Speck: accelerating GPU sparse matrix-matrix multiplication through lightweight analysis. In: Proceedings of the 25th ACM SIGPLAN Symposium on Principles and Practice of Parallel Programming, pp. 362–375. PPoPP 2020, Association for Computing Machinery, New York, NY, USA (2020). https://doi.org/10.1145/3332466.3374521

22. Satish, N., Harris, M., Garland, M.: Designing efficient sorting algorithms for many-core GPUs. In: 2009 IEEE International Symposium on Parallel Distributed Processing, pp. 1–10 (2009). https://doi.org/10.1109/IPDPS.2009.5161005

23. Schaub, M.T., Trefois, M., van Dooren, P., Delvenne, J.C.: Sparse matrix factorizations for fast linear solvers with application to laplacian systems. SIAM J. Matrix Anal. Appl. **38**(2), 505–529 (2017). https://doi.org/10.1137/16m1077398

24. Xie, Z., Tan, G., Liu, W., Sun, N.: IA-SPGEMM: an input-aware auto-tuning framework for parallel sparse matrix-matrix multiplication. In: Proceedings of the ACM International Conference on Supercomputing, pp. 94–105. ICS 2019, Association for Computing Machinery, New York, NY, USA (2019). https://doi.org/10.1145/3330345.3330354

25. Yang, W., Li, K., Mo, Z., Li, K.: Performance optimization using partitioned SPMV on GPUs and multicore CPUs. IEEE Trans. Comput. **64**(9), 2623–2636 (2015). https://doi.org/10.1109/TC.2014.2366731

Data Caching Based Transfer Optimization in Large Scale Networks

Xinxin Han[1,2], Guichen Gao[1,2], Yang Wang[1], Hing-Fung Ting[3],
and Yong Zhang[1(✉)]

[1] Shenzhen Institute of Advanced Technology, Chinese Academy of Sciences,
Shenzhen, China
{xx.han,gc.gao,yang.wang1,zhangyong}@siat.ac.cn
[2] University of Chinese Academy of Sciences, Beijing, China
[3] Department of Computer Science, The University of Hong Kong,
Pok Fu Lam, Hong Kong, China
hfting@cs.hku.hk

Abstract. With the tremendous increased scale of networks, data transferring plays a very important role in high performance computing. Traditionally, when a task is completed, the data associated with the task will be released to free the space for the following tasks. However, the following task may require the same data, which will be fetched back and the efficiency will be affected too. In this paper, we implement a data caching method so as to improve the performance of the system. When a task is completed, the data will be stored for a longer time. If the previous used data is requested before releasing, there will be no transferring cost. With the help of data caching, the performance of the network in handling real-time tasks is significantly improved while the traffic jamming is also decreased. We extend the previous single data model to multiple type of data among servers. Tasks arrive in an online fashion, and when handling a task on some network server, we have no information about the future tasks. Assuming the storage cost is submodular, which is well adopted in reality, we propose an online algorithm to minimize the total transfer and caching costs of the system. Via theoretical analysis, the competitive ratio of the algorithm is proved to be a constant.

Keywords: High performance computing · Data caching · Cloud computing · Competitive algorithm

1 Introduction

In high performance computing, processors or servers are often widely spread and each task might be assigned to any feasible processor. The data in corresponding

This work is supported by National Key R&D Program of China (2018YFB0204005), NSFC 12071460, Shenzhen research grant (KQJSCX20180330170311901, JCYJ20180 305180840138 and GGFW2017073114031767), Hong Kong GRF-17208019.

© Springer Nature Switzerland AG 2021
Y. Zhang et al. (Eds.): PDCAT 2020, LNCS 12606, pp. 330–340, 2021.
https://doi.org/10.1007/978-3-030-69244-5_29

to the task will be located at the server. Otherwise, the data will be transferred from other node and leads to some transfer cost. During these years, the size of networks is increased tremendously, and thus data transferring affects the performance of high performance computing a lot.

With the help of wired and wireless network, servers can be easily connected with each other. It can be used to realize resource sharing on the networks by data caching. What's more, storing data makes the content closer to the users. Therefore, caching data in advance on the server can reduce the jamming on the network and decrease the delay of providing the requested data for tasks execution. Data jamming is also an extensively studied issue [22]. In addition it is also cheaper to avoid downloading data from the remote cloud and it improves users' service satisfaction to some extent.

Since a few decades ago there has been widely studied work [1,3,8,13,14] for data caching problem which is a very significant issue in edge and cloud computing scheduling [7,19–21]. [14] studied the data caching problem with one data in the network. They assume that all the requests proposed by users can be finished by only one data which is limited in many practical applications. [13] consider the problem of minimizing the cost with a content which can be transferred to satisfy the request of the users. They present polynomial time optimal caching strategies to minimize the total monetary cost of all the service requests on a high-speed network following the caching paradigm.

It is common in classic caching problems to consider its homogeneous model. Based on the caching model with a content, in [13] they study data caching problem with constraint heterogeneous model in a fully connected topology to minimize the total cost of transfer. They present a dynamic programming to solve the problem. Meanwhile, they consider the homogeneous model where the caching cost and the transfer cost are the identical for the servers at the same time. Following the cost and network models in [13], Wang et al. [15] extend the analysis to multiple data items. In their homogeneous cost model, they give an offline optimal algorithm to minimize the total cost. In particular, it is sufficient to serve all the user request with a single copy when the ratio between the transfer cost and storage cost is low. What's more, based on different forms with the constraints on the number of copies and the volume of communications, they propose a polynomial optimal solutions. In addition, they also discuss data caching problem with a heterogeneous cost model and present a 2-approximation greedy algorithm to the general case. More details are considered in [17] about the homogeneous model of the data caching problem with different forms.

Submodularity is also considered in data caching issues. In [6,12], the authors consider the problem of minimizing the delay. They prove the model is NP-complete and show the problem is submodular. Based on the property, they give an approximation algorithm about the content placement problem. They extend their work to the case of different path delay in [12]. [9] study the problem of collaborative caching in cellular network. They show that the offline model has a submodular property with knowing the popularity of the contents, and they present an online algorithm about unlimited cache and limited cache respectively.

There also exists extensively research work in different forms for data caching involved in online model and offline model. In 1995, [10] proposed a greedy heuristic strategy to minimize the total cost of data caching. The heterogeneous model, similar with the Rectilinear Steiner Arborescence problem, is NP-hard [2, 11], to some extent. Based on the homogeneous model, Veeravalli et al. [13] proposed a polynomial time algorithm to minimize the total cost of all requests. Wang et al. [15] improved its time complexity to $O(m^2n)$. Meanwhile, they proposed a 2-approximate greedy algorithm for heterogeneous model with some practical limitations.

[4] presents an online competition algorithm to minimize the total caching cost. For the same objective, [5] presented an online algorithm for the collaborative cache model in multi-cluster collaborative systems. Wang et al. [16] presented a random competition algorithm under the condition that the server could be migrated. Based on a new homogeneous cache model, the transfer cost and storage cost of the data between servers are fixed respectively, Wang et al. [17] presented a dynamic programming algorithm with time complexity to minimize total cost. They also investigated the online model of the problem, using the idea of pre-caching to prove an approximation algorithm with a competitive ratio of 3. For the same data caching problem, [18] studied the semi-heterogeneous model, that is, the transfer cost of the data is the same, but the storage cost on the server is different, and they designed a 2-competitive algorithm.

In this paper, we consider data caching problem with multiple data items on high-speed networks (execution time of the requests and the transfer time of the data are ignored). Our objective is to minimize the total cost which contains the transmission cost and the storage cost. With this purpose, the servers can collaborate with each other to cache the data and provide them to the request. For the homogeneous model in which the storage cost of the data is a constant unit time and the transmission cost between two servers is fixed, we assume the storage cost doesn't increase when the new data is added into the server. In the second model, we suppose that the storage cost of each server is submodular for the data set. Upon these assumptions, our contributions in this paper can be summarized as follows.

- We propose an online algorithm for the homogeneous model, which can achieve competitive ratio equal to 3.
- Assuming that the storage cost of the server per unit time has a submodular property, we give a modified online competitive algorithm with a constant performance ratio.

The rest of the paper is organized as follows. In Sect. 2, we introduce the problem and some notations. We present our online algorithm for the homogeneous model and submodular model in Sect. 3 and in Sect. 4, respectively. Finally, we concludes the paper and give our views about future work which can be studied in Sect. 5.

2   Problem Statement

In this section, we first define some basic concepts. Then, we describe the data caching problem with k data items based on homogeneous model and submodular model. Main notations used in this paper are showed in Table 1.

Suppose that the set of n servers is $S = \{S_1, S_2, \cdots, S_n\}$ in a network. And there are m data items denoted as $M = \{1, 2, \cdots, m\}$. The servers are connected to each other through links. When a new request arrives at a server, the server will check whether there exists the data needed by the request. If it exists, the request will be satisfied immediately. Otherwise, the data will be transferred from the other servers where the data exists directly. The system model is shown in Fig. 1. We assume that any data transfer between two servers incurs a cost. In addition, there is storage cost if the data is stored on the server. We consider two online models in this paper. In our first model, we assume that the storage cost of the data is unrelated with the server and the number of data. That is to say, the cost of storing multiple data is the same as one data. Meanwhile, the transfer cost is denoted as λ. We call this model as homogeneous cost model. Then when the data is stored on the server, the holding cost unit time is represented as μ. In the second model, we consider that the storage cost of the data is submodular. Moreover, we present a modified online algorithm with a constant performance ratio.

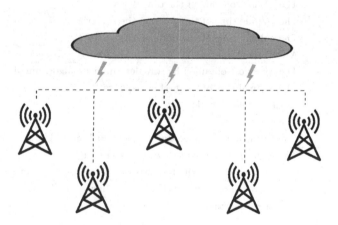

Fig. 1. The system model.

Definition 1. *Let T is a finite ground set, and f be a set function $f : 2^T \to \mathbb{R}$. f is a submodular function if for any $A \subseteq B \subseteq T$ we have:*

(1) $f(A) \leq f(B)$.
(2) $f(A \cup \{i\}) - f(A) \geq f(B \cup \{i\}) - f(B)$, for any $i \notin B$.

The submodularity of the storage function is reasonable in reality, the cost of data stored on the server increases as the data increases. Moreover, as the data set grows, the possible benefits of adding a data to the set are diminishing.

About two models, we decide whether the data is stored on the server and how long time to stored when every request is finished. The objective is to provide a caching strategy to minimize the total cost which contains the transfer cost and the storage cost. In the following sections, first, we propose an online competitive algorithm for the homogeneous model. Then, when the storage function is submodular we also give an online algorithm.

Table 1. Notations and parameters

Parameter	Definition/value
S	The server set
M	The data set
λ	The transfer cost between any two servers
μ	The storage cost of any data stored on the server S_i per unit time
S_j	The j-th server
D_j	The storge data set on the j-th server
$r^i_{jk} = (S_j, t^i_j, k)$	The i-th request on server S_j which comes at $t = t^i_j$ need data k
C_{jk}	A decision variable, $C_{jk} \in \{0, 1\}$
C_k	The number of copy of the data k
Δt	The holding time of the data on the server S_j
Δt	The holding time of the data on the server S_j
Δ	A very small interval
$C_{\mathcal{A}}$	The total cost obtained by any deterministic algorithm $\mathcal{A}$
OPT	The optimal cost of the first n requests
$f(\cdot)$	A submodular function
t_e	The current time
t^k_s	The first coming time of the request which need data k
$|r_{jk}|$	The number of the request which need data k
$|r_j|$	The total number of the requests on the server j up to time t_e
c	A positive constant which is not less than $min f(k)/\lambda$
l	A positive constant

3   Homogeneous Model Analysis

In this section, we give an 3-competitive algorithm for the online version of this problem. Without loss of generality, we assume all the data are copied on the server S_1 at the initial. $r^i_{jk} = (S_j, t^i_j, k)$ is the i-th request on server S_j which comes at $t = t^i_j$ need data k. Let $C_{jk} = 1$ if data k is on the server S_j and $C_{jk} = 0$ otherwise for every $k \in M, S_j \in S$. Thus $C_{1k} = 1$ for every $k \in M$ when $t = 0$, and whenever every data should be stored on at least one server, i.e. $C_k = \sum_{j=1}^{n} C_{jk} \geq 1$ for every $k \in M$. The homogeneous online algorithm we give operates as follows:

step 1: For each request r_{jk}^i, if there exists data k on server S_j when $t = t_j^i$, i.e. $C_j = 1$, the data is available; else, transfer data k from one of servers where the data is kept to server S_j and set $C_{jk} \leftarrow 1$, $D_j = D_j \cup \{k\}$.

step 2: Consider the holding time Δt of request r_{jk}^i, where $\Delta t = \lambda/\mu$. If the next request is coming when $t \leq t_j^i + \Delta t$, then $i \leftarrow i+1$ and go to step 1; else if there is no new request coming and $C_k > 1$ for $k \in D_j$ when $t = t_j^i + \Delta t$, then delete all the data from S_j, and set $C_{jk} \leftarrow 0$ for every $k \in M$; else if there exists a $C_l = 1$ for $l \in D_j$ when $t = t_j^i + \Delta t$, then S_j is the only server holding the data l, we need to keep all the data which is in D_j on S_j one more Δt: and if the next request is coming when $t \leq t_j^i + 2\Delta t_j$, then $i \leftarrow i+1$ and go to step 1; else, we need to transfer those data which only have a copy (denoted as D_j') to the server i which has the most data and continue to keep the data on the server until a new request comes, let $C_{jk} \leftarrow 0$, $D_i \leftarrow D_i \cup D_j'$, $D_j \leftarrow \emptyset$, $C_{ik} \leftarrow 1$ for $k \in D_i$, and then go to step 1.

Theorem 1. *The homogeneous online algorithm is a 3-competitive algorithm.*

Proof. Let $C_A(r_{jk}^i)$ define the cost of request r_{jk}^i obtained by the homogeneous online algorithm, $C'(r_{jk}^i)$ is the optimal cost of request r_{jk}^i gained by the offline algorithm. First, we analysis the ratio of request r_{jk}^i by considering the following cases.

case 1: If r_{jk}^i is the first request on server S_j for $1 \leq j \leq n$, without loss of generality, we assume the first request arrives at the server S_1, we have $C_A(r_{1k}^1) = \mu t_1$, $C'(r_{1k}^1) = \mu t_1$. And $C_A(r_{jk}^1) = \lambda$, $C'(r_{jk}^1) = \lambda$. Thus $C_A(r_{jk}^1)/C'(r_{jk}^1) = 1 < 2$.

case 2: Considering the cost of the i-th($i > 1$) request on server S_j, we discuss four cases in detail.

a. If request r_{jk}^i is coming during holding time Δt of request r_{jl}^{i-1} and

- If data $k \in D_j$, $C_A(r_{jk}^i) = C'(r_{jk}^i) \leq \mu\Delta \leq \lambda$, then $C_A(r_{jk}^i)/C'(r_{jk}^i) = 1 < 2$.
- If data $k \notin D_j$, $C_A(r_{jk}^i) = \mu\Delta + \lambda$, $C'(r_{jk}^i) = \lambda$, then $C_A(r_{jk}^i)/C'(r_{jk}^i) = (\mu\Delta + \lambda)/\lambda < 2$.

b. If request r_{jk}^i is coming after holding Δt_j and $C_k > 1$ for $k \in D_j$, delete all the data from S_j, $C_A(r_{jk}^i) = \mu\Delta t + \lambda = 2\lambda$, $C'(r_{jk}^i) = \lambda$, then $C_A(r_{jk}^i)/C'(r_{jk}^i) = 2$.

c. If there is no new request r_{jk}^i coming and there exists a data which has no copy on any server other than S_j when holding Δt, according to the algorithm, the data in set D_j should be stored on S_j one more Δt.

- If the new request is coming during the second holding time Δt and the data k is in D_j, then $\lambda < C_A(r_{jk}^i) \leq 2\mu\Delta t \leq 2\lambda, \lambda \leq C'(r_{jk}^i) \leq 2\lambda$, then $C_A(r_{jk}^i)/C'(r_{jk}^i) < 2$.

– If the new request is coming during the second holding time Δt and the data k is not kept in D_j, then $C_A(r^i_{jk}) = \mu\Delta t + \mu\Delta + \lambda \le 3\lambda, C'(r^i_j) = \lambda$, then $C_A(r^i_{jk})/C'(r^i_{jk}) < 3$. However, if there are h subcases simultaneously shown in Fig. 2, the optimal strategy is to store all the data on the same server. Then $C_A(\cdot)_h = h(\lambda + \mu\Delta) + h\lambda, C'(\cdot)_h \ge (h-1)\lambda + \lambda + \mu\Delta + (h-1)\lambda$, then $C_A(\cdot)_h/C'(\cdot)_h \le 3h/(2h-1) \le 2$ (when $h \ge 2$).

d. If there is no new request r^i_{jk} coming and has some data no copy on any server other than S_j when holding Δt, according to the algorithm, the data should be stored on S_j one more Δt, the new request is still not coming during the second holding time Δt, then transfer the data of D_j to server S_i. And we assume that multiple data packages for transmission, so the cost of transfering multiple data is no less than the cost of one data. After time Δ a new request is coming, $C_A(r^i_{jk}) = \mu \cdot 2\Delta t + 2\lambda + \mu\Delta = 2\lambda + 2\lambda + \mu\Delta, C'(r^i_{jk}) = \min\{2\lambda + \mu\Delta, \lambda + \lambda + \mu(2\Delta t + \Delta)\}$, then $C_A(r^i_{jk})/C'(r^i_{jk}) < 2$. And there might be h substructure simultaneously among servers. $C_A(\cdot)_h = 2h\lambda + k\lambda + \mu t + (h-1)\lambda, C'(\cdot)_h = k\lambda + (h-1)\lambda + 2\lambda + \mu t$, where $k \ge h$, then $C_A(\cdot)_h/C'(\cdot)_h \le (3h-1)/(h+1) \le 2$ (when $h \le 3$).

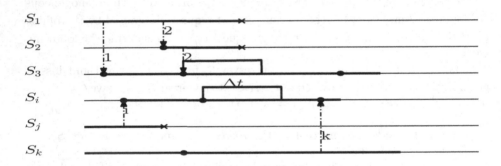

Fig. 2. Case c in the homogeneous data caching model.

Suppose C_A is the total cost of the first n requests obtained by the homogeneous online algorithm, OPT is the total optimal cost of the first n requests obtained by an offline algorithm. We have

$$C_A/OPT = \left(\sum_{1 \le j \le n} C_A(r_j) + m\lambda \right) / \sum_{1 \le j \le n} C'(r_j) < \left(\sum_{1 \le j \le n} 3C'(r^i_j) + m\lambda \right) / \sum_{1 \le j \le n} C'(r^i_j)),$$

Therefore

$$\lim_{n \to \infty} C_A/OPT = 3.$$

■

To better explain the algorithm 1, as shown in the Fig. 3, we give an example of the homogeneous data caching problem solved by our algorithm 1. Set $\mu = 2$, $\lambda = 6$. The black bold line indicates data is saved, and the dotted line represents the data transfer. Black dots indicate requests, and the cross indicates the deletion of data. According to the algorithm, we know that $\Delta t = 3$. The total cost obtained by algorithm 1 is $28 + 44 + 24 + 12 + 16 = 124$, where $C(S_1) = 2 \times 8 + 6 + 2 \times 2 = 28$, $C(S_2) = 6 + 2 \times 3 + 6 + 2 \times 3 + 6 + 2 \times 3 + 2 \times 4 = 44$, $C(S_3) = 6 + 2 \times 3 + 6 + 2 \times 3 = 24$, $C(S_4) = 6 + 2 \times 3 = 12$, $C(S_5) = 6 + 2 \times 5 = 16$. And the optimal total cost is $7 \times 6 + 14 \times 2 + 2 \times 2 + 6 = 80$.

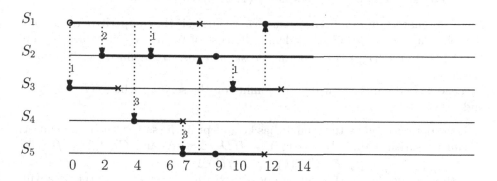

Fig. 3. An example of algorithm 1 for the homogeneous data caching model.

4   Submodular Model Analysis

We present another online algorithm about data caching problem with submodular model in this section. The worst performance ratio of the homogeneous online algorithm is poor when the storage cost on the server is submodular for data. When we analyze the performance of the algorithm, we consider an adversary in this paper which deliberately choose difficult data to maximize the ratio of the algorithm.

Supposed that $f(\cdot)$ is the unit cost of the data storged on the server j and it is a submodular function $f : 2^M \rightarrow \mathbb{R}$. We modify the second step of the homogeneous online algorithm as follows which is called the modified submodular algorithm(MS Algorithm for short).

step 2: If $f(D_j)\Delta_t \leq \lambda$ and there is a new request coming, then go to step 1. Else if $f(D_j)\Delta_t \geq \lambda$, delete the data set A such that $A = max[f(D_j) - f(B)]$, where the remaining data set B contains the data which has no copy other than server S_j or $(t_e - t_s^k)/|r_{jk}| < t_e/|r_j|$. Then set $D_j \leftarrow B$ consider $f(D_j)\Delta_t \leq \lambda$ one more time at $t = t_e$, if the request is coming, then go to step 1; else transfer the data which has no copy to the cheapest server such that $min[f(D_i \cup D) - f(D_i)]$, where D is the data subset of D_j which has no copy other than server S_j, then set $D_i \leftarrow D_i \cup D$ and $D_j \leftarrow \emptyset$.

Where t_e is current time, t_s^k is the first coming time of the request which need data k, $|r_{jk}|$ is the number of the request which need data k and $|r_j|$ is the total number of the requests on the server j up to time t_e. The formula $t_e/|r_j|$ represents the average time of the data. If data is needed frequently or if new data is needed recently, these data should be stored as much as possible. Thus we show this condition by formula $(t_e - t_s^k)/|r_{jk}| < t_e/|r_j|$. We analysis the competitive ratio of the MS Algorithm next. And we assume that the ratio between the smallest cost of storing one data for unit time on a server and the transfer cost is not less than a constant c.

Theorem 2. *The MS Algorithm is a $O(1)$-competitive algorithm.*

Proof. Similar to the homogeneous model, we analysis the competitive ratio from the following cases. Let $C_A(r_{jk}^i)$ define the cost of request r_{jk}^i obtained by the MS Algorithm, $C'(r_{jk}^i)$ is the optimal cost of request r_{jk}^i gained by the offline algorithm.

   case 1: If there is a new request coming on the server j when $f(D_j)\Delta_t \leq \lambda$, and

- if the data needed by the request has been kept in the server, then the request can be satisfied directly. $C_A(r_{jk}^i) = f(D_j)\Delta \leq \lambda$, and $C'(r_{jk}^i) = f(k)\Delta$. $C_A(r_{jk}^i)/C'(r_{jk}^i) \leq f(D_j)\Delta/f(k)\Delta \leq \lambda/f(k)\Delta \leq lc$.
- if the data needed by the request doesn't exist in the server, then the data will be transferred from other servers to the server j. $C_A(r_{jk}^i) = f(D_j)\Delta + \lambda \leq 2\lambda$, and $C'(r_{jk}^i) = \lambda$. So we have $C_A(r_{jk}^i)/C'(r_{jk}^i) \leq 2$.

   case 2: If there is no new request coming on the server j when $f(D_j)t = \lambda$, and we still consider whether the needed data is kept in the server. The two cases are the same as the case 1.

   Suppose C_A is the total cost of the first n requests obtained by the MS online algorithm when request r_n is coming, OPT is the total optimal cost of the first n requests obtained by an offline algorithm. We have

$$C_A/OPT = (\sum_{1 \leq j \leq n} C_A(r_j) + m\lambda)/\sum_{1 \leq j \leq n} C'(r_j) < (\sum_{1 \leq j \leq n} lcC'(r_j^i) + m\lambda)/\sum_{1 \leq j \leq n} C'(r_j^i)),$$

Therefore

$$\lim_{n \to \infty} C_A/OPT = O(1).$$

∎

5   Conclusion and Future Work

Caching the data in advance on the server reduces the traffic on the network, which can improve users' service satisfaction. In this paper, we consider two models of data caching problem. First, we studied the homogeneous model and proposed an 3-competitive online algorithm. In addition, when the storage function of the data is submodular, we presented a modified online algorithm with the performance ratio is a constant.

In the future research work, we can consider the heterogeneous model of the data caching model. When the transfer cost of the data between two servers not only related to the data but also the server, the model will be more challenging.

References

1. Cao, P., Irani, S.: Cost-aware WWW proxy caching algorithms. In: Proceedings of the USENIX Symposium on Internet Technologies and Systems, p. 18 (1997)
2. Charikar, M., Halperin, D., Motwani, R.: The dynamic servers problem. In: Proceedings of the Ninth Annual ACM-SIAM Symposium on Discrete Algorithms (SODA 1998), pp. 410–419 (1998)
3. Fan, X., Cao, J., Mao, H., Wu, W., Zhao, Y., Xu, C.: Web access patterns enhancing data access performance of cooperative caching in IMANETs. In: Proceedings of the 17th IEEE International Conference on Mobile Data Management, pp. 50–59 (2016)
4. Gharaibeh, A., Khreishah, A., Khalil, I.: An O(1)-competitive online caching algorithm for content centric networking. In: IEEE INFOCOM 2016–35th Annual IEEE International Conference on Computer Communications (2016)
5. Gharaibeh, A., Khreishah, A., Ji, B., Ayyash, M.: A provably efficient online collaborative caching algorithm for multicell-coordinated systems. IEEE Trans. Mob. Comput. **15**(8), 1863–1876 (2016)
6. Golrezaei, N., Shanmugam, K., Dimakis, A., Molisch, A., Caire, G.: FemtoCaching: wireless video content delivery through distributed caching helperss. In: IEEE INFOCOM, pp. 1107–1115 (2012)
7. Li, F., Yu, D., Yang, H., Yu, J., Holger, K., Cheng, X.: Multi-armed bandit based spectrum scheduling algorithms in wireless networks: a survey. IEEE Wireless Commun. Mag. **27**(1), 24–30 (2020)
8. Karger, D., et al.: Web caching with consistent hashing. In: Proceedings of the 8th International Conference on World Wide Web, pp. 1203–1213 (1999)
9. Ostovari, P., Wu, J., Khreishah, A.: Efficient online collaborative caching in cellular networks with multiple base stations. In: 2016 IEEE 13th International Conference on Mobile Ad Hoc and Sensor Systems (MASS), Brasilia, pp. 136–144 (2016)
10. Papadimitriou, C., Ramanathan, S., Rangan, P., et al.: Multimedia information caching for personalized video-on-demand. Comput. Commun. **18**(3), 204–216 (1995)
11. Papadimitriou, C.H., Ramanathan, S., Venkat Rangan, P.: Optimal information delivery. In: Staples, J., Eades, P., Katoh, N., Moffat, A. (eds.) ISAAC 1995. LNCS, vol. 1004, pp. 181–187. Springer, Heidelberg (1995). https://doi.org/10.1007/BFb0015422
12. Shanmugam, K., Golrezaei, N., Dimakis, A., Molisch, A., Caire, G.: FemtoCaching: wireless content delivery through distributed caching helpers. IEEE Trans. Inf. Theory **59**(12), 8402–8413 (2013)
13. Veeravalli, B.: Network caching strategies for a shared data distribution for a predefined service demand sequence. IEEE Trans. Knowl. Data Eng. **15**(6), 1487–1497 (2003)
14. Wang, Y., He, S., Fan, X., Xu, C., Sun, X.: On cost-driven collaborative data caching: a new model approach. IEEE Trans. Parallel Distrib. Syst. **30**(3), 662–676 (2018)

15. Wang, Y., Veeravalli, B., Tham, C.: On data staging algorithms for shared data accesses in clouds. IEEE Trans. Parallel Distrib. Syst. **24**(4), 825–838 (2013)
16. Wang, Y., Shi, W., Hu, M.: Virtual servers co-migration for mobile accesses: online versus off-line. IEEE Trans. Mob. Comput. **14**(12), 2576–2589 (2015)
17. Wang, Y., He, S., Fan, X., Xu, C., Sun, X.: On cost-driven collaborative data caching: a new model approach. IEEE Trans. Parallel Distrib. Syst. **30**(3), 662–676 (2019)
18. Wang, Y., et al.: Cost-driven data caching in the cloud: an algorithmic approach. Accepted by INFOCOM 2021 (2021)
19. Yu, L., Cai, Z.: Dynamic scaling of virtualized networks with bandwidth guarantees in cloud datacenters. In: The 35th Annual IEEE International Conference on Computer Communications (INFOCOM 2016) (2016)
20. Zheng, X., Cai, Z., Li, J., Gao, H.: A study on application-aware scheduling in wireless networks. IEEE Trans. Mob. Comput. **16**(7), 1787–1801 (2017)
21. Zhu, T., Shi, T., Li, J., Cai, Z., Zhou, X.: Task scheduling in deadline-aware mobile edge computing systems. IEEE Internet Things J. **6**(3), 4854–4866 (2019)
22. Zou, Y., et al.: Fast distributed backbone construction despite strong adversarial jamming. In: INFOCOM 2019 (2019)

Second-Order Convolutional Neural Network Based on Cholesky Compression Strategy

Yan Li[1,2], Jing Zhang[1], and Qiang Hua[1(✉)]

[1] Key Laboratory of Machine Learning and Computational Intelligence of Hebei Province, College of Mathematics and Information Science, Hebei University, Baoding 071002, China
huaq@hbu.cn
[2] Research Center for Applied Mathematics and Interdisciplinary Sciences, Beijing Normal University at Zhuhai, Beijing 519087, China

Abstract. In the past few years, Convolution Neural Network (CNN) has been successfully applied to many computer vision tasks. Most of these networks can only extract first-order information from input images. The second-order statistical information refers to the second-order correlation obtained by calculating the covariance matrix, the fisher information matrix, or the vector outer product operation on the local feature group according to the channels. It has been shown that using second-order information on facial expression datasets can better capture the distortion of facial area features, while at the same time generate more parameters which may cause much more computational cost. In this article we propose a new CNN structure including layers which can (i) incorporate first-order information into the covariance matrix; (ii) use eigenvalue vectors to measure the importance of feature channels; (iii) reduce the bilinear dimensionality of the parameter matrix; and (iv) perform Cholesky decomposition on the positive definite matrix to complete the compression of the second-order information matrix. Due to the incorporation of both first-order and second-order information and the Cholesky compression strategy, our proposed method reduces the number of parameters by half of the SPDNet model, and simultaneously achieves better results in facial expression classification tasks than the corresponding first-order model and the reference second-order model.

Keywords: Convolution Neural Network · First-order information · Second-order pooling · Cholesky decomposition

1 Introduction

In recent years, Convolution Neural Network (CNN) is very effective in classification tasks [1–3]. It has been proved that as the network layers deepens, the expressive ability of the network will be stronger [2]. At the same time, the number of parameters is rapidly increasing, and the phenomenon of gradient disappear or explosion may also appear. The use of grouped convolutional layers and small-size filters can achieve large-size filter receptive fields while reducing the number of parameters [3]. In order to reduce the over-fitting of the fully connected layer, the regularization method of dropout was proposed [1]. The above

© Springer Nature Switzerland AG 2021
Y. Zhang et al. (Eds.): PDCAT 2020, LNCS 12606, pp. 341–352, 2021.
https://doi.org/10.1007/978-3-030-69244-5_30

mentioned CNNs are good representative deep neural network models in image classification problems. However, these traditional CNNs only extract first-order statistical information such as do not consider the relationship between channels from the input image, which do not incorporate second-order statistical information such as covariance.

Inspired by the extraction process of manual features, a type of method was proposed to replace the fully connected layer with a second-order pooling strategy, mainly using the covariance descriptor in CNN [4–7]. Among them, [4] proposed a bilinear model containing two feature extractors, and the output was multiplied by the outer product and pooled to obtain the image description operator. [5] gave the theoretical basis for matrix back propagation of deep networks. To improve the performance of large-scale classification, [6] proposed a method of matrix power normalization covariance, and the forward and backward propagation formulas of nonlinear matrix functions were also given. [7] encoded the covariance matrix calculated by the covariance descriptor from the local image features. In [8, 9], a bilinear pooling compression strategy is proposed to capture the pairwise correlation between feature channels, which can effectively perform fine-grained recognition tasks, but it is non-parametric or only designed for specific classification forms that limits the network's representation ability and applicability. By incorporating second-order statistical information, these proposed models can capture more fine-grained features in the input data, thus improving the performance of CNN models. However, most of current work has a common problem of generating large number of parameters and being easy to overfit. In this paper, we propose a second-order compression strategy to reduce the number of parameters without decreasing the accuracy rate noticeably. The method is developed on the good mathematical properties of the second-order information matrix which is a symmetric positive definite matrix. We find a decomposition method of this matrix to reduce the number of parameters based on the symmetry property, and at the same time preserve most of the useful information. The effectiveness of our method is verified on the facial expression dataset through conducting extensive experiments.

2   Related Work

Recent years, in the imagenet visual recognition challenge, the appearance of alexnet pushed deep learning to be the research focus [10]. One of the characteristics of deep learning is that as the network structure gets deeper and deeper, there will be gradient disappearance or explosion [10–12]. Several optimization strategies were given to tackle the problem [13–16]. In these networks, although non-linear operations are performed through activation, pooling or other strategies, the resulting operations still cannot extract second-order statistics, and they do not involve higher-order statistical information such as covariance matrix.

The second-order statistics in the era of manual features, especially the Regional Covariance Descriptor (RCD) [15, 17–20], can effectively handle visual recognition tasks. However, only part of these early work combined RCD with deep learning models. More recently, in [21], a method for saliency calculation tasks with RCD as input was developed. In 2017, Yu K et al. proposed a second-order compression method for bilinear dimensionality reduction of the covariance matrix and parameter vectorization output [22]. In 2018, a statistical analysis method was used to compress the second-order information matrix [23]. In [24], the popular SPDNet was developed mainly using bilinear dimensionality reduction, eigenvalue decomposition rectification, and the riemannian manifold structure was mapped to the Euclidean space through the logarithmic transformation. Based on SPDNet, a method was proposed in [25] which combined covariance matrix with a BiRe block by bilinear dimensionality reduction and eigenvalue rectification, and achieved good results in facial expression classification. [26] proposed a second-order statistical information channel attention mechanism, which adaptively rescaled features by considering feature statistics higher than the first-order. This channel attention mechanism enables the network model to focus on more feature information and enhance the ability of discriminative learning. To summarize, it can be seen that the combination of second-order statistical information and deep learning for image classification has achieved preliminary results in recent years [9, 23–26]. Our work proposes a new method based on the network structure of [24] and [25]. Firstly, we use eigenvalue vector to measure the importance of the channels by eigenvalue decomposition of the information matrix. Secondly, the Cholesky decomposition is used to compress the second-order information matrix to reduce the number of parameters.

3 Methodology

The traditional CNN structure is mainly composed of convolutional layer, maximum or average pooling layer, activation function layer and fully connected layer. These layers only capture first-order statistics. The second-order statistical information is the second-order correlation of the feature group obtained by calculating the covariance matrix, the fisher information matrix or the vector outer product operation based on the channel.

3.1 Basic SPDNet Model

Since SPDNet [24] is a representative model which incorporates second-order statistical information, we firstly introduce SPDNet briefly in this section as the basic model to be compared with our proposed model. The basic structure is shown in Fig. 1. The components and design ideas of each module is explained as follows.

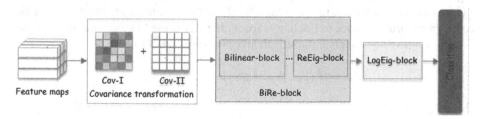

Fig. 1. Basic structure of SPDNet.

Covariance Transformation [24]. A set of feature maps is given as $F = \{f_1, f_2, \ldots, f_n\} \in R^d$, where $n = w * h$ represents the number of pixels in a feature map, and d the channel dimension. The covariance matrix C can be expressed as follows (**Cov-I**):

$$C = \frac{1}{(n-1)} \sum_{i=1}^{n} (f_i - \mu)(f_i - \mu)^T \tag{1}$$

where μ is the mean vector of F, i.e., $\mu = \frac{1}{n} \sum_{i=1}^{n} f_i$.

Matrix C is a positive semi-definite matrix, which can be regularized by adding trace C^+ (**Cov-II**) to ensure the positive definite property. γ is the regularization coefficient.

$$C^+ = C + \gamma \, trace(C)I \tag{2}$$

Bilinear Transformation [22]. Given a matrix M of size dxd as input, the introduced parameter matrix W is used to compute the output matrix Y:

$$Y = WMW^T \tag{3}$$

where $W \in R^{dxd'}$ is trainable, and d' controls the size of the output matrix Y. In order to preserve the rank of the output covariance matrix, the orthogonal constraint is enforced on the parameter matrix W.

Eigenvalue Rectification (ReEig) [24]. The eigenvalue rectification layer is similar to the nonlinear unit layer of relu activation function in traditional convolutional neural networks. If Y_{K-1} is the input SPD matrix, Y_K is the output matrix and $\in$ is the eigenvalue rectification threshold. The k-th eigenvalue rectification layer can be expressed as:

$$Y_K = f_r^k(Y_{K-1})U_{K-1}max(\in I, \sigma_{K-1})U_{K-1}^T \tag{4}$$

where Y_{K-1} can be decomposed into U_{K-1} and Σ_{k-1} by eigenvalues as $Y_{K-1} = U_{K-1}\Sigma_{K-1}U_{K-1}^T$, and the max operation is applied at the matrix element level. The BiRe-block is obtained by combining the Bilinear-block and ReEig-block (see Fig. 1).

Logarithm of Eigenvalue (LogEig) [24]. Since the SPD matrix is located on the riemannian manifold, the LogEig layer maps the elements in the riemannian manifold with the lie group structure back to Euclidean space. If Y_{K-1} is used as the input matrix, the f_l^k of the LogEig layer applied to the k-th layer is expressed as follows:

$$Y_K = f_l^k(Y_{K-1}) = U_{K-1}log(\Sigma_{K-1})U_{K-1}^T \tag{5}$$

where Y_{K-1} can be decomposed into U_{K-1} and Σ_{K-1} as $Y_{K-1} = U_{K-1}\Sigma_{K-1}U_{K-1}^T$, and then the logarithm of Y_{K-1} can be obtained by taking logarithm of the eigenvalue diagonal matrix elements.

3.2   The Proposed Second-Order CNN Model

Based on the SPDNet architecture, a new structure of CNN is proposed in this section focusing on modeling second-order statistics, which mainly includes three parts. Firstly, we perform linear operations on the second-order and the first-order to compute the information matrix M^* containing both the two levels of feature information, and then perform eigenvalue decomposition on M^*. The obtained eigenvalue p is then used as a weight vector to measure the importance of the channel. Secondly, to reduce the size of parameters, Cholesky decomposition is implemented on the positive definite matrix M^*. After that, we take the logarithm of the diagonal elements of the lower triangular matrix H obtained by the decomposition to obtain H^*, which is vectorized to the classifier. Finally, we give the formula to determine the size of the bilinear rectification transformation weight matrix empirically.

Although the components of the proposed CNN are similar to the those of SPDNet, the main difference is that the new CNN model combines both first-order and second-order information, and a compress strategy is proposed to reduce the size of information matrix. In the following, we explain the main process of the proposed model.

(1)   **The combination of covariance transformation with first-order information**
   Given a set of feature maps, we firstly perform global average pooling on the feature maps by channel to represent first-order information, and then the symmetric positive definite matrix C^+ calculated by formula 2 is simply summed with the first-order information. It is defined as follows (**Cov-IV**):

$$M^* = C^+ + \mu\mu^T \tag{6}$$

   where $\mu \in R^d$ is the mean vector of the feature maps. Since C^+ is the positive definite form of the covariance matrix C, this linear representation can reflect both the second-order information and the first-order information.

(2)   **Channel weighting** Given the current input be the second-order information matrix M^* computed by formula (6), and the eigenvalue or singular value of M^* is decomposed into an eigenvector matrix U and the eigenvalue matrix V^* as in (7):

$$M^* = UV^*U^T \tag{7}$$

   Let σ be the rectification threshold for the positive small eigenvalue in V^*, the weighting vector p can be expressed as:

$$p = (\lambda_1, \lambda_2, \sigma, \cdots, \lambda_i, \cdots, \lambda_n)^T \tag{8}$$

where $\lambda_i, i = 1, 2, \ldots, n$ is the eigenvalues of M^*.

The above two steps of the proposed CNN model can be demonstrated by the following Fig. 2. Let $F = \{f_1, f_2, \cdots, f_n\}$ represent the feature map of the last layer of convolution, which is the input of the eigenvalue transformation. Vector p computed based on formula (8) is used as the weight vector and the feature map F is point-multiplied as the output feature map: $F = \{\lambda_1 f_1, \lambda_2 f_2, \sigma f_3, \ldots\ldots, \lambda_n f_n\}$.

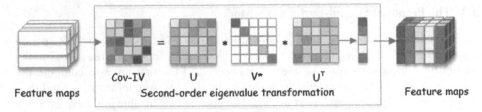

Fig. 2. The process of second-order eigenvalue transformation

(3) **Cholesky compression** Suppose the input matrix $M^* = [m_{ij}^*]$ (i,j = 1,2,...,n) is a symmetric positive definite matrix. We then perform Cholesky decomposition of matrix M^*. Firstly, M^* can be expressed as the product of a lower triangular matrix $H=[h_{ij}](h_{ii} > 0)$ and H^T with positive diagonal elements, and the diagonal elements are also eigenvalues of H.

$$M^* = H * H^T \tag{9}$$

The logarithm of the eigenvalue of the lower triangular matrix H can be expressed as.

Cholesky Decomposition Logarithmic Transformation (Cholesky-DLT):

$$H^* = \begin{cases} h_{11}^* = log\left(\sqrt{m_{11}^*}\right), h_{i1}^* = \frac{m_{i1}^*}{h_{11}^*}, (i = 2, 3, \ldots, n) \\ h_{kk}^* = log\left(\sqrt{m_{kk}^* - \sum_{i=1}^{k-1} \left(h_{ki}^*\right)^2}\right), (k = 2, 3, \ldots, n) \\ h_{ik}^* = \frac{1}{h_{kk}^*}\left(m_{ik}^* - \sum_{j=1}^{k-1} h_{ij}^* h_{ki}^*\right), (i = k+1, \ldots, n) \end{cases} \tag{10}$$

(4) **BiRe_ block** When $\omega = 1/2^n$ is the scaling factor that controls the output matrix, we can always merge BiRe blocks into one BiRe_ block, where n is the number of BiRe blocks, and the size of the parameter matrix W is fixed at ω d × d. Compared with multiple BiRe blocks, the parameter amount is obviously reduced.

The overall architecture and main steps of the proposed second-order CNN model can be shown in Fig. 3. By combing the first-order and second-order information in Cov-IV, the model can capture more useful features in the input data, and Cholesky decomposition is then used to reduce the number of parameters. Finally the vectorized H* is output to the classifier for label prediction.

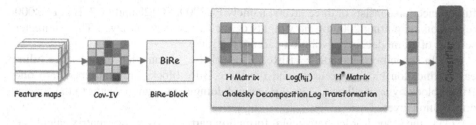

Fig. 3. The basic steps of our model

4 Experimental Results and Analyses

In this section, to validate the effectiveness the proposed CNN model, it is used on some benchmark facial expression recognition datasets. The results are compared with other sever standard first-order and second-order models and their different variations.

4.1 Datasets

We have used the wild natural facial expression SFEW2.0 [27, 28] dataset and the real-world emotional facial RAF dataset [29]. The SFEW 2.0 [27] and AFEW [28] datasets use landmarks and alignment images obtained from a hybrid model based on parts [30], containing 1394 images of which 958 are for training and 436 are for test. The RAF dataset [29] was prepared by collecting images from various search engines, and the author provided manually annotated facial landmarks. The RAF dataset contains 15,331 images, which are labeled with seven basic emotion categories, of which 12,271 are used as the training set and 3068 are used as the test set. The facial expression dataset contains 7 categories, including neutral, disgust, angry, fear, happy, sad, and surprise. Certainly, there are other image-based datasets, such as EmotioNet [31] and FER2013 [32]. EmotioNet is the largest facial expression recognition dataset available, but its tags are not complete. FER-2013 contains relatively small image size and does not contain RGB information.

4.2 Experimental Setup

(1) Preprocessing part: In our experiment, the data is preprocessed by flipping, cropping and rotating, and the image size is cropped to 100 * 100 pixels. In formula 2, the coefficient for regularizing the covariance matrix (Cov-II) is set as $\gamma = 10^{-4}$. In formula 4, the rectification threshold is set to $\in = 10^{-4}$. In formula 8, the eigenvalue is replaced by the positive value $\sigma = 10^{-4}$.

(2) In first-order models: The basic first-order model of Baseline [25], Conv (3×3, $O = 64$, $S = 1$, $P =$ same) is used in convolution layers, where the convolution kernel size is 3×3; the number of output channels O is 64; the step size S is 1, and the adaptive padding input feature map is consistent with the output feature map. The maximum pooling is then adopted and the pooling window is 2×2 with step size 2. The number of channels of the output feature map is 256 passing through the relu layer. Finally, the

full connection consists of three layers, namely FC2000, FC128 and FC7. E.g, FC2000 represents outputting 2000 of the weights of the full connection layer. The parameter settings of all models in convolution part are the same.

(3) In second-order models: 1BiRe, 3BiRe, BiRe_ means that one or three bilinear rectification blocks are used, and the three BiRe blocks are combined into one BiRe_ block by ω coefficient. $256 + p$ means adding the weight p vector to the current convolution layer.

(4) In the second-order statistical information part, the covariance matrix calculated by convolution at the last layer is $C \times C$, and the same experimental setup is adopted according to [25], namely, BiRe1 weight matrix [C/2, C], BiRe2 weight matrix [C/4, C/2], and BiRe3 weight matrix [C/8, C/4]. It can be seen that the number of parameters in second-order part is relatively large. Model5 synthesizes Bire1–Bire3 into BiRe_ according to BiRe-block, the weight matrix W has the size of [C/8, C], and the number of parameters is reduced relative to Bire1–Bire3. The p vector is the weight vector of the feature maps output by convolution at the last layer.

4.3   The Used Models for Comparisons

Table 1 lists eight models applied in our four groups of comparative experiments. In the first group, Baseline [25] and Model1 adopt standard first-order convolution neural network [25], which distinguishes the importance of channel by adding p weight vector in the last convolution layer of Model1.The other three groups are second-order models, the common difference is that LogEig and Cholesky-DLT transform methods are used in each group of comparative experiments. Cholesky-DLT not only realizes mapping but also compresses.

4.4   Analysis of Results

The eight models mentioned in the previous section are implemented on the three image datasets introduced in Sect. 4.1. The classification performance with respect to accuracy is shown in Table 2.

The accuracy of Baseline [25] is the lowest among the eight models since it only adopts the first-order VGG network. Model1 adds the weight vector to the channel, which improves the accuracy a little (say, by 0.61% on RAF data) compared with Baseline [25]. The accuracy of the three groups of second-order models on the RAF and SFEW datasets of our Models 3–5 and the compared models have been improved, and the accuracy of the RAF dataset is more than 81%. Models 3–5 achieve much better classification performance due to the incorporation of second-order information and Model4 achieves the best results in average on the three datasets. For example, the accuracies are improved more than 1% on all datasets using Model4 than those using Model2. Overall, it is found that our proposed second-order CNN model is more suitable for feature learning of image data. It is worth noting that in Models 3–5, while improving the accuracy, the Cholesky decomposition is used to replace the original matrix with a lower triangular matrix to reduce the full connection parameter by half. On some datasets, the effect of applying the second-order model is not ideal. This may because the second-order covariance matrix or the number of parameters is too large. Even this happens, the result is still acceptable.

Table 1. The used first and second order network models

Model	First, Second order layer	Fc layer
Baseline [25]	Conv64, 96, 128, 128, 256, 256	Fc2000, 128, 7
Model1	Conv64, 96, 128, 128, 256, 256 + p	Fc2000, 128, 7
Model2 [25]	Cov-II, 1BiRe, LogEig	Fc2000, 128, 7
Model3	Cov-IV, 1BiRe, Cholesky-DLT	Fc2000, 128, 7
SPDNet3 [24]	Cov-II, 3BiRe, LogEig	Fc2000, 128, 7
Model4	Cov-IV, 3BiRe, Cholesky-DLT	Fc2000, 128, 7
SPDNet1 [24]	Cov- II, BiRe_, LogEig	Fc2000, 128, 7
Model5	Cov-IV, BiRe_, Cholesky-DLT	Fc2000, 128, 7

Table 2. On the accuracy of the RAF, SFEW2.0 and AFEW facial expression datasets. AFEW data is feature data, and the first-order part does not participate in training.

Datasets	RAF-DB	SFEW-DB	AFEW-DB
Baseline [25]	79.54%	45.30%	–
Model1	80.15%	46.50%	–
Model2 [25]	80.03%	48.00%	48.113%
Model3	**81.15%**	**48.90%**	49.120%
SPDNet3 [24]	80.62%	47.32%	50.120%
Model4	**81.79%**	**48.30%**	**50.223%**
SPDNet1 [24]	80.51%	47.48%	**50.269%**
Model5	**81.09%**	**48.50%**	**50.594%**

Table 3. The trainable parameters in second-order CNN models

Model	Model2 [25]	SPDNet3 [24]	SPDNet1 [24]	Model3	Model4	Model5
SO-params	32k	43k	8k	32k	43k	8k
FC-params	33027k	2307k	2307k	16771k	1315k	1315k
Total-params	33059k	2350k	2315k	16803k	1358k	1323k

Table 3 shows the number of trainable parameters. In convolution part, the number of parameters of all models are the same. In the second-order part, the vectorized output of lower triangular matrix obtained by the Cholesky decomposition is used to reduce the full connection parameters of Models 3–5. Therefore, compared with Model2, Models 3–5 have much less parameters. Especially Models 4–5 can achieve superior classification performance with the least performance cost of parameters among the six second-order models.

Fig. 4 shows the confusion matrix of model5 on RAF and AFEW datasets, clearly depicting the probability value of each category predicting to other categories.

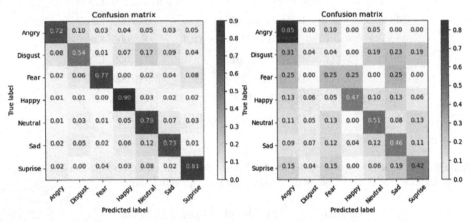

Fig. 4. Confusion matrix of the RAF dataset (left) and the AFEW dataset (right)

5   Conclusions

In this paper, our purpose is to explore the second-order matrix transformation method and parameter compression strategy for CNN structure, thus improving the image classification performance without increasing parameters noticeably. Specifically, our work has three main contributions: First, first-order information is used to linearly transform the second-order information matrix which could capture more information of input data; second, the eigenvalue decomposition is implemented on the second-order information matrix, and then the eigenvalue vector is used to measure the importance of the convolutional neural network channels; third, using Cholesky to decompose a symmetric positive definite matrix for effectively reducing the number of parameters. The proposed CNN structure is applied to the task of facial expression classification. The experimental results show that our framework can outperform the representative first-order network and some basic second-order benchmark models. Using the models with our proposed structure, the classification accuracy on three benchmark datasets has been improved with less parameters. Future work includes studying the combination of second-order statistical information and the attention mechanism which could be more effective in the task of image segmentation or image classification.

Acknowledgements. This work is supported by NSF of Hebei Province (No. F2018201096), NSF of Guangdong Province (No. 2018A0303130026), the Key Science and Technology Foundation of the Educational Department of Hebei Province (ZD 2019021), and NSFC (No. 61976141).

References

1. Krizhevsky, A., Sutskever, I., Hinton, G.: ImageNet classification with deep convolutional neural networks. In: NIPS, pp. 1106–1114 (2012)
2. He, K., Zhang, X., Ren, S, Sun, J.: Deep residual learning for image recognition. In: IEEE Conference on Computer Vision and Pattern Recognition, pp. 770–778 (2016)
3. Simonyan, K., Zisserman, A.: Very deep convolutional networks for large-scale image recognition. In ICLR, pp. 340–352 (2015)
4. Lin, T., RoyChowdhury, A., Maji, S.: Bilinear CNN models for fine-grained visual recognition. In: ICCV, pp. 1449–1457 (2015)
5. Ionescu, C., Vantzos, O., Sminchisescu, C.: Matrix backpropagation for deep networks with structured layers. In: IEEE International Conference on Computer Vision, pp. 990–1002 (2015)
6. Li, P., Xie, J., Wang, Q., Zuo, W.: Is second-order information helpful for large-scale visual recognition? In: ICCV, pp. 1205–1213 (2017)
7. Tuzel, O., Porikli, F., Meer, P.: Region Covariance: A Fast Descriptor for Detection and Classification. In: Leonardis, A., Bischof, H., Pinz, A. (eds.) ECCV 2006. LNCS, vol. 3952, pp. 589–600. Springer, Heidelberg (2006). https://doi.org/10.1007/11744047_45
8. Gao, Y., Beijbom, O., Zhang, N., Darrell, T.: Compact bilinear pooling. In: International Conference on Computer Vision and Pattern Recognition, pp. 317–326 (2016)
9. Kong, S., Fowlkes, C.: Low-rank bilinear pooling for fine-grained classification. In: Conference on Computer Vision and Pattern Recognition, pp. 880–890 (2017)
10. Krizhevsky, A., Sutskever, I., Hinton, G.: ImageNet classification with deep convolutional neural networks. Adv. Neural Inf. Process. Syst. **25**(2) (2012)
11. He, K., Zhang, X., Ren, S., Sun, J.: Deep residual learning for image recognition. In Proceedings of the IEEE Conference on Computer Vision and Pattern Recognition, pp. 770–778 (2016)
12. Simonyan, K., Zisserman, A.: Very deep convolutional networks for large-scale image recognition. In: ICLR, pp. 553–572 (2015)
13. Duchi, J., Hazan, E., Singer, Y.: Adaptive Subgradient Methods for Online Learning and Stochastic Optimization. J. Mach. Learn. Res. **12**(7), 257–269 (2011)
14. Kingma, D., Ba, J.: Adam: A method for stochastic optimization. Computer Science, pp. 1135–1142 (2015)
15. Srivastava, N., Hinton, G., Krizhevsky, A., Salakhutdinov, R.: Dropout - a simple way to prevent neural networks from overfitting. J. Mach. Learn. Res. **15**(1), 1929–1958 (2014)
16. Zeiler, M.: ADADELTA: An Adaptive Learning Rate Method. arXiv.org (2012)
17. Tuzel, O., Porikli, F., Meer, P.: Pedestrian detection via classification on riemannian manifolds. In: IEEE TPAMI, pp. 1980–1991 (2008)
18. Pennec. X., Fillard, P., Ayache, N.: A riemannian framework for tensor computing. In: IJCV, pp. 990–1112 (2006)
19. Ha, M., San-Biagio, M., Murino, V.: Log-Hilbert-Schmidt metric between positive definite operators on Hilbert spaces. In NIPS, pp. 1124–1134 (2014)
20. Sra, S.: A new metric on the manifold of kernel matrices with application to matrix geometric means. In: NIPS, pp. 2010–2023 (2012)
21. Xu, X., Mu, N., Zhang, X.: Covariance descriptor based convolution neural network for saliency computation in low contrast images. In: International Joint Conference on Neural Networks, pp. 1220–1229 (2016)
22. Yu, K., Salzmann, M.: Second-order convolutional neural networks. In: Computer Vision and Pattern Recognition, pp. 1305–1316 (2017)

23. Yu, K., Salzmann, M.: Statistically-Motivated Second-Order Pooling. In: Ferrari, V., Hebert, M., Sminchisescu, C., Weiss, Y. (eds.) ECCV 2018. LNCS, vol. 11211, pp. 621–637. Springer, Cham (2018). https://doi.org/10.1007/978-3-030-01234-2_37
24. Huang, Z., Van Gool, L.: A Riemannian network for SPD matrix learning. In: Internaltional Conference on Computer Vision and Pattern Recognition, pp. 2036–2042 (2017)
25. Acharya, D., Huang, Z., Paudel, D.: Covariance pooling for facial expression recognition. In: IEEE Computer Society Conference on Computer Vision and Pattern Recognition Workshops, pp. 480–487 (2018)
26. Dai, T., Cai, J., Zhang, Y., Xia, S., Zhang, L.: Second-order attention network for single image super-resolution. In: International Conference on Computer Vision and Pattern Recogintion, pp. 1123–1135 (2019)
27. Dhall, A., et al.: Collecting large, richly annotated facial expression databases from movies. IEEE Multimedia 19(3), 34–41 (2012)
28. Dhall, A., et al.: Emotion recognition in the wild challenge 2014: Baseline, data and protocol. In: ACM ICMI (2014)
29. Li, S., Deng, W, Du, J.: Reliable crowdsourcing and deep locality-preserving learning for expression recognition in the wild. In: The IEEE Conference on Computer Vision and Pattern Recognition, pp. 89–96 (2017)
30. Zhu, X., Ramanan, D.: Face detection, pose estimation and landmark estimation in the wild. In: IEEE Conference on Computer Vision and Pattern Recognition, pp. 3431–3445 (2012)
31. Benitez-Quiroz, C.F., Srinivasan, R., Martinez, A.M.: Emotionet: an accurate, real-time algorithm for the automatic annotation of a million facial expressions in the wild. In: IEEE Conference on Computer Vision and Pattern Recognition, pp. 5562–5570 (2016)
32. Goodfellow, I.J.: Challenges in representation learning. Neural Netw. 64(C), 59–63 (2015)

Submodular Maximization with Bounded Marginal Values

Ruiqi Yang[1], Suixiang Gao[1], Changjun Wang[2(✉)], and Dongmei Zhang[3]

[1] School of Mathematical Sciences, University of Chinese Academy Sciences,
Beijing 100049, People's Republic of China
{yangruiqi,sxgao}@ucas.ac.cn
[2] Department of Operations Research and Information Engineering, Beijing
University of Technology, Beijing 100124, People's Republic of China
wcj@bjut.edu.cn
[3] School of Computer Science and Technology, Shandong Jianzhu University,
Jinan 250101, People's Republic of China
zhangdongmei@sdjzu.edu.cn

Abstract. We study the problem of maximizing non-monotone submodular functions subject to a p-independence system constraint. Although the submodularity ratio has been well-studied in maximizing set functions under monotonic scenario, the defined parameter may bring hardness of approximation for the maximization of set functions in the non-monotonic case. In this work, utilizing a lower bound for the marginal values, we investigate the Repeated Greedy introduced by (Feldman et al. 2017) and obtain a parameterized performance guarantee for the above constrained submodular maximization problem.

Keywords: Submodular maximization · Independence systems · Approximation algorithms

1 Introduction

Submodularity is a ubiquitous concept in many fields such as operations research, industry engineering, computer science, and economics, among others. Constrained submodular function maximization has received much attentions recently, including cardinality, knapsack, and matroid constraints. Nemhauser et al. [24] provide a greedy procedure, which plays an important role in developing algorithms for submodular maximization problems. The greedy chooses elements with maximum marginal gain in every iteration until the cardinality constrained is met and gets a $(1 - e^{-1})$-approximation for the cardinality constrained submodular maximization. On the other hand, no algorithm can yield an approximation guarantee with $(1 - e^{-1} + \varepsilon)$ for any $\varepsilon > 0$, under the assumption P$\neq$ NP [9].

We follow with interest the parameterized submodular maximization under the non-monotone setting in this work. A parameter *curvature* α of submodular

© Springer Nature Switzerland AG 2021
Y. Zhang et al. (Eds.): PDCAT 2020, LNCS 12606, pp. 353–361, 2021.
https://doi.org/10.1007/978-3-030-69244-5_31

functions introduced in [7] is used to study the worst case of the performance guarantee for the submodular maximization. They presented a $(1 - e^{-\alpha})/\alpha$-approximation for the cardinality constrained submodular maximization. Das and Kempe [8] introduce a parameter of *submodularity ratio* for any nonnegative set function, which is used to measure the magnitude of submodularity of the utility function. For maximizing a cardinality constrained set function characterized by the submodularity ratio γ, they [8] get a $(1 - e^{-\gamma})$-approximation. Bian et al. [2] introduce the *generalized curvature* α. Together with the submodularity ratio γ, they present a tight $(1 - e^{-\alpha\gamma})/\alpha$-approximation for the former cardinality constrained maximization. Wang et al. [27] study a matroid constrained maximization of set functions by introducing a parameter of *elemental curvature*. Gong et al. [15] introduce a parameter of *generic submodular ratio* $\hat{\gamma}$, and provide a $\hat{\gamma}(1 - e^{-1})(1 - e^{-\hat{\gamma}} - O(\varepsilon))$-approximation for the above matroid constrained set function maximization. Feige and Izsak [10,13] introduce a parameter of *supermodular degree* $\mathcal{D}_f^+$ and offer a $1/(p(\mathcal{D}_f^+ + 1) + 1)$-approximation for the set function maximization with more generalized p-extendible system constraint. In addition, Bogunovic et al. [3] study the maximization of set functions which combine with parameters of *supermodularity ratio, inverse generalized curvature*, and *subadditivity ratio* and so on. The more work for the parameterized submodular maximization can be found in [1,20,21]. To estimate the utility value in an effective type, Horel and Singer [19] introduce the ε-approximately submodular maximization and give a $(1 - e^{-1} - O(k\varepsilon))$-approximation for the cardinality constrained maximization problem. Utilizing the parameter of the submodular ratio γ, Qian et al. [25] provide a $(1 - \varepsilon)(1 - e^{-\gamma})/(1 + \varepsilon)$-approximation for the above cardinality constrained maximization problem. For the p-matroid constrained approximately submodular maximization, Gölz and Procaccia [14] present a $1/(p + 1 + 4r\varepsilon/(1 - \varepsilon))$-approximation, where r is the maximum size of feasible solutions. Singer and Hassidim [18] provide a $1/(p + 1)$-approximation for the above p-matroid constrained approximately problem.

1.1   Related Work

Unlike the monotone setting of the submodular maximziation, there exists a natural obstacle in the study of non-monotonic case. Feige et al. [11] present a series of algorithms for the non-monotone unconstrained submodular maximization. By randomly choosing elements, they show the approximation ratio gets 0.25. By utilizing local search add and delete operations, they present a 1/3-approximation. Further, they provide a 0.4-approximation by randomizing the above local search. If the utility function reduces to symmetric, the above randomized algorithm approaches to 0.5-approximation. In last, they show there does not exist $(0.5 + \varepsilon)$-approximation under the assumption of P $\neq$ NP, where $\varepsilon > 0$ is any give constant. In addition, Based on local search technique, Buchbinder et al. [6] provide a randomized linear time 0.5-approximation. The latest work introduced in [4], gets a deterministic 0.5-approximation for this unconstrained submodular maximization problem.

We briefly investigate the maximization of submodular functions with complex constraints as follows. In [4], they get a deterministic $1/e$-approximation for the cardinality constrained submodular maximization. For the maximization of submodular functions with matroid constraint, Buchbinder and Feldman [5] derive a randomized 0.385-approximation algorithm. Recently, Han et al. [17] present a deterministic 0.25-approximation with time complexity of $O(nr)$, where r denotes the rank of the matroid. Moreover, Lee et al. [22] derive a $1/(p+1+1/(p-1)+\varepsilon)$-approximation for the submodular maximization with p-matroid. For the general p-independence system constraint, Gupta et al. [16] yield an almost $1/(3p)$-approximation, which has time complexity of $O(n\ell p)$, where ℓ denotes the maximum size of any feasible solutions. Keeping the same of the time complexity, Mirzasoleiman et al. [23] improve the approximation ratio to about $1/(2p)$. Feldman et al. [12] provide a $1/(p+\sqrt{p})$-approximation with an improved time complexity of $O(n\ell\sqrt{p})$.

The former defined parameters usually prefer to the monotone setting, but may fail in the non-monotone scenario. In this paper, we consider the non-monotonic submodualr maximization problem subject to a more general p-independence system constraint. By introducing a lower bound of marginal values, we investigate the Repeated Greedy and get a parameterized performance guarantee for this problem.

The rest of the paper is organized as follows. Section 2 gives necessary preliminaries and definitions. Section 3 investigates the Repeated Greedy and the theoretical analysis are summarized in Sect. 4. In last, Sect. 5 offers a conclusion for our work.

2   Preliminaries

Given a ground set $\mathcal{V} = \{u_1, ..., u_n\}$, and a utility function $f : 2^{\mathcal{V}} \to \mathbb{R}_+$. The utility function f is submodular, if

$$f(A) + f(B) \geq f(A \cup B) + f(A \cap B), \forall A, B \subseteq \mathcal{V}.$$

For any $A, B \subseteq \mathcal{V}$, denote $f_A(B) = f(A \cup B) - f(A)$ as the amount of change by adding B to A. For the sake of brevity and readability, we set $f_A(u) = f(A \cup \{u\}) - f(A)$ for any singleton element $u \in \mathcal{V}$. The set function f is monotonic (non-decreasing) if $f_A(u) \geq 0$. The submodularity ratio measures how close of f being submodular and we restate it as following.

Definition 1. *([8]) Given an integer k, the* submodularity ratio *of a non-negative set function f with respect to $\mathcal{V}$ is*

$$\gamma_{\mathcal{V},k}(f) = \min_{A \subseteq \mathcal{V}, B:|B| \leq k, A \cap B = \emptyset} \frac{\sum_{u \in B} f_A(u)}{f_A(B)}.$$

Let k be the maximum size of any feasible solution, and omit $k, \mathcal{V}$ and f for clarity. Bian et al. [2] introduce an equivalent formulation of submodularity ratio by the largest scalar γ such that

$$\sum_{u \in B} f_A(u) \geq \gamma \cdot f_A(B), \forall A, B \subseteq \mathcal{V}, A \cap B = \emptyset.$$

Based on the definition of submodularity ratio, there are some properties which can be concluded by the observation below.

Observation 1 *[2]. If f is a non-decreasing set function, then*

1. $\gamma \in [0,1]$ and
2. f is submodular iff $\gamma = 1$.

Similarity, we restate a parameter of the generic submodularity ratio introduced by [15], which measures how many times the marginal gains of joining an element to a set than that of a superset.

Definition 2. *[15] The* generic submodularity ratio *of any increasing set function f is defined as the largest scalar γ such that*

$$f_S(u) \geq \gamma \cdot f_T(u), \forall S \subseteq T \subseteq \mathcal{V}, u \in \mathcal{V} \setminus T.$$

It immediately deserves the following properties.

Observation 2 *[15]. If f is an increasing set function, then*

1. $\gamma \in (0,1]$,
2. f is submodular iff $\gamma = 1$, and
3. $\sum_{u \in B} f_A(u) \geq \gamma \cdot f_A(B), \forall A, B \subseteq \mathcal{V}.$

In this work, we study the maximization of submodular functions losing monotonicity. Under the non-monotonic case, the marginal value $f_A(u)$ may be negative and then the submodularity ratio can not be bounded. We consider a cut maximum example depicted as in Fig. 1. By the stated example, it follows that the submodularity ratio $\gamma = -1(< 0)$. Now we investigate the setting of all the marginal values are lower bounded by a parameter. Assume $\theta \in \mathbb{R}_+$, we have $f_A(u) \geq -\theta$ for any $u \in \mathcal{V}, A \subseteq \mathcal{V}$.

We now recall the p-independence system. Let $\mathcal{I} = \{A_i\}_i$ be a finite collection of subsets chosen from $\mathcal{V}$. The tuple $(\mathcal{V}, \mathcal{I})$ is defined as an *independence system* if for any $A \in \mathcal{I}$, $A' \subseteq A$ implies that $A' \in \mathcal{I}$. The sets of $\mathcal{I}$ are called the *independent sets* of the independence system. An independent set B contained in a subset $X \subseteq \mathcal{V}$ is a *base* (basis) of X if no other independent set $A \subseteq X$ strictly contains B. By these terminologies, the p-independence system is defined as follows.

Definition 3. *([12]) An independence system $(\mathcal{V}, \mathcal{I})$ is a p-independence system if, for every subset $X \subseteq \mathcal{V}$ and for any two bases B_1, B_2 of X, we have $|B_1|/|B_2| \leq p$.*

In last, we also assume that there exist utility function value oracle and independence oracle, respectively. That is to say, for any $A \subseteq \mathcal{V}$, one can query the value of $f(A)$ and query if A is in $\mathcal{I}$ or not in constant time.

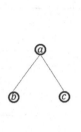

B \ A	∅	{a}	{b}	{c}	{a,b}	{a,c}	{b,c}	{a,b,c}
∅	1	1	1	1	1	1	1	1
{a}	1	1	1	1	1	1	1	1
{b}	1	1	1	1	1	1	1	1
{c}	1	1	1	1	1	1	1	1
{a,b}	3	1	1	-1	1	1	1	1
{a,c}	3	1	-1	1	1	1	1	1
{b,c}	1	1	1	1	1	1	1	1
{a,b,c}	$+\infty$	1	-1	-1	1	1	1	1

Fig. 1. We consider the maximization of cut functions with graph $G = (\mathcal{V}, E)$, where $\mathcal{V} = \{a, b, c\}$, $\mathcal{E} = \{(a, b), (a, c)\}$, and $f(S) = |\{(u, v) \in \mathcal{E} : |S \cap \{u, v\}| = 1\}|$ for any $S \subseteq \mathcal{V}$. For any $A, B \subseteq \mathcal{V}$, we present the ratios of $\sum_{u \in B} f_A(u)$ and $f_A(B)$ by the above table. In addition, for some subsets, the marginal value may be zero, we set $1 = \frac{0}{0}$ for clarity.

3 Algorithm

In this section, we applied the Repeated Greedy introduced by [12] to the maximization of submodular functions with bounded marginal values. The main pseudo-codes are summarized as Algorithm 1.

The algorithm processes r rounds and firstly runs a standard greedy procedure at the beginning of every round $i \in [r]$. Denote $\mathcal{V}_i$ as the ground set of elements and set S_i as the output returned by the greedy algorithm with respect to round i. Based on the output set S_i, the algorithm calls on any sub-procedure, which is used to the unconstrained submodular scenario. Let S_i' be the output returned by the sub-procedure according to S_i. For clarity, we denote Greedy and ALG as the above greedy and unconstrained sub-procedure, respectively. The algorithm keeps the solution sets $\{S_i, S_i'\}_i$ in all rounds and outputs the set with the maximum utility value.

4 Theoretical Analysis

As the feasibility of returned solution T can be derived by Observation 1 [12], we omit it here. We consider the approximation of the Repeated Greedy applied to the setting of maximizing submodular functions with bounded marginal values. Denote O as an optimal solution for maximizing f over $\mathcal{V}$ under p-independence system $\mathcal{I}$, i.e., $O = \arg \max_{S \subseteq \mathcal{V}, S \in \mathcal{I}} f(S)$. We study the properties of the greedy procedure applied to the above setting by the following lemma. Assume S be set of elements returned by Greedy according to their added order, and set $S = \{e_1, ..., e_k\}$. Let S^j be the set of first j elements added to S, i.e., $S^j = \{e_1, ..., e_j\}$. Denote $\delta_j = f(S^{j-1} + e_j) - f(S^{j-1})$, by the assumption of marginal values, we have $\delta_j \geq -\theta$. In addition, ALG represents as an α-approximation for

---

Algorithm 1. [12] Repeated Greedy($\mathcal{V}, f, \mathcal{I}, r$)

1: **Initialization** $i = 1, \mathcal{V}_1 \leftarrow \mathcal{V}$
2: **repeat**
3:     $S_i \leftarrow$ Greedy($\mathcal{V}_i, f, \mathcal{I}$)
4:     $S_i' \leftarrow$ ALG(S_i, f)
5:     $\mathcal{V}_{i+1} \leftarrow \mathcal{V}_i \setminus S_i$
6: **until** $i = r$
7: **return** $T \leftarrow \arg\max_{i \in [r]} \{f(S_i), f(S_i')\}$

---

the unconstrained submodular maximization, we summarize it as the following lemma.

Lemma 1. *[12] For any round i, we have*

1. $f(S_i) \geq \frac{1}{p+1} \cdot f(S_i \cup (O \cap \mathcal{V}_i))$, *and*
2. $f(S_i') \geq \alpha \cdot f(S_i \cap O)$.

Before presenting the average value of the union $\{S_i\}_i \in [r]$ and the optimum, by following lemma, we give an extended property of submodular functions.

Lemma 2. *Let $g : 2^\mathcal{V} \to \mathbb{R}_+$ be a submodular function and assume $g_S(e) \geq -\theta$ for any subset S and element e. Let S be a random subset chosen from $\mathcal{V}$, where each element appears with probability at most q. Then,*

$$\mathbb{E}[g(S)] \geq g(\emptyset) - \theta \cdot q|S|.$$

Proof. Let $S = \{e_1, ..., e_{|S|}\}$ be a subset with an arbitrary order and S_i be the first i elements in S according to this order. Denote X_i as the indicator for the event of $e_i \in S$. By the basic knowledge of probability theory, we yield

$$\mathbb{E}[g(S)] = \mathbb{E}[g(\emptyset) + \sum_{i=1}^{|S|} X_i \cdot g_{S_{i-1} \cap S}(e_i)]$$

$$\geq g(\emptyset) + \sum_{i=1}^{|S|} \mathbb{E}[X_i] \cdot g_{S_{i-1}}(e_i)$$

$$\geq g(\emptyset) - \theta \cdot \sum_{i=1}^{|S|} \Pr[X_i]$$

$$\geq g(\emptyset) - \theta \cdot q|S|,$$

where the first inequality follows from the submodularity, the last two inequalities are obtained by the assumptions of the marginal values and probability, respectively. $\square$

We uniformly choose S form $\{S_i\}_{i \in [r]}$ with probability $1/r$ and assume the size of each of greedy solutions is bounded by k. Further, let $g(S) = f(S \cup O)$ for any $S \subseteq V$, then g is non-negative and submodular. By Lemma 2, we obtain

$$\frac{1}{r} \cdot \sum_{i=1}^{r} f(S_i \cup O) = \mathbb{E}[f(S \cup O)] \geq g(\emptyset) - \theta \cdot q|S| \geq f(O) - \frac{k\theta}{r}. \tag{1}$$

We restate another property of submodular functions by the following lemma.

Lemma 3. *[12] Given any non-negative submodular function f. For any three subsets A, B, C, we have*

$$f(A \cup B) \leq f(A \cup (B \cap C)) + f(B \setminus C).$$

Now, we summarize our main result by the following theorem.

Theorem 1. *Assume ALG is an α-approximation and the marginal values are lower bound $-\theta$, Algorithm 1 returns a solution T such that*

$$f(T) \geq \frac{f(O)}{p + 1 + (r-1)/2\alpha} - \frac{k\theta}{(p+1)r + r(r-1)/2\alpha}.$$

Proof. Since $O \setminus V_i$ can be represented as $\cup_{j=1}^{i-1}(O \cap S_j)$ for any round $i \in [r]$, by inequality (1), we obtain

$$r \cdot f(O) - k\theta \leq \sum_{i=1}^{r} f(S_i \cup O)$$

$$\leq \sum_{i=1}^{r} f(S_i \cup (O \cap V_i)) + \sum_{i=1}^{r} f(O \setminus V_i)$$

$$\leq \sum_{i=1}^{r} f(S_i \cup (O \cap V_i)) + \sum_{i=1}^{r} \sum_{j=1}^{i-1} f(O \cap S_j)$$

$$\leq (p+1) \cdot \sum_{i=1}^{r} f(S_i) + (1/\alpha) \cdot \sum_{i=1}^{r} \sum_{j=1}^{i-1} f(S_j')$$

$$\leq [(p+1)r + r(r-1)/2\alpha] \cdot f(T).$$

The above second and fourth inequalities are obtained by Lemmas 3 and 1, respectively. The third inequality follows by submodularity. By rearranging the above inequality, we get

$$f(T) \geq \frac{f(O)}{p + 1 + (r-1)/2\alpha} - \frac{k\theta}{(p+1)r + r(r-1)/2\alpha}.$$

$\square$

Corollary 1. *If θ is reduced to 0 and $r = 1$, Algorithm 1 gets a $1/(p+1)$-approximation for the submodular maximization with p-independence system constraint.*

5 Conclusion

In this paper, we study the p-independence system constrained problem of maximizing submodular functions. It is different from the monotonic scenario, in which one can learn the properties of set function with the help of parameters of the submodular ratio, generalized curvature, etc. Since the above defined parameters can not be bounded in the non-monotonic setting, we introduce a parameter θ to denote the lower bound of the maraginal values. By investigating the Repeated Greedy introduced by [12], we yield a parameterized performance guarantee for the stated problem.

Acknowledgements. The third author is supported by Scientific Research Project of Beijing Municipal Education Commission (No. KM201910005012) and the National Natural Science Foundation of China (No. 11971046). The fourth author is supported by the National Natural Science Foundation of China (No. 11871081).

References

1. Bai, W., Bilmes, J.: Greed is still good: maximizing monotone submodular + supermodular (BP) functions. In: Proceedings of the 35th International Conference on Machine Learning, pp. 314–323 (2018)
2. Bian, A.A., Buhmann, J.M., Krause, A., Tschiatschek, S.: Guarantees for greedy maximization of non-submodular functions with applications. In: Proceedings of the 34th International Conference on Machine Learning 498–507 (2017)
3. Bogunovic, I., Zhao, J., Cevher, V.: Robust maximization of non-submodular objectives. In: Proceedings of the 21th International Conference on Artificial Intelligence and Statistics, pp. 890–899 (2018)
4. Buchbinder, N., Feldman, M.: Deterministic algorithms for submodular maximization problems. ACM Trans. Algorithms **14**(3), 32 (2018)
5. Buchbinder, N., Feldman, M.: Constrained submodular maximization via a nonsymmetric technique. Math. Oper. Res. **44**(3), 988–1005 (2019)
6. Buchbinder, N., Feldman, M., Seffi, J., Schwartz, R.: A tight linear time 1/2-approximation for unconstrained submodular maximization. SIAM J. Comput. **44**(5), 1384–1402 (2015)
7. Conforti, M., Cornuéjols, G.: Submodular set functions, matroids and the greedy algorithm: tight worst-case bounds and some generalizations of the Rado-Edmonds theorem. Discrete Appl. Math. **7**(3), 251–274 (1984)
8. Das, A., Kempe, D.: Submodular meets spectral: greedy algorithms for subset selection, sparse approximation and dictionary selection. In: Proceedings of the 28th International Conference on Machine Learning, pp. 1057–1064 (2011)
9. Feige, U.: A threshold of $\ln n$ for approximating set cover. J. ACM **45**(4), 634–652 (1998)
10. Feige, U., Izsak, R.: Welfare maximization and the supermodular degree. In: Proceedings of the 4th Conference on Innovations in Theoretical Computer Science, pp. 247–256 (2013)
11. Feige, U., Mirrokni, V.S., Vondrák, J.: Maximizing non-monotone submodular functions. SIAM J. Comput. **40**(4), 1133–1153 (2011)

12. Feldman, M., Harshaw, C., Karbasi, A.: Greed is good: near-optimal submodular maximization via greedy optimization. In: Proceedings of the 30th Annual Conference on Learning Theory, pp. 758–784 (2017)
13. Feldman, M., Izsak, R.: Constrained monotone function maximization and the supermodular degree. In: Proceedings of the 17th International Workshop on Approximation Algorithms for Combinatorial Optimization Problems and the 18th International Workshop on Randomization and Computation, pp. 160–175 (2014)
14. Gölz, P., Procaccia, A.D.: Migration as submodular optimization. In: Proceedings of the 33rd AAAI Conference on Artificial Intelligence, pp. 549–556 (2019)
15. Gong, S., Nong, Q., Liu, W., Fang, Q.: Parametric monotone function maximization with matroid constraints. J. Global Optim. **75**(3), 833–849 (2019). https://doi.org/10.1007/s10898-019-00800-2
16. Gupta, A., Roth, A., Schoenebeck, G., Talwar, K.: Constrained non-monotone submodular maximization: Offline and secretary algorithms. In: Proceedings of the 6th International Conference on Internet and Network Economics, pp. 246–257 (2010)
17. Han, A., Cao, Z., Cui, S., Wu, B.: Deterministic approximation for submodular maximization over a matroid in nearly linear time, 2020, to appear in NeurIPS
18. Hassidim, A., Singer, Y.: Optimization for approximate submodularity. In: Proceedings of the 32nd Annual Conference on Neural Information Processing Systems, pp. 396–407 (2018)
19. Horel, T., Singer, Y.: Maximization of approximately submodular functions. In: Proceedings of the 30th International Conference on Neural Information Processing Systems, pp. 3053–3061 (2016)
20. Jiang, Y., Wang, Y., Xu, D., Yang, R., Zhang, Y.: Streaming algorithm for maximizing a monotone non-submodular function under d-knapsack constraint. Optim. Lett. **13**(82), 1–14 (2019)
21. Kuhnle, A., Smith, J.D., Crawford, V.G., Thai, M.T.: Fast maximization of non-submodular, monotonic functions on the integer lattice. In: Proceedings of the 35th International Conference on Machine Learning 2791–2800 (2018)
22. Lee, J., Sviridenko, M., Vondrák, J.: Submodular maximization over multiple matroids via generalized exchange properties. Math. Oper. Res. **35**(4), 795–806 (2010)
23. Mirzasoleiman, B., Badanidiyuru, A., Karbasi, A.: Fast constrained submodular maximization: personalized data summarization. In: Proceedings of the 33rd International Conference on Machine Learning, pp. 1358–1366 (2016)
24. Nemhauser, G.L., Wolsey, L.A., Fisher, M.L.: An analysis of approximations for maximizing submodular set functions-I. Math. Program. **14**(1), 265–294 (1978)
25. Qian, C., Shi, J., Yu, Y., Tang, K., Zhou, Z.: Subset selection under noise. In: Proceedings of the 31st Annual Conference on Neural Information Processing Systems, pp. 3560–3570 (2017)
26. Sviridenko, M., Vondrák, J., Ward, J.: Optimal approximation for submodular and supermodular optimization with bounded curvature. Math. Oper. Res. **42**(4), 1197–1218 (2017)
27. Wang, Z., Moran, B., Wang, X., Pan, Q.: Approximation for maximizing monotone non-decreasing set functions with a greedy method. J. Combin. Optim. **31**(1), 29–43 (2014). https://doi.org/10.1007/s10878-014-9707-3

A Streaming Model for Monotone Lattice Submodular Maximization with a Cardinality Constraint

Zhenning Zhang[1], Longkun Guo[2(✉)], Linyang Wang[1], and Juan Zou[3]

[1] Department of Operations Research and Information Engineering,
Beijing University of Technology, Beijing 100124, People's Republic of China
zhangzhenning@bjut.edu.cn, wangly0908@emails.bjut.edu.cn
[2] School of Computer Science and Technology, Qilu University of Technology
(Shandong Academy of Sciences), Jinan 250353, People's Republic of China
longkun.guo@gmail.com
[3] School of Mathematics and Science,
Qufu Normal University, Qufu 273165, People's Republic of China
zoujuanjn@163.com

Abstract. In this paper, we consider a streaming model of maximizing monotone lattice submodular function with a cardinality constraint on the integer lattice. As (lattice) submodularity does not imply the diminishing return property on the integer lattice, we introduce the Sieve-Streaming algorithm combining with a modified binary search subroutine to solve the problem. We also show it is with an approximation ratio $1/2 - \epsilon$, a memory complexity $O(\epsilon^{-1}k \log k)$, and a query complexity $O(\epsilon^{-2} \log^2 k)$ per element.

Keywords: Streaming algorithm · Lattice submodular · Integer lattice · Memory complexity · Cardinality

1 Introduction

Submodular maximization on the integer lattice $\mathbb{Z}_+^E$, which arises from real scenarios such as sensor placement and optimal budget allocation, has received much attention. Here, E can be considered as a ground set with size n. Comparing with the classical submodular maximization of set functions, each element is allowed to have positive integer copies for submodular maximization on the integer lattice. It is well known that submodularity and diminishing return submodularity (DR-submodularity) are equivalent for set functions. However, for a lattice function $f : \mathbb{Z}_+^E \to \mathbb{R}_+$, this property does not hold any more. A function $f : \mathbb{Z}_+^E \to \mathbb{R}_+$ is lattice submodular, if $f(\boldsymbol{x}) + f(\boldsymbol{y}) \geq f(\boldsymbol{x} \vee \boldsymbol{y}) + f(\boldsymbol{x} \wedge \boldsymbol{y})$ holds for any vector $\boldsymbol{x}, \boldsymbol{y} \in \mathbb{Z}_+^E$, where $\boldsymbol{x} \vee \boldsymbol{y}$ and $\boldsymbol{x} \wedge \boldsymbol{y}$ are the coordinate-wise maximum and minimum, that is, $(\boldsymbol{x} \vee \boldsymbol{y})(e) = \max\{\boldsymbol{x}(e), \boldsymbol{y}(e)\}$ and $(\boldsymbol{x} \wedge \boldsymbol{y})(e) = \min\{\boldsymbol{x}(e), \boldsymbol{y}(e)\}$, $e \in E$. A function $f : \mathbb{N}^E \to \mathbb{R}_+$ is diminishing return submodular (DR-submodular), if $f(\boldsymbol{x} + \chi_e) - f(\boldsymbol{x}) \geq f(\boldsymbol{y} + \chi_e) - f(\boldsymbol{y})$

© Springer Nature Switzerland AG 2021
Y. Zhang et al. (Eds.): PDCAT 2020, LNCS 12606, pp. 362–370, 2021.
https://doi.org/10.1007/978-3-030-69244-5_32

for any $\boldsymbol{x} \leq \boldsymbol{y}$ and $e \in E$, where χ_e is a unit standard vector with 1 at the coordinate e and 0 elsewhere. The lattice submodularity is a weaker condition than the DR-submodularity on the integer lattice. That is, from the lattice submodularity, $f(\boldsymbol{x} + \chi_e) - f(\boldsymbol{x}) \geq f(\boldsymbol{x} + 2\chi_e) - f(\boldsymbol{x} + \chi_e)$ may not hold.

In this paper, we study a streaming model for maximizing the monotone non-negative lattice submodular function with a cardinality k over the integer lattice, which can be characterized as

$$\max_{\boldsymbol{x}(E) \leq k} f(\boldsymbol{x}), \tag{1}$$

where f is a lattice submodular function, $\boldsymbol{x}(E) = \sum_{e \in E} \boldsymbol{x}(e)$ and $\boldsymbol{x}(e)$ is the copies of e in vector $\boldsymbol{x}$. For a streaming model, the element $e \in E$ is supposed to arrive one by one with its copies $\boldsymbol{b}(e)$. We focus on devising a one-pass algorithm with approximation ratio, memory complexity and query complexity.

Recently, there are many work [2,5,6,8–10] considering the offline models of the maximization of lattice submodular functions over the integer lattice. Soma et al. [8] provide a pseudo-polynomial time algorithm for maximizing a monotone lattice submodular function under a knapsack constraint. Soma and Yoshida [10] develop a polynomial time algorithm for lattice submodular maximization under a cardinality constraint. Gottschalk and Peis [2] give a 1/3-approximation algorithm for non-monotone lattice submodular maximization. Nong et al. [6] consequently improve the approximation ratio to 1/2. However, few work investigates the streaming model for maximizing a lattice submodular function, although the online model is more prevalent in real applications. Thus, we study the streaming algorithms for lattice submodular maximization on the integer lattice.

However, streaming algorithms [1,3,4,7,11] for the maximization of set submodular functions have been well developed. The streaming model can not preserve all the elements and its copies at the same time. It is supposed that the element $e \in E$ arrives one by one with its copies $\boldsymbol{b}(e)$. We not only have to decide whether the element is preserved but also give the level of the preserved element. We aim to design a one-pass algorithm and show its approximation ratio, memory complexity and query complexity.

The main idea is to combine the Sieve-Streaming algorithm [1] with a modified binary search subroutine to identify the amount of each reserved element. Firstly, we give an offline algorithm by the known estimation of optimal value. Secondly, we introduce a two-passes algorithm by estimating the optimal value. Finally, we state the one-pass streaming algorithm. The algorithm is established by estimating the interval of the optimal value according to the arriving elements and refining the interval. Furthermore, we also analyze the performance of the streaming algorithms.

2   Preliminaries

We first introduce some notations and definitions which will be used throughout this paper.

We use $E = \{e_1, \ldots, e_n\}$ to denote the ground set with size n, and $\boldsymbol{x} \in \mathbb{Z}_+^E$ a n-dimensional vector, where $\boldsymbol{x}(e_i)$ is the component of coordinate e_i ($e_i \in E$) of $\boldsymbol{x}$. Let χ_{e_i} be the unit standard vector with coordinate e_i being 1 and 0 elsewhere. For a vector $\boldsymbol{x}$ and a subset $S \subseteq E$, we have $\boldsymbol{x}(S) = \sum_{i \in S} \boldsymbol{x}(i)$. The set $\mathrm{supp}^+(\boldsymbol{x}) = \{e \in E | \boldsymbol{x}(e) > 0\}$.

Let $f : \mathbb{Z}_+^E \to \mathbb{R}_+$ be a function defined on the integer lattice. The monotonicity of the function f is $f(\boldsymbol{x}) \leq f(\boldsymbol{y})$, for any $\boldsymbol{x}, \boldsymbol{y} \in \mathbb{Z}_+^E$ with coordinate-wise $\boldsymbol{x} \leq \boldsymbol{y}$. The non-negative of the function f means $f(\boldsymbol{x}) \geq 0$ for all $\boldsymbol{x} \in \mathbb{Z}_+^E$.

We introduce the definitions of submodularity and diminishing return submodularity of the function f on the integer lattice.

Definition 1. *A function $f : \mathbb{Z}_+^E \to \mathbb{R}_+$ is diminishing return submodular (DR-submodular) if*

$$f(\boldsymbol{x} + \chi_e) - f(\boldsymbol{x}) \geq f(\boldsymbol{y} + \chi_e) - f(\boldsymbol{y}) \tag{2}$$

for any $\boldsymbol{x} \leq \boldsymbol{y}$ and $e \in E$, where χ_e is a unit standard vector with 1 at the coordinate e and 0 elsewhere.

Definition 2. *A function $f : \mathbb{Z}_+^E \to \mathbb{R}_+$ is (lattice) submodular if*

$$f(\boldsymbol{x}) + f(\boldsymbol{y}) \geq f(\boldsymbol{x} \vee \boldsymbol{y}) + f(\boldsymbol{x} \wedge \boldsymbol{y}) \tag{3}$$

for any $\boldsymbol{x}, \boldsymbol{y} \in \mathbb{Z}_+^E$, where $\boldsymbol{x} \vee \boldsymbol{y}$ and $\boldsymbol{x} \wedge \boldsymbol{y}$ are the coordinate-wise maximum and minimum, that is, $(\boldsymbol{x} \vee \boldsymbol{y})(e) = \max\{\boldsymbol{x}(e), \boldsymbol{y}(e)\}$ and $(\boldsymbol{x} \wedge \boldsymbol{y})(e) = \min\{\boldsymbol{x}(e), \boldsymbol{y}(e)\}$, $e \in E$.

In general, the lattice submodularity of f does not imply the diminishing return property. It is a weaker condition than the DR-submodularity when the domain is the integer lattice. However, a weaker version of the diminishing return property nevertheless holds as stated in the following lemma.

Lemma 1. *Let f be a monotone lattice submodular function. We have*

$$f(\boldsymbol{x} \vee l\chi_e) - f(\boldsymbol{x}) \geq f(\boldsymbol{y} \vee l\chi_e) - f(\boldsymbol{y}) \tag{4}$$

for any $e \in E$, $l \in \mathbb{N}$, $\boldsymbol{x}, \boldsymbol{y}$ with $\boldsymbol{x} \leq \boldsymbol{y}$.

Lemma 2. *Let f be a lattice submodular function. For any $\boldsymbol{x}, \boldsymbol{y}$, we receive*

$$f(\boldsymbol{x} \vee \boldsymbol{y}) - f(\boldsymbol{x}) \leq \sum_{e \in \mathrm{supp}^+(\boldsymbol{y} - \boldsymbol{x})} \left(f(\boldsymbol{y}(e)\chi_e \vee \boldsymbol{x}) - f(\boldsymbol{x}) \right), \tag{5}$$

where $\mathrm{supp}^+(\boldsymbol{y} - \boldsymbol{x}) = \{e \in E | (\boldsymbol{y} - \boldsymbol{x})(e) > 0\}$.

Let $\mathcal{F}_b$ denote the set of all non-negative monotone lattice submodular functions with $f(\boldsymbol{0}) = 0$ and defined on a domain $\{\boldsymbol{x} \in \mathbb{Z}_+^E : \boldsymbol{x} \leq \boldsymbol{b}\}$. In this paper, we design stream algorithms for maximization of lattice submodular functions $f \in \mathcal{F}_b$ with a cardinality constraint, i.e. in the form of (1).

For $f \in \mathcal{F}_b$, and vectors $\boldsymbol{x}$, $\boldsymbol{y}$, we use $f(\boldsymbol{x}|\boldsymbol{y}) = f(\boldsymbol{x} + \boldsymbol{y}) - f(\boldsymbol{y})$ for brevity. Meanwhile, we use $\boldsymbol{x}^*$ and OPT to denote an optimal solution vector and the optimal value of the problem (1), respectively.

3 A Streaming Algorithm with Known OPT

First, we introduce an offline algorithm for problem (1) under the assumption of a known parameter v with $\gamma\text{OPT} \leq v \leq \text{OPT}$, where $\gamma \in [0,1]$. Similar to the algorithm in [1], we first introduce a threshold greedy algorithm. However, the difference with [1] is that we should determine the amount of the arriving element so that the average marginal increment exceeds the threshold. Unfortunately, binary search algorithm [10] doesn't work for the lattice submodular function anymore, since we can not bound the marginal increment $f((\boldsymbol{x}^* - \boldsymbol{x}) \vee \boldsymbol{0}|\boldsymbol{x})$, where $\boldsymbol{x}$ is the final solution returned by the algorithm. To overcome this difficulty, by utilizing the idea in [10], we introduce a modified binary search algorithm in Algorithm 2. Lemma 3 states the properties of the amount l returned by Algorithm 2. Meanwhile, the approximation of Algorithm 1 is arrived by these properties.

Lemma 3. *Suppose that the element e arrives with its $\boldsymbol{b}(e)$ copies associated with a given vector $\boldsymbol{x}$ and a threshold η. Algorithm 2 has the following properties.*

Algorithm 1: Streaming-OPT

Input: $f \in \mathcal{F}_b$, cardinality k, ground set E, and v such that $\gamma\text{OPT} \leq v \leq \text{OPT}$
Output: a vector $\boldsymbol{x} \in \mathbb{Z}_+^E$

1   $\boldsymbol{x} \leftarrow 0$;
2   **for** *each $e \in E$* **do**
3   **if** $\boldsymbol{x}(E) < k$ **then**
4         $l \leftarrow$ BinarySearchLattice $\left(f, \boldsymbol{x}, \boldsymbol{b}, e, k, \frac{v}{2k}\right)$;
5         $\boldsymbol{x} \leftarrow \boldsymbol{x} + l\chi_e$;
6   **end**
7   **end**
8   **return** $\boldsymbol{x}$

Algorithm 2: BinarySearchLattice($f,\boldsymbol{x},e,\boldsymbol{b}(e),k,\eta,\epsilon$)

Input: $f \in \mathcal{F}_b$, $\boldsymbol{x} \in \mathbb{Z}_+^E$, $e \in E$, $\boldsymbol{b}(e) \in \mathbb{Z}_+$, cardinality k, $\eta \in \mathbb{R}_+$, and $\epsilon \in \mathbb{R}_+$
Output: $l \in \mathbb{Z}_+$

1   $l_s \leftarrow 1$, $l_t \leftarrow \min\{k - \boldsymbol{x}(E), \boldsymbol{b}(e)\}$;
2   **if** $f(l_t\chi_e|\boldsymbol{x}) < 0$ **then**
3      $l = 0$;
4   **end**
5   **if** $f(l_s\chi_e|\boldsymbol{x}) > 0$ **then**
6      $l_{\min} \leftarrow 1$, $l_{\max} \leftarrow \min\{k - \boldsymbol{x}(E), \boldsymbol{b}(e)\}$;
7      $l \leftarrow$ Threshold($f,e,\boldsymbol{x},l_{\min}, l_{\max},\epsilon$);
8   **end**
9   **else**
10     $l_{\min} \leftarrow$ Binary($f,e,\boldsymbol{x},l_s,l_t,0$), $l_{\max} \leftarrow \min\{k - \boldsymbol{x}(E), \boldsymbol{b}(e)\}$;
11     $l \leftarrow$ Threshold($f,e,\boldsymbol{x},l_{\min}, l_{\max},\epsilon$);
12   **end**

Algorithm 3: Binary($f,e,\boldsymbol{x},l_s,l_t,\theta$)

Input: $f \in \mathcal{F}_b$, $e \in E$, and $\boldsymbol{x} \in \mathbb{N}^E$, l_s, l_t, θ
Output: l_t
1 **while** $l_t \neq l_s + 1$ **do**
2 $\quad$ $m = \lfloor (l_t + l_s)/2 \rfloor$;
3 $\quad$ **if** $f(m\chi_e|\boldsymbol{x}) > \theta$ **then**
4 $\quad\quad$ | $\quad l_t = m$
5 $\quad$ **end**
6 $\quad$ **else**
7 $\quad\quad$ | $\quad l_s = m$
8 $\quad$ **end**
9 $\quad$ **return** l_t
10 **end**

Algorithm 4: Threshold($f,e,\boldsymbol{x},l_{\min}, l_{\max},\epsilon$)

Input: $f \in \mathcal{F}_b$, $l_{\min}$, $l_{\max}$, $\boldsymbol{x} \in \mathbb{N}^E$, $\eta \in \mathbb{Z}_+$, $\epsilon \in \mathbb{Z}_+$
Output: l
1 **for** ($h = f(l_{\max}\chi_e|\boldsymbol{x})$; $h \geq (1-\epsilon)f(l_{\min}\chi_e|\boldsymbol{x})$; $h = (1-\epsilon)h$) **do**
2 $\quad$ $l \leftarrow$ Binary($f,e,\boldsymbol{x},l_{\min},l_{\max},h$);
3 $\quad$ **if** $f(l\chi_e|\boldsymbol{x}) \geq (1-\epsilon)l\eta$ **then**
4 $\quad\quad$ | $\quad$ **return** l
5 $\quad$ **end**
6 **end**
7 **return** $l = 0$

---

1. If there exists $0 < l^* \leq l_{\max}$ such that $f(l^*\chi_e|\boldsymbol{x}) \geq l^*\eta$, then there is a l returned by Algorithm 2 with $l_{\min} \leq l \leq l_{\max}$ such that $f(l\chi_e|\boldsymbol{x}) \geq (1-\epsilon)l\eta$. And for any $l \leq l' \leq l_{\max}$, we have $f(l'\chi_e|\boldsymbol{x}) < \max\{\frac{f(l\chi_e|\boldsymbol{x})}{1-\epsilon}, l'\eta\}$, where $l_{\min}$ is the smallest positive integer to make $f(l_{\min}\chi_e|\boldsymbol{x}) > 0$, and $l_{\max} = \min\{k - \boldsymbol{x}(E), \boldsymbol{b}(e)\}$.

2. The running time of Algorithm 2 is $O\left(\epsilon^{-1}\log\left(\frac{f(l_{\max}\chi_e|\boldsymbol{x})}{f(l_{\min}\chi_e|\boldsymbol{x})}\right) \cdot \log(l_{\max})\right)$.

Lemma 4. Let $\boldsymbol{x}_{i-1}$ be the initial vector of the i-th iteration, e_i be the element exactly arrived at the i-th iteration of Algorithm 1. The amount l_i returns from Algorithm 2. Then, for each iteration i of Algorithm 1, we have

$$f(\boldsymbol{x}_{i-1} + l_i\chi_{e_i}) \geq (1-\epsilon)\frac{v(\boldsymbol{x}_{i-1}(E) + l_i)}{2k}, \tag{6}$$

for $i = 1, \ldots, n$.

Lemma 5. For the i-th iteration, element e_i is arriving with its copies $\boldsymbol{b}(e_i)$ and the initial vector is $\boldsymbol{x}_{i-1}$. If there is a positive integer $l_i > 0$ produced by Algorithm 2, it obviously satisfies

$$\frac{f(l_i\chi_{e_i}|\boldsymbol{x}_{i-1})}{l_i} \geq (1-\epsilon)\frac{v}{2k}. \tag{7}$$

Algorithm 5: Two-Passes-Streaming

> **Input:** $f \in \mathcal{F}_b$, cardinality k, ground set E, α such that $\alpha = \max_{e \in E} f(\chi_e)$,
> and β such that $\beta = \max_{e \in E} f(b(e)\chi_e)$, $\epsilon \in (0,1)$
> **Output:** a vector $x \in \mathbb{Z}_+^E$

1   $\mathcal{H}_\epsilon = \{(1 + \epsilon)^i | i \in \mathbb{Z}, \alpha/(1 + \epsilon) \le (1 + \epsilon)^i \le 2k \cdot \beta\}$;
2   For each $v \in \mathcal{H}_\epsilon$, set $x^v \leftarrow 0$;
3   **for** $i = 1, \ldots, n$ **do**
4   **for** $v \in \mathcal{H}_\epsilon$ **do**
5   **if** $x^v(E) < k$ **then**
6            $l \leftarrow$ BinarySearchLattice $\left(f, x^v, b, e_i, k, \frac{v}{2k}\right)$;
7            $x^v \leftarrow x^v + l\chi_{e_i}$;
8   **end**
9   **end**
10 **end**
11 **return** $\arg\max_{v \in \mathcal{H}_\epsilon} f(x^v)$

Let $y = (x^* - x) \vee 0$, where x^* is an optimal solution and x is the final solution returned by Algorithm 1. Finally, we have the following conclusion:

$$f(y(e_i)\chi_{e_i}|x) \le \max \left\{ \frac{\epsilon}{1 - \epsilon} f(l_i \chi_{e_i}|x_{i-1}), (\epsilon l_i + y(e_i)) \frac{v}{2k} \right\}, \qquad (8)$$

for $i = 1, \ldots, n$.

Theorem 1. *For a given $\gamma \in [0, 1]$, Algorithm 1 gives a one-pass algorithm with an approximation ratio $\left(\frac{1}{2} - \epsilon\right)\gamma$, a memory complexity k, and a per-element query complexity $O(\epsilon^{-1} \log k)$.*

4 Two-Pass Streaming Algorithm

Since the optimal value of (1) is actually unknown, we employ the following lemma to estimate the OPT.

Lemma 6. *By introducing $\alpha = \max_{e \in E} f(\chi_e)$ and $\beta = \max_{e \in E} f(b(e)\chi_e)$, we have $\mathcal{H}_\epsilon = \{(1 + \epsilon)^i | i \in \mathbb{Z}, \alpha/(1 + \epsilon) \le (1 + \epsilon)^i \le k \cdot \beta\}$, in which there exists a $v \in \mathcal{H}_\epsilon$, such that $(1 - \epsilon)\mathrm{OPT} \le v \le \mathrm{OPT}$.*

Thus, the pseudocode is formally described in Algorithm 5. In the first pass, we get the maximum values $\alpha = \max_{e \in E} f(\chi_e)$ and $\beta = \max_{e \in E} f(b(e)\chi_e)$; while in the second, we estimate the value v in Algorithm 1 by refining the domain $[\alpha, k\beta]$. Finally, we return the vector with the maximum function value.

Theorem 2. *With $\epsilon \in (0, 1)$, Algorithm 5 gives a two-pass algorithm that deserves an approximation ratio of $\left(\frac{1}{2} - \epsilon\right)$ and consumes $O(\epsilon^{-1} k \log k)$ memory complexity and $O(\epsilon^{-2} \log^2 k)$ update time per element.*

5   Sieve-Steaming Algorithm for Lattice Submodular Maximization

To get a one-pass Sieve-Steaming algorithm for lattice submodular maximization, we have to estimate the values α and β accompanied with the arriving element. Thus, the estimation interval $\mathcal{H}_\epsilon^i = \{(1+\epsilon)^s | \alpha_i/(1+\epsilon) \leq (1+\epsilon)^s \leq 2k\beta_i\}$ of the optimal value changes with the arriving element e_i, $i = 1, \ldots, n$, where $\alpha_i = \max\{\alpha_{i-1}, f(\chi_{e_i})\}$ and $\beta_i = \max\{\beta_{i-1}, f(\boldsymbol{b}(e_i)\chi_{e_i})\}$. For every interval $\mathcal{H}_\epsilon^i$, if the parameter v is first appeared in this interval, the initial iterative vector $\boldsymbol{x}^v$ is assigned to be zero. Meanwhile, for the parameter v which doesn't belong to $\mathcal{H}_\epsilon^i$ any more, the corresponding vector $\boldsymbol{x}^v$ is deleted. Finally, the estimated interval $\mathcal{H}_\epsilon^n$ is equivalent to $\mathcal{H}_\epsilon$ in the two passes algorithm (Algorithm 5). Thus, Sieve-Steaming algorithm returns the solution $\boldsymbol{x}$ with the maximum value $\max_{v \in \mathcal{H}_\epsilon^n} f(\boldsymbol{x}^v)$. The pseudocode is stated in Algorithm 6.

Lemma 7. *For any parameter $v \in \mathcal{H}_\epsilon^n$, let $\boldsymbol{x}^v$ be the finial solution returned by the 14-th step of Algorithm 6 and $\boldsymbol{y}^v = (\boldsymbol{x}^* - \boldsymbol{x}^v) \vee \boldsymbol{0}$. Suppose that v is first appeared in $\mathcal{H}_\epsilon^{i(v)}$. Thus, we have*

$$f(\boldsymbol{y}^v(e_j)\chi_{e_j}|\boldsymbol{x}^v) < \boldsymbol{y}^v(e_j) \cdot \frac{v}{2k}, \tag{9}$$

for $j = 1, \ldots, i(v) - 1$, and

$$f(\boldsymbol{y}^v(e_j)\chi_{e_j}|\boldsymbol{x}^v) \leq \max\left\{\frac{\epsilon}{1-\epsilon}f(l_j\chi_{e_j}|\boldsymbol{x}_{j-1}^v), (\epsilon l_i + \boldsymbol{y}^v(e_j))\frac{v}{2k}\right\}, \tag{10}$$

for $j = i(v), \ldots, n$.

Algorithm 6: Sieve-Streaming-Algorithm/Lattice Submodular

Input: $f \in \mathcal{F}_{\boldsymbol{b}}$, $\boldsymbol{b} \in \mathbb{Z}_+^E$, cardinality k, ground set E, and $\epsilon \in (0, 1)$
Output: a vector $\boldsymbol{x} \in \mathbb{Z}_+^E$
1 $\mathcal{H} = \{(1+\epsilon)^i | i \in \mathbb{Z}_+\}$;
2 For each $v \in \mathcal{H}$, set $\boldsymbol{x}^v \leftarrow 0$;
3 $\alpha \leftarrow 0$, $\beta \leftarrow 0$;
4 **for** $i = 1, \ldots, n$ **do**
5 $\quad$ $\alpha \leftarrow \max\{\alpha, f(\chi_{e_i})\}$, $\beta \leftarrow \max\{\beta, f(\boldsymbol{b}(e)\chi_{e_i})\}$;
6 $\quad$ $\mathcal{H}_\epsilon^i = \{(1+\epsilon)^s | \alpha/(1+\epsilon) \leq (1+\epsilon)^s \leq 2k\beta\}$;
7 $\quad$ Delete all $\boldsymbol{x}^v$, where $v \notin \mathcal{H}_\epsilon^i$;
8 $\quad$ **for** $v \in \mathcal{H}_\epsilon^i$ **do**
9 $\quad\quad$ **if** $\boldsymbol{x}^v(E) < k$ **then**
10 $\quad\quad\quad$ $l \leftarrow$ BinarySearchLattice $\left(f, \boldsymbol{x}^v, \boldsymbol{b}, e_i, k, \frac{v}{2k}\right)$;
11 $\quad\quad\quad$ $\boldsymbol{x}^v \leftarrow \boldsymbol{x}^v + l\chi_{e_i}$;
12 $\quad\quad$ **end**
13 $\quad$ **end**
14 **end**
15 **return** $\arg\max_{v \in \mathcal{H}_\epsilon^n} f(\boldsymbol{x}^v)$

Theorem 3. *With $\epsilon \in (0,1)$, Algorithm 6 gives a one pass $(\frac{1}{2}-\epsilon)$-approximation algorithm. Meanwhile, the memory complexity of Algorithm 6 is $O(\epsilon^{-1}k \log k)$, and the queries complexity is $O(\epsilon^{-2} \log^2 k)$ per element.*

The proofs of all the above Lemmas and Theorems will be given in the journal version of this paper due to length limitation.

6 Conclusion

In the paper, we observe that lattice submodularity and diminishing return property are not equivalent on the integer lattice as lattice submodularity is a weaker version because it can not conclude $f(\chi_e + \boldsymbol{x}) - f(\boldsymbol{x}) \geq f(2\chi_e + \boldsymbol{x}) - f(\chi_e + \boldsymbol{x})$. We investigate the more complicated case of submodular maximization with a cardinality constraint on the integer lattice and devise streaming algorithms. First, we give an offline algorithm with a known estimation of OPT. Second, we introduce a two-pass algorithm to give the estimation interval of OPT. Finally, we give the real one-pass streaming algorithm by estimating the interval along with the arriving elements, together with analysis of its approximation ratio, memory complexity and query complexity.

Acknowledgements. The first author is supported by National Natural Science Foundation of China (No. 12001025) and Science and Technology Program of Beijing Education Commission (No. KM201810005006). The second author is supported by Natural Science Foundation of China (No. 61772005) and Natural Science Foundation of Fujian Province (No. 2017J01753). The third author is supported by Beijing Natural Science Foundation Project No. Z200002 and the National Natural Science Foundation of China (No. 11871081). The fourth author is supported by Natural Science Foundation of China (No. 11801310).

References

1. Badanidiyuru, A., Mirzasoleiman, B., Karbasi, A., Krause, A.: Streaming submodular maximization: massive data summarization on the fly. In: Proceedings of SIGKDD, pp. 671–680 (2014)
2. Gottschalk, C., Peis, B.: Submodular function maximization on the bounded integer lattice. In: Proceedings WAOA, pp. 133–144 (2015)
3. Huang, C., Kakimura, N.: Improved streaming algorithms for maximizing monotone submodular functions under a knapsack constraint. In: Proceedings of WADS, pp. 438–451 (2019)
4. E. Kazemi, E., Mitrovic, M., Zadimoghaddam, M., Lattanzi, S., Karbasi, A.: Submodular streaming in all its glory: tight approximation, minimum memory and low adaptive complexity. In: Proceedings of ICML, pp. 3311–3320 (2019)
5. Kuhnle, A., Smith, J.D., Crawford, V.G., Thai, M.T.: Fast maximization of nonsubmodular, monotonic functions on the integer lattice. In: Proceedings of ICML, pp. 2791–2800 (2018)
6. Nong, Q., Fang, J., Gong, S., Du, D., Feng, Y., Qu, X.: A 1/2-approximation algorithm for maximizing a non-monotone weak-submodular function on a bounded integer lattice. J. Comb. Optim. 39, 1208–1220 (2020)

7. Norouzi-Fard, A., Tarnawski, J., Mitrovic, S., Zandieh, A., Mousavifar, A., Svensson, O.: Beyond 1/2-approximation for submodular maximization on massive data streams. In: Proceedings of ICML, pp. 3829–3838 (2018)

8. Soma, T., Kakimura, N., Inaba, K., Kawarabayashi, K.: Optimal budget allocation: theoretical guarantee and efficient algorithm. In: Proceedings of ICML, pp. 351–359 (2014)

9. Soma, T., Yoshida, Y.: A generalization of submodular cover via the diminishing return property on the integer lattice. In: Proceedings of NIPS, pp. 847–855 (2015)

10. Soma, T., Yoshida, Y.: Maximizing monotone submodular functions over the integer lattice. Math. Program. **172**, 539–563 (2018)

11. Wang, Y., Xu, D., Wang, Y., Zhang, D.: Non-submodular maximization on massive data streams. J. Global Optim. **76**(4), 729–743 (2019). https://doi.org/10.1007/s10898-019-00840-8

The Prize-Collecting k-Steiner Tree Problem

Lu Han[1], Changjun Wang[2], Dachuan Xu[2], and Dongmei Zhang[3(✉)]

[1] Academy of Mathematics and Systems Science, Chinese Academy of Sciences,
Beijing 100190, People's Republic of China
hanlu@amss.ac.cn

[2] Department of Operations Research and Information Engineering,
Beijing University of Technology, Beijing 100124, People's Republic of China
{wcj,xudc}@bjut.edu.cn

[3] School of Computer Science and Technology, Shandong Jianzhu University,
Jinan 250101, People's Republic of China
zhangdongmei@sdjzu.edu.cn

Abstract. In this paper, we study the prize-collecting k-Steiner tree problem (PC k-ST), which is an interesting generalization of both the k-Steiner tree problem (k-ST) and the prize-collecting Steiner tree problem (PCST). In the PC k-ST, we are given an undirected connected graph $G = (V, E)$, a subset $R \subseteq V$ called terminals, a root vertex $r \in V$ and an integer k. Every edge has a non-negative edge cost and every vertex has a non-negative penalty cost. We wish to find an r-rooted tree F that spans at least k vertices in R so as to minimize the total edge costs of F as well as the penalty costs of the vertices not in F. As our main contribution, we propose two approximation algorithms for the PC k-ST with ratios of 5.9672 and 5. The first algorithm is based on an observation of the solutions for the k-ST and the PCST, and the second one is based on the technique of primal-dual.

Keywords: Prize-collecting · Steiner tree · Approximation algorithm

1 Introduction

We propose and focus on the prize-collecting k-Steiner tree problem (PC k-ST) in this paper. In the PC k-ST, we are given an undirected connected graph $G = (V, E)$, where V denotes all the vertices and E denotes all the edges, a subset $R \subseteq V$ called terminals, a root vertex $r \in V$ and an integer k. Every edge $e \in E$ is associated with a non-negative edge cost c_e. Every vertex $v \in V$ is associated with a non-negative penalty cost π_v. We wish to find an r-rooted tree F that spans at least k vertices in R so as to minimize the total costs, which is the total edge costs of F as well as the penalty costs of the vertices not in F. Without loss of generality, assume that $k \leq |R|$. For a tree F, we use $V(F)$ and $E(F)$ to denote the vertices and edges of F.

Extensive applications can be found in the PC k-ST, since it generalizes both the well-known NP-hard k-Steiner tree problem (k-ST) and prize-collecting Steiner tree problem (PCST).

© Springer Nature Switzerland AG 2021
Y. Zhang et al. (Eds.): PDCAT 2020, LNCS 12606, pp. 371–378, 2021.
https://doi.org/10.1007/978-3-030-69244-5_33

(1) The k-ST: In the k-ST, we are given an undirected connected graph $G = (V, E)$, where V are the vertices and E are the edges, a subset $R \subseteq V$ called terminals, a root vertex $r \in V$ and an integer k. Every edge has a non-negative edge cost c_e. Our aim is to find an r-rooted tree F that spans at least k vertices in R so as to minimize the total costs, which is the total edge costs of F. When we set $\pi_v = 0$ for every vertex $v \in V$, the PC k-ST simplifies to the k-ST.

(2) The PCST: In the PCST, we are given an undirected connected graph $G = (V, E)$, where V denotes all the vertices and E denotes all the edges, and a root vertex $r \in V$. Every edge $e \in E$ is associated with a non-negative edge cost c_e. Every vertex $v \in V$ is associated with a non-negative penalty cost π_v. We wish to find an r-rooted tree F with minimized total costs, which is the total edge costs of F and the penalty costs of the vertices not in F. When we set $R = \emptyset$ and $k = 0$, the PC k-ST simplifies to the PCST.

Many researchers are interested in designing approximation algorithms for the NP-hard problems [1,5,7,12,13,16–18]. For the k-ST, Chudak et al. [7] give a 5-approximation algorithm via the primal-dual technique and Lagrangean relaxation. Ravi et al. [16] claim that a ρ-approximation for the k-minimum spanning tree problem (k-MST) can yield a 2ρ-approximation algorithm for it. In the k-MST, an undirected connected graph $G = (V, E)$, a root vertex $r \in V$ and an integer k are given. Every edge has a non-negative edge cost and we want to find an r-rooted tree F that spans at least k vertices with minimized total edge costs. Since Garg [10]gives the current best approximation ratio of 2 for the k-MST, there exists an approximation ratio of 4 for the k-ST. Besides the state-of-art result for the k-MST, a great deal of other algorithms also has been presented [2–4,6–9,15,16].

The first constant approximation ratio of 3 for the PCST can be received from adopting the LP-rounding based idea proposed by Bienstock et al. [5]. Later, Goemans and Williamson [11] propose an elegant primal-dual based 2-approximation algorithm and Archer et al. [1] offer the current best ratio of 1.9672.

As our main contribution, we first propose a 5.9672-approximation algorithm for the PC k-ST, which is inspired by the work of Matsuda and Takahashi [14]. The idea of the algorithm based on an observation that for any PC k-ST instance, constructing some essential k-ST and PCST instances and combining their feasible solutions can provide a feasible solution for the PC k-ST. Then, based on the work of Chudak et al. [7] on the k-ST, we design a more complicated algorithm and successfully improve the ratio to 5.

The remainder of the paper is organized as follows. Some preliminaries are given in Sect. 2. The first 5.9672-approximation algorithm is presented in Sect. 3. Section 4 shows the improved 5-approximation algorithm. Section 5 offers some discussions.

2   Preliminaries

For any PC k-ST instance $\mathcal{I} = (G, R, r, k, \{c_e\}_{e \in E}, \{\pi_v\}_{v \in V})$, by getting rid of the penalty cost π_v for every $v \in V$, we obtain a k-ST instance $\mathcal{I}_{kST} = (G, R, r, k, \{c_e\}_{e \in E})$. Besides, for the PC k-ST instance $\mathcal{I}$, if we throw out the terminals R and the integer k, we get a PCST instance $\mathcal{I}_{PCST} = (G, r, \{c_e\}_{e \in E}, \{\pi_v\}_{v \in V})$. For any PC k-ST instance $\mathcal{I}$, let $F_{\mathcal{I}}$ denote its optimal solution and let $OPT_{\mathcal{I}}$ denote

the corresponding optimal total costs, i.e., $OPT_{\mathcal{I}} = \sum_{e \in E(F_{\mathcal{I}})} c_e + \sum_{v \in V \setminus V(F_{\mathcal{I}})} \pi_v$.
Likewise, denote $F_{\mathcal{I}_{kST}}$ and $F_{\mathcal{I}_{PCST}}$ as the optimal solutions of the corresponding instances $\mathcal{I}_{kST}$ and $\mathcal{I}_{PCST}$, and let $OPT_{\mathcal{I}_{kST}}$ and $OPT_{\mathcal{I}_{PCST}}$ be their total costs, i.e., $OPT_{\mathcal{I}_{kST}} = \sum_{e \in E(F_{\mathcal{I}_{kST}})} c_e$ and $OPT_{\mathcal{I}_{PCST}} = \sum_{e \in E(F_{\mathcal{I}_{PCST}})} c_e + \sum_{v \in V \setminus V(F_{\mathcal{I}_{PCST}})} \pi_v$.

The following lemma bounds $OPT_{\mathcal{I}_{kST}}$ in terms of $OPT_{\mathcal{I}}$.

Lemma 1. *The total cost of the optimal tree $F_{\mathcal{I}_{kST}}$ for the k-ST instance $\mathcal{I}_{kST}$ is no more than the total cost of the optimal tree $F_{\mathcal{I}}$ for the PC k-ST instance $\mathcal{I}$, i.e.,*

$$\sum_{e \in E(F_{\mathcal{I}_{kST}})} c_e \leq \sum_{e \in E(F_{\mathcal{I}})} c_e + \sum_{v \in V \setminus V(F_{\mathcal{I}})} \pi_v.$$

Proof. Since the tree F_{kST} is the optimal tree for the k-ST instance $\mathcal{I}_{kST}$, its total edge costs is at most the total edge cost of the tree $F_{\mathcal{I}}$, which is a feasible solution for instance $\mathcal{I}_{kST}$. Thus,

$$\sum_{e \in E(F_{\mathcal{I}_{kST}})} c_e \leq \sum_{e \in E(F_{\mathcal{I}})} c_e$$

$$\leq \sum_{e \in E(F_{\mathcal{I}})} c_e + \sum_{v \in V \setminus V(F_{\mathcal{I}})} \pi_v.$$

The following lemma bounds $OPT_{\mathcal{I}_{PCST}}$ in terms of $OPT_{\mathcal{I}}$.

Lemma 2. *The total cost of the optimal tree $F_{\mathcal{I}_{PCST}}$ for the PCST instance $\mathcal{I}_{PCST}$ is no more than the total cost of the optimal tree $F_{\mathcal{I}}$ for the PC k-ST instance $\mathcal{I}$, i.e.,*

$$\sum_{e \in E(F_{\mathcal{I}_{PCST}})} c_e + \sum_{v \in V \setminus V(F_{\mathcal{I}_{PCST}})} \pi_v \leq \sum_{e \in E(F_{\mathcal{I}})} c_e + \sum_{v \in V \setminus V(F_{\mathcal{I}})} \pi_v.$$

This lemma obviously holds since the optimal tree $F_{\mathcal{I}}$ for the PC k-ST instance $\mathcal{I}$ is also a feasible solution for the PCST instance $\mathcal{I}_{PCST}$.

3   The First Algorithm

In this section, we propose the first 5.9672-approximation algorithm for the PC k-ST. The idea of the algorithm is based on an observation that for any PC k-ST instance $\mathcal{I}$, combining the feasible solutions of its corresponding instances $\mathcal{I}_{kST}$ and $\mathcal{I}_{PCST}$ can derive its own feasible solution.

The first algorithm, which has three main steps, is formally shown in Algorithm 1. In the first two steps, we construct the k-ST instance $\mathcal{I}_{kST}$ and PCST instance $\mathcal{I}_{PCST}$ from the PC k-ST instance $\mathcal{I}$, then, we solve the two constructed instances in order to obtain their solutions F_1 and F_2. In the last step, we construct a feasible solution F for the PC k-ST instance $\mathcal{I}$ from the solutions F_1 and F_2.

The main result of Algorithm 1 is as follows.

Algorithm 1:

Input: An PC k-ST instance $\mathcal{I} = (G, R, r, k, \{c_e\}_{e \in E}, \{\pi_v\}_{v \in V})$.

Output: An r-rooted tree F which spans at least k vertices in R.

Step 1 Construct the k-ST instance $\mathcal{I}_{kST}$ and solve it to obtain a solution F_1.

For the PC k-ST instance $\mathcal{I} = (G, R, r, k, \{c_e\}_{e \in E}, \{\pi_v\}_{v \in V})$, get rid of the penalty cost π_v for every $v \in V$ and obtain a k-ST instance $\mathcal{I}_{kST} = (G, R, r, k, \{c_e\}_{e \in E})$. Solve $\mathcal{I}_{kST}$ with the current best α-approximation algorithm for the k-ST and get a feasible solution F_1 for $\mathcal{I}_{kST}$ where $\alpha = 4$.

Step 2 Construct the PCST instance $\mathcal{I}_{PCST}$ and solve it to obtain a solution F_2.

For the PC k-ST instance $\mathcal{I} = (G, R, r, k, \{c_e\}_{e \in E}, \{\pi_v\}_{v \in V})$, get rid of the terminals R and the integer k to obtain a PCST instance $\mathcal{I}_{PCST} = (G, r, \{c_e\}_{e \in E}, \{\pi_v\}_{v \in V})$. Solve $\mathcal{I}_{PCST}$ with the current best β-approximation algorithm for the PCST and get a feasible solution F_2 for $\mathcal{I}_{PCST}$ where $\beta = 1.9672$.

Step 3 Construct a feasible solution F for the PC k-ST instance $\mathcal{I}$.

Find the minimum spanning tree F of $(V(F_1) \cup V(F_2), E(F_1) \cup E(F_2))$ and output it as the feasible solution for $\mathcal{I}$.

Theorem 1. *For any instance $\mathcal{I}$ of the PC k-ST, Algorithm 1 can output a feasible solution F where the total costs of F is within a factor of 5.9672 the total cost of the optimal solution for $\mathcal{I}$.*

Proof. Since F_1 is a feasible solution for the k-ST instance $\mathcal{I}_{kST}$, we have that $|V(F_1) \cap R| \geq k$. Recalling that F is the minimum spanning tree F of $(V(F_1) \cup V(F_2), E(F_1) \cup E(F_2))$. Thus, $|V(F)| = |V(F_1)| + |V(F_2)|$. Therefore,

$$|V(F) \cap R| \geq |V(F_1) \cap R| \geq k.$$

The above inequality implies that F is a feasible solution for the PC k-ST instance $\mathcal{I}$.

From Lemma 1 and Lemma 2 and the fact that $E(F) \subseteq E(F_1) \cup E(F_2)$, we have that

$$\sum_{e \in E(F)} c_e + \sum_{v \in V \setminus V(F)} \pi_v$$

$$\leq \sum_{e \in E(F_1)} c_e + \sum_{e \in E(F_2)} c_e + \sum_{v \in V \setminus V(F_2)} \pi_v$$

$$\leq \alpha \cdot OPT_{\mathcal{I}_{kST}} + \beta \cdot OPT_{\mathcal{I}_{PCST}}$$

$$\leq (\alpha + \beta) \cdot OPT_{\mathcal{I}}.$$

Recalling $\alpha = 4$ and $\beta = 1.9672$, the approximation ratio becomes 5.9672.

4   The Improved Algorithm

In this section, we present the main idea of the improved 5-approximation algorithm for the PC-k-ST, which is based on the technique of primal-dual and Lagrangean relaxation.

We show the integer program and the dual program that are fundamental for the 5-approximation algorithm here. For any subset $S \subseteq V \setminus \{r\}$, let $\pi(S)$ be the total penalty costs of the vertices in S. Denote $\delta(S)$ as all the edges with only one endpoint in S. The PC k-ST can be formulated as the following integer program, where variable x_e indicates whether edge e is chosen, and variable z_S indicates whether subset S is penalized.

$$\min \sum_{e \in E} c_e x_e + \sum_{S \subseteq V \setminus \{r\}} \pi(S) z_S$$

$$\text{s.t.} \sum_{e \in \delta(S)} x_e + \sum_{T:T \supseteq S} z_T \geq 1, \qquad \forall S \subseteq V \setminus \{r\},$$

$$\sum_{S \subseteq V \setminus \{r\}} |S \cap R| z_S \leq |R| - k,$$

$$x_e \in \{0, 1\}, \qquad \forall e \in E,$$

$$z_S \in \{0, 1\}, \qquad \forall S \subseteq V \setminus \{r\}.$$

By relaxing all the variables in the integer program, the LP relaxation can be gained.

$$\min \sum_{e \in E} c_e x_e + \sum_{S \subseteq V \setminus \{r\}} \pi(S) z_S$$

$$\text{s.t.} \sum_{e \in \delta(S)} x_e + \sum_{T:T \supseteq S} z_T \geq 1, \qquad \forall S \subseteq V \setminus \{r\},$$

$$\sum_{S \subseteq V \setminus \{r\}} |S \cap R| z_S \leq |R| - k,$$

$$x_e \geq 0, \qquad \forall e \in E,$$

$$z_S \geq 0, \qquad \forall S \subseteq V \setminus \{r\}.$$

From the duality theory and introducing dual variables, the dual program can be obtained.

$$\max \sum_{S \subseteq V \setminus \{r\}} y_S - \lambda(|R| - k)$$

$$\text{s.t.} \sum_{S \subseteq V \setminus \{r\}: e \in \delta(S)} y_S \leq c_e, \qquad \forall e \in E, \qquad (1)$$

$$\sum_{T:T \subseteq S} y_T \leq \pi(S) + \lambda |S \cap R|, \qquad \forall S \subseteq V \setminus \{r\},$$

$$y_S \geq 0, \qquad \forall S \subseteq V \setminus \{r\},$$

$$\lambda \geq 0.$$

Note that the dual program (1) can be viewed as a parametric version of the dual program of the PCST, if we set the penalty cost $\pi_v = \pi_v + \lambda$ for every vertex $v \in R$ and set $\pi_v = \pi_v$ for every vertex $v \in V \setminus R$. The following are the dual program of the

PCST.

$$\max \sum_{S \subseteq V \setminus \{r\}} y_S$$

$$\text{s. t.} \sum_{S \subseteq V \setminus \{r\}: e \in \delta(S)} y_S \leq c_e, \qquad \forall e \in E,$$

$$\sum_{T:T \subseteq S} y_T \leq \pi(S), \qquad \forall S \subseteq V \setminus \{r\},$$

$$y_S \geq 0, \qquad \forall S \subseteq V \setminus \{r\},$$

$$\lambda \geq 0.$$

For the PCST, Goemans and Williamson [11] give the GW 2-approximation algorithm with Lagrangean multiplier preserving property. The following theorem show the main result of the GW algorithm.

Theorem 2. *For any instance $\mathcal{I}_{PCST}$ of the PCST, GW algorithm can output a feasible solution F_{GW} and a feasible dual solution $\{y_S^{GW}\}_{S \subseteq V \setminus \{r\}}$ that satisfies*

$$\sum_{e \in E(F_{GW})} c_e + \left(2 - \frac{1}{n-1}\right) \sum_{v \in V \setminus V(F_{GW})} \pi_v \leq \left(2 - \frac{1}{n-1}\right) \sum_{S \subseteq V \setminus \{r\}} y_S^{GW},$$

where n is the number of vertices in V.

Let $c_{\min}$ be the smallest non-zero edge cost . Now, we are ready to present our second algorithm, which is formally shown in Algorithm 2 and utilizes the GW algorithm as a subroutine. The details of the GW algorithm are omitted here.

The main result of Algorithm 2 is as follows.

Theorem 3. *For any instance $\mathcal{I}$ of the PC k-ST, Algorithm 2 can output a feasible solution F where the total costs of F is within a factor of 5 the total cost of the optimal solution for $\mathcal{I}$.*

Due to space constraint, the proofs of Theorem 2 are removed but will further appear in a full version.

5   Discussions

In this paper, we propose the PC k-ST, which is a meaningful generalization of the k-ST and PCST, and offer approximation algorithms with ratio of 5.9672 and 5. For further study, we are interested in improving the approximation ratio of 5, since the gap between the upper and lower bound for the PC k-ST is not as narrow as being satisfied.

Acknowledgements. The first author is supported by the National Natural Science Foundation of China (No. 12001523). The second author is supported by Scientific Research Project of Beijing Municipal Education Commission (No. KM201910005012) and the National Natural Science Foundation of China (No. 11971046). The third and fourth authors are supported by the National Natural Science Foundation of China (No. 11871081). The third author is also supported by Beijing Natural Science Foundation Project No. Z200002.

Algorithm 2:

Input: An PC k-ST instance $\mathcal{I} = (G, R, r, k, \{c_e\}_{e \in E}, \{\pi_v\}_{v \in V})$.

Output: An r-rooted tree F which spans at least k vertices in R.

Step 1 Run the GW algorithm to find two suitable value of λ.

  Step 1.1 Set $\lambda_1 := 0$ and $\lambda_2 := \sum_{e \in E} c_e$. Run GW algorithm with λ_1 and λ_2 in order to obtain solution $F_1, \{y_S^1\}_{S \subseteq V \setminus \{r\}}$ and $F_2, \{y_S^1\}_{S \subseteq V \setminus \{r\}}$. Set $k_1 = |V(F_1) \cap R|$ and $k_2 = |V(F_2) \cap R|$.

  Step 1.2 **While** $\lambda_1 - \lambda_2 > \frac{c_{\min}}{2n^2}$ **do**
  set $\lambda_m := \frac{\lambda_1 + \lambda_2}{2}$. Run GW algorithm with λ_m in order to get solution F_m, $\{y_S^m\}_{S \subseteq V \setminus \{r\}}$. Set $k_m = |V(F_m) \cap R|$.
  If $\lambda_m = k$, update $F := F_m$ and stop the algorithm.
  If $\lambda_m > k$, update $F_2 := F_m$, $\lambda_2 := \lambda_m$ and
  $\{y_S^2\}_{S \subseteq V \setminus \{r\}} := \{y_S^m\}_{S \subseteq V \setminus \{r\}}$.
  If $\lambda_m < k$, update $F_1 := F_m$, $\lambda_1 := \lambda_m$ and
  $\{y_S^1\}_{S \subseteq V \setminus \{r\}} := \{y_S^m\}_{S \subseteq V \setminus \{r\}}$.

Step 2 Construct a feasible solution F for the PC k-ST instance $\mathcal{I}$.

  Step 2.1 Double all the edges in $E(F_2)$ and find its Euler tour T_2.

  Step 2.2 Find a tour T_s with only the vertices in $V(F_2)$.

  Step 2.3 Find the cheapest path P with exactly $k - k_1$ vertices in the T_s.

  Step 2.4 Connect the closest vertex in P to the root vertex r.

  Step 2.5 Combine F_1, P and the edge in Step 2.4 to produce and output F.

References

1. Archer, A., Bateni, M.H., Hajiaghayi, M.T., Karloff, H.: Improved approximation algorithms for prize-collecting Steiner tree and TSP. SIAM J. Comput. **40**, 309–332 (2011)
2. Arora, S., Karakostas, G.: A $2 + \varepsilon$ approximation algorithm for the k-MST problem. In: Proceedings of the 11th Annual ACM-SIAM Symposium on Discrete Algorithms, pp. 754–759 (2000)
3. Arya, S., Ramesh, H.A.: 2.5-factor approximation algorithm for the k-MST problem. Inf. Process. Lett. **65**, 117–118 (1998)
4. Awerbuch, B., Azar, Y., Blum, A., Vempala, S.: Improved approximation guarantees for minimum-weight k-trees and prize-collecting salesmen. SIAM J. Comput. **28**, 254–262 (1999)
5. Bienstock, D., Goemans, M.X., Simchi-Levi, D., Williamson, D.P.: A note on the prize collecting traveling salesman problem. Math. Program. **59**, 413–420 (1993)
6. Blum, A., Ravi, R., Vempala, S.: A constant-factor approximation algorithm for the k-MST problem. In: Proceedings of the 28th Annual ACM Symposium on Theory of Computing, pp. 442–448 (1996)
7. Chudak, F.A., Roughgarden, T., Williamson, D.P.: Approximate k-MSTs and k-Steiner trees via the primal-dual method and Lagrangean relaxation. Math. Program. **100**, 411–421 (2004)
8. Fischetti, M., Hamacher, H.W., Jørnsten, K., Maffioli, F.: Weighted k-cardinality trees: complexity and polyhedral structure. Networks **24**, 11–21 (1994)
9. Garg, N.: A 3-approximation for the minimum tree spanning k vertices. In: Proceedings of the 37th Annual Symposium on Foundations of Computer Science, pp. 302–309 (1996)
10. Garg, N.: Saving an epsilon: a 2-approximation for the k-MST problem in graphs. In: Proceedings of the 37th Annual ACM Symposium on Theory of Computing, pp. 396–402 (2005)

11. Goemans, M.X., Williamson, D.P.: A general approximation technique for constrained forest problems. SIAM J. Comput. **24**, 296–317 (1995)
12. Han, L., Xu, D., Du, D., Wu, C.: A 5-approximation algorithm for the k-prize-collecting Steiner tree problem. Optim. Lett. **13**, 573–585 (2019)
13. Han, L., Xu, D., Du, D., Wu, C.: A primal-dual algorithm for the generalized prize-collecting Steiner forest problem. J. Oper. Res. Soc. China **5**, 219–231 (2017)
14. Matsuda, Y., Takahashi, S.: A 4-approximation algorithm for k-prize collecting Steiner tree problems. Optim. Lett. **13**, 341–348 (2019)
15. Rajagopalan, S., Vazirani, V.V.: Logarithmic approximation of minimum weight k trees. Unpublished manuscript (1995)
16. Ravi, R., Sundaram, R., Marathe, M.V., Rosenkrantz, D.J., Ravi, S.S.: Spanning trees short or small. SIAM J. Discrete Math. **9**, 178–200 (1996)
17. Xu, Y., Xu, D., Du, D., Wu, C.: Improved approximation algorithm for universal facility location problem with linear penalties. Theor. Comput. Sci. **774**, 143–151 (2019)
18. Xu, Y., Xu, D., Du, D., Zhang, D.: Approximation algorithm for squared metric facility location problem with non uniform capacities. Discrete Appl. Math. **264**, 208–217 (2019)

Optimal Algorithm of Isolated Toughness for Interval Graphs

Fengwei Li[1], Qingfang Ye[1], Hajo Broersma[2], and Xiaoyan Zhang[3(✉)]

[1] College of Basic Science, Ningbo University of Finance and Economics,
Ningbo 315327, Zhejiang, People's Republic of China
fengwei.li@hotmail.com, fqy-y@163.com
[2] Faculty of EEMCS, University of Twente,
P.O. Box 217, 7500 AE Enschede, The Netherlands
h.j.broersma@utwente.nl
[3] School of Mathematical Science and Institute of Mathematics,
Nanjing Normal University, Nanjing 210023, Jiangsu, People's Republic of China
zhangxiaoyan@njnu.edu.cn

Abstract. Factor and fractional factor are widely used in many fields related to computer science. The isolated toughness of an incomplete graph G is defined as $i\tau(G) = \min\{\frac{|S|}{i(G-S)} : S \in C(G), i(G-S) > 1\}$. Otherwise, we set $i\tau(G) = \infty$ if G is complete. This parameter has a close relationship with the existence of factors and fractional factors of graphs. In this paper, we pay our attention to computational complexity of isolated toughness, and present an optimal polynomial time algorithm to compute the isolated toughness for interval graphs, a subclass of co-comparability graphs.

Keywords: Isolated toughness · Factor · Fractional factor · Interval graph · Polynomial time algorithm

1 Introduction

Throughout this paper, we use Bondy and Murty [1] for terminology and notations not defined here and consider finite simple undirected graphs only. The vertex set of a graph G is denoted by V and the edge set of G is denoted by E. For $X \subseteq V(G)$, let $\omega(G - X)$ and $i(G - X)$, respectively, denote the number of components, the number of components which are isolated vertices in $G - X$. We use $\delta(G)$ and $\kappa(G)$ to denote the minimum degree and connectivity of G, respectively. For any $X \subseteq V$, denote $G[X]$ to be the subgraph of G induced by X. Let $\kappa(G)$ denotes the connectivity of graph G. A subset $X \subseteq V$ is a *cutset* of a graph $G = (V, E)$ if $G - X$ has more than one component. Note that $X = \emptyset$ is a cutset of G if and only if G is disconnected. We let $C(G)$ denote the set of all cutsets of G. A *clique* of a graph is an induced subgraph that is a complete graph.

© Springer Nature Switzerland AG 2021
Y. Zhang et al. (Eds.): PDCAT 2020, LNCS 12606, pp. 379–388, 2021.
https://doi.org/10.1007/978-3-030-69244-5_34

The study of the factor and fractional factor of graphs is a new problem raised in recent years [21]. Let g and f be two non-negative integer-valued functions defined on $V(G)$ such that $g(x) \leq f(x)$ for each $x \in V(G)$. A (g, f)-factor of G is a spanning subgraph H of G such that $g(x) \leq d_H(x) \leq f(x)$ holds for each $x \in V(G)$. Similarly, H is an f-factor of G if $g(x) = f(x)$ for each $x \in V(G)$.

Fractional factors can be considered as the rationalization of the traditional factors by replacing integer-valued function by a more generous "fuzzy" function (i.e., a $[0, 1]$-valued indicator function). Fractional factors have wide-ranging applications in areas such file transfer problems in computer networks, timetable problems and scheduling, etc.

In 1973, Chvátal [5] introduced the notion of toughness for studying Hamiltonian cycles and regular factors in graphs. The *toughness* [5] of an incomplete connected graph G is defined as

$$\tau(G) = \min\{\frac{|S|}{\omega(G - S)} : S \in C(G), \omega(G - S) > 1\}.$$

This parameter has become an important graph invariant for studying various fundamental properties of graphs. In particular, Chvátal conjectured that k-toughness implies a k-factor in graphs and this conjecture was confirmed positively by Enomoto et al. [6].

Motivated from Chvátal's toughness by replacing $\omega(G - X)$ with $i(G - X)$ in the above definition, Ma and Liu [15] introduced the *isolated toughness*, $i\tau(G)$, as a new parameter to investigate and discuss the necessary and sufficient condition for a graph to have a (fractional) factor of the graph.

Definition 1 [15]. The *isolated toughness* of an incomplete connected graph G is defined as

$$i\tau(G) = \min\{\frac{|S|}{i(G - S)} : S \in C(G), i(G - S) > 1\},$$

where the maximum is taken over all the cutsets of G. Especially, for complete graph K_n, define $i\tau(K_n) = \infty$.

The following result is basic in fractional factor theory.

Theorem 1 [21]. *A graph G has a fractional 1-factor iff $i(G - X) \leq |X|$ for any $X \subseteq V(G)$.*

Thus, we can easily get the following theorem which provides a characterization for the existence of fractional 1-factors in terms of $i\tau(G)$.

Theorem 2 [22]. *Let G be a graph of order $n \geq 2$. Then G has a fractional 1-factor iff $i\tau(G) \geq 1$.*

Ma and Liu [15] proved that graph G has a fractional 2-factor if $i\tau(G) \geq 2$ and $\delta(G) \geq 2$. Furthermore, they showed that graph G has a fractional k-factor if $i\tau(G) \geq k$ and $\delta(G) \geq k$, and if $\delta(G) \geq i\tau(G) \geq a - 1 + \frac{a}{b}$, then G has a fractional $[a, b]$-factor, where $a < b$ are two positive integers [16]. Ma and Yu [19] proved that if G is a graph with $\delta(G) \geq a$, $i\tau(G) \geq a - 1 + \frac{a-1}{b}$, and $G - S$ has no $(a - 1)$-regular component for any subset $S \subseteq V(G)$, then G has an $[a, b]$-factor. For more results about (isolated) toughness condition for existence of (fractional) factor in graphs we refer to [17,18].

In this paper, we discuss the computational complexity of isolated toughness in graphs and we give a polynomial time algorithm to compute isolated toughness for interval graphs.

2   Preliminaries

In this section, we recall some definitions, notations and lemmas which will be used throughout the paper.

First, we define the minimal cutset.

Definition 2 [11]. A subset $X \subseteq V$ of G is called an a, b-*cutset* for nonadjacent vertices a and b of graph G if the removal of X separates a and b in distinct connected components. If no proper subset of X is an a, b-cutset of graph G, then X is called a *minimal a, b-cutset* of G. A *minimal cutset* X of G is a set of vertices such that X is a minimal a, b-cutset for some nonadjacent vertices a and b.

The following Lemma provides an easy test of whether or not a given vertex set X is a minimal cutset [12].

Lemma 1 [12]. *Let X be a cutset of the graph $G = (V, E)$. Then X is a minimal cutset if and only if there are at least two different connected components of $G - X$ such that every vertex of X has a neighbor in both of these components.*

For $k \in \{0, 1, 2 \ldots, n\}$ we define $i_k(G)$ as the maximum number of isolated vertices the graph G can obtain after accurately removing k vertices from G, i.e., $i_n(G) = 0$ and for $k < n$

$$i_k(G) = max\{i(G - S) : S \subseteq V, |S| = k\}.$$

It is easy to see that for any incomplete graph G, we have

$$i\tau(G) = \min\{\frac{k}{i_k(G)} : i_k(G) > 1\}.$$

The following theorem give a formula to compute the $i_k(G)$ for $k \in \{0, 1, 2 \ldots, n\}$.

Theorem 3. *Let G be an incomplete graph and let $k \in \{0, 1, 2 \ldots, n\}$. If $i_k(G) > 1$ then*

$$i_k(G) = \max_{|X^*| \leq k} \max_{0 \leq r_{h+1}, r_{h+2}, \ldots, r_p \leq n} \left\{ \sum_{j=h+1}^{p} i_{r_j}(G[C_j]) + h : \sum_{j=h+1}^{p} r_j = k - |X^*| \right\},$$

where the maximum is taken over all minimal cutsets X^* of the graph G, and over all nonnegative integer vectors $(r_{h+1}, r_{h+2}, \ldots, r_p)$. Furthermore, $C_1, C_2, \ldots, C_h$ are the components of $G - X^*$ which are isolated vertices, and $C_{h+1}, C_{h+2}, \ldots, C_p$ are the connected components of $G - X^*$ which are not isolated vertices.

Proof. First let X be a cutset of G with $|X| = k$ and $i(G - X) = i_k(G) > 1$. Let X^* be a minimal cutset of G that is a subset of X, we suppose $C_1, C_2, \ldots, C_h$ be the components of $G - X^*$ which are isolated vertices, and let $C_{h+1}, C_{h+2}, \ldots, C_p$ be the connected components of $G - X^*$ which are not isolated vertices. Then $C_1, C_2, \ldots, C_h$ are also the components of $G - X$, so, we consider the sets $X_j = X \cap C_j$, $j \in \{h+1, h+2, \ldots, p\}$. Then, we know that

$$i_k(G) = i(G - X) = \sum_{j=h+1}^{p} i(G[C_j - X_j]) + h \leq \sum_{j=l+1}^{p} i_{|X_j|}(G[C_j]) + h$$

$$\leq \max_{|X^*| \leq k} \max_{0 \leq r_{h+1}, r_{h+2}, \ldots, r_p \leq n} \left\{ \sum_{j=h+1}^{p} i_{r_j}(G[C_j]) + h : \sum_{j=h+1}^{p} r_j = k - |X^*| \right\}.$$

On the other hand, let X^* be a minimal cutset of G. Furthermore let $C_1, C_2, \ldots, C_h$ be the components of $G - X^*$ which are isolated vertices, and let $C_{h+1}, C_{h+2}, \ldots, C_p$ be the connected components of $G - X^*$ which are not isolated vertices. Let $(r_{h+1}, r_{h+2}, \ldots, r_p)$ be a vector making the right hand side of the above formula to be maximal. For every $j \in \{h+1, h+2, \ldots, p\}$, we choose a set X_j of $G[C_j]$ such that $|X_j| = r_j$ and $i_{r_j}(G[C_j]) + h = i(G[C_j - X_j]) + h$. Thus, $X = X^* \cup (\cup_{j=h+1}^{p} X_j)$ is a subset of G and

$$|X| = |X^*| + \sum_{j=h+1}^{p} |X_j| = |X^*| + \sum_{j=h+1}^{p} r_j = k.$$

Furthermore, we have

$$\max_{|X^*| \leq k} \max_{0 \leq r_{h+1}, r_{h+2}, \ldots, r_p \leq n} \left\{ \sum_{j=h+1}^{p} i_{r_j}(G[C_j]) + h : \sum_{j=h+1}^{p} r_j = k - |X^*| \right\}$$

$$= \sum_{j=h+1}^{p} i_{r_j}(G[C_j]) + h = i(G[C_j - X_j]) + h \leq i_k(G).$$

This completes the proof. ∎

Theorem 4. *Let G be an incomplete graph, let X^* be a minimal cutset of G and let $C_1, C_2, \ldots, C_h$ are the components of $G - X^*$ which are isolated vertices, and $C_{h+1}, C_{h+2}, \ldots, C_p$ are the connected components of $G - X^*$ which* *are not isolated vertices. For every $j \in \{h+1, h+2 \ldots, p\}$, let the list H_j be* $(i_0(G[C_j]), i_1(G[C_j]), \ldots, i_{|C_j|}(G[C_i]))$. *There is an algorithm computing*

$$\max_{0 \leq r_{h+1}, r_{h+2}, \ldots, r_p \leq n} \left\{ \sum_{j=h+1}^{p} i_{r_j}(G[C_j]) + h : \sum_{j=h+1}^{p} r_j = k - |X^*| \right\},$$

for every $k \geq |X^*|$ from the list $(H_{h+1}, H_{h+2}, \ldots, H_p)$ in time $O(n^3)$.

Proof. Let X^* be a minimal cutset of G, we suppose $C_1, C_2, \ldots, C_h$ be the components of $G - X^*$ which are isolated vertices, and let $C_{h+1}, C_{h+2}, \ldots, C_p$ be the connected components of $G - X^*$ which are not isolated vertices. let $i_j^{(r)}(G - X^*)$ $(h + 1 \leq r \leq p)$ to be the largest number of isolated vertices of the graph $G[\cup_{j=h+1}^{p} C_j]$ after the deleting of $k - |X^*|$ vertices in $\cup_{j=h+1}^{r} C_j$. Thus, $i_j^{(p)}(G - X^*)$ is exactly the largest number of isolated vertices of $G[\cup_{j=h+1}^{p} C_j]$ can have after the removal of $k - |X^*|$ vertices in $\cup_{j=h+1}^{p} C_j$ which is exactly

$$\max_{0 \leq r_{h+1}, r_{h+2}, \ldots, r_p \leq n} \left\{ \sum_{j=h+1}^{p} i_{r_j}(G[C_j]) + h : \sum_{j=h+1}^{p} r_j = k - |X^*| \right\},$$

Let the list $L^{(r)}$ be

$$(i_0^{(r)}(G - X^*), i_1^{(r)}(G - X^*), \ldots, i_{|\cup_{j=h+1}^{t} C_j|}^{(r)}(G - X^*)),$$

$h + 1 \leq r \leq p$. Then $L^{(h+1)} = H_{h+1}$.

Furthermore, the algorithm iteratively computes for $r = h + 2, h + 3, \ldots, p$ the list $L^{(r)}$ from $L^{(r-1)}$ and H_r by using

$$i_k^{(r)}(G - X^*) = \max \left\{ i_a^{(r-1)}(G - X^*) + i_b^{(r)}(G[C_r]) : a + b = k \right\}.$$

The calculation of an entry $i_k^{(r)}(G - X^*)$ can be completed in time $O(n)$. Hence, we can compute all $O(n^2)$ entries in time $O(n^3)$ by the algorithm. This completes the proof. ∎

3   Isolated Toughness for Interval Graphs

An undirected graph G is called an *interval graph* if its vertices can be put into one to one correspondence with a set of intervals ℓ of a linearly ordered set (like the real line) such that two vertices are connected by an edge if and only if their corresponding intervals have nonempty intersection [9].

Interval graphs are a well-known family of perfect graphs with plenty of nice structural properties [4,7,9,10,20]. Kratsch et al. [13] computed the toughness and the scattering number for interval and other graphs. Li and Li [14] proved the problem of computing the neighbor scattering number of an interval graph can be solved in polynomial time. Broersma et al. [3] gave linear-time algorithms for computing the scattering number and Hamilton-connectivity of interval graphs. In this section, we prove that there exists polynomial time algorithm for computing isolated toughness of an interval graph.

The following lemmas give some useful properties of interval graphs.

Lemma 2 [9]. *Any induced subgraph of an interval graph is an interval graph.*

Lemma 3 [2]. *Any interval graph with order n and size m can be recognized in $O(m + n)$ time.*

Lemma 4 [8]. *A graph G is an interval graph if and only if the maximal cliques of G can be linearly ordered, such that, for every vertex v of G, the maximal cliques containing v occur consecutively.*

We call such a linear ordering of the maximal cliques of an interval graph a *consecutive clique arrangement*. Booth and Lueker [2] give a linear time *PQ*-tree algorithm for interval graphs, meanwhile, this algorithm can compute a consecutive clique arrangement of the interval graph too.

The following lemma determine the minimal cutsets of an interval graph.

Lemma 5 [13]. *Let G be an interval graph and let $L_1, L_2, \cdots, L_t$, $t \leq n$, be a consecutive clique arrangement of G. Then the set of all minimal cutsets of G consists of vertex set $C_s = L_s \cap L_{s+1}$, $s \in \{1, 2, \cdots, t-1\}$.*

From Lemma 5, we know that an interval graph G of order n possess at most n minimal cutsets.

Definition 3 [13]. Let G be an interval graph with consecutive clique arrangement $L_1, L_2, \cdots, L_t$. We define $L_0 = L_{t+1} = \emptyset$. For all l, r with $1 \leq l \leq r \leq t$ we define $\mathfrak{P}(l, r) = (\cup_{i=l}^{r} L_i) - (L_{l-1} \cup L_{r+1})$. A set $\mathfrak{P}(l, r)$, $1 \leq l \leq r \leq t$, is said to be a *piece* of G if $\mathfrak{P}(l, r) \neq \emptyset$ and $G[\mathfrak{P}(l, r)]$ is connected. Furthermore, $V = \mathfrak{P}(1, t)$ is a piece of G (even if G is disconnected).

It is obvious that cliques in $G[\mathfrak{P}(l, r)]$ are listed in the same order as that they are listed in graph G.

Lemma 6 [13]. *Let X be a minimal cutset of connected subgraph $G[\mathfrak{P}(l, r)]$, $1 \leq l \leq r \leq t$. Then there exists a minimal cutset C_s of G, $1 \leq s \leq r$, such that $X = C_s \cap \mathfrak{P}(l, r) = C_s - (L_{l-1} \cup L_{r+1})$. Moreover, every connected component of $G[\mathfrak{P}(l, r) - X]$ is a piece of G.*

From the definition of piece of G, any interval graph contains two kind of pieces. A piece is named *complete* if it induces a complete graph. Otherwise, we call it *incomplete*. For every complete piece $G[\mathfrak{P}(l, r)]$, $l \leq r$, holds

$$i_k(G[\mathfrak{P}(l, r)]) = \begin{cases} 0, if \ k \in \{0, 1, 2, \ldots, |G[\mathfrak{P}(l, r)]| - 2\} \\ 1, if \ k = |G[\mathfrak{P}(l, r)]| - 1 \end{cases} \tag{1}$$

The incomplete piece $G[\mathfrak{P}(l,r)]$, $1 \le l \le r \le t$, has minimal cutsets, and for every $k \in \{\kappa(G[\mathfrak{P}(l,r)]), \ldots, |G[\mathfrak{P}(l,r)]| - 2\}$, the following equality holds

$$i_k(G[\mathfrak{P}(l,r)]) = \max \sum_{j=h+1}^{p} i_{r_j}(G[C_i]) + h, \qquad (2)$$

where the maximum is taken over all $C_s \cap \mathfrak{P}(l,r)$, $s \in \{l+1, l+2, \cdots, r-1\}$, that are minimal cutsets of $G[\mathfrak{P}(l,r)]$, satisfying the condition that $|C_s \cap \mathfrak{P}(l,r)| \le k$ and over all nonnegative integer vectors $(r_{l+1}, r_{l+2}, \ldots, r_p)$ fulfilling the condition that $\sum_{j=h+1}^{p} r_j = k - |C_s \cap \mathfrak{P}(l,r)|$. $C_1, C_2, \cdots, C_h$ are the components of $G[\mathfrak{P}(l,r) - C_s]$ which are isolated vertices, and $C_{h+1}, C_{h+2}, \cdots, C_p$ are the connected components of $G[\mathfrak{P}(l,r) - C_s]$ which are not isolated vertices.

Let G be a complete interval graph. Then $i\tau(G) = \infty$. Otherwise, based on Theorem 3, the isolated toughness of incomplete interval graphs can be computed by Algorithm 1.

In the following theorem, we prove the correctness of Algorithm 1 and make clear that the algorithm can be executed in polynomial time.

Theorem 5. *Algorithm 1 outputs the isolated toughness for an input interval graph G of order n within time complexity $O(n^6)$.*

Proof. The correctness of the algorithm can be deduced from the Theorem 3 and Lemma 6. It is easy to see that the steps at lines 2–3 can be performed in $O(1)$ time. The steps at lines 5–11 and 19 can be executed in time $O(n^4)$ in a straightforward manner.

In the steps at line 12 and lines 16–18, an $O(n+m)$ algorithm can be used to test connectedness and calculation components for up to n^2 graphs $G[\mathfrak{P}(l,r)]$. If $G[\mathfrak{P}(l,r)]$ is disconnected and Q_j is a component, then $Q_j = \mathfrak{P}(l_j, r_j)$ with $l(j) = \min\{l(v) : v \in Q_j\}$ and $r(j) = \max\{l(v) : v \in Q_j\}$ which can be computed in time $O(n)$. Hence, the steps at line 12 and lines 16–18 can be executed in time $O(n^4)$.

The steps at lines 13–15 can be executed for at most n^3 triples (s,l,r) with $l \le s \le r$. If $\mathfrak{P}(l,r) - C(s) \neq \emptyset$, then the components of $G[\mathfrak{P}(l,r) - C(s)]$ are computed as indicated in Lemma 6, by using the marks of (l,s) and $(s+1,r)$, namely, if the mark is 'complete' or 'incomplete', then (l,s) and $(s+1,r)$, respectively, are stored, and if the mark is 'disconnected', then the corresponding linked list is added. Thus the linked list of (s,l,r) can be computed in time $O(n)$. From Lemma 1 we know that $\mathfrak{P}(l,r) \cap C(s)$ is a minimal cutset of $G[\mathfrak{P}(l,r) - C(s)]$ if and only if there are at least two components in the list of (s,l,r) such as every vertex of $\mathfrak{P}(l,r) \cap C(s)$ has a neighbour in them. From the consecutive clique arrangement, it suffices to check the two components Q_j of $G[\mathfrak{P}(l,s)]$ with the two largest values of r_j and the two components Q_j of $G[\mathfrak{P}(s+1,r)]$ with the two smallest values of l_j, this can be done in time $O(n)$. Hence, the steps at lines 13–15 can be executed in time $O(n^4)$.

The steps on lines 20–22 require that the right side of the Eq. (2) be calculated for each $k \in \{\kappa(G[\mathfrak{P}(l, l + d)]), \ldots, |\mathfrak{P}(l, l + d)| - 2\}$. The list $H_j =$

Algorithm 1: Algorithm Isolated Toughness

Input: An interval graph G with consecutive clique arrangement $L_1, L_2, \cdots, L_t$.
Output: Isolated toughness $i\tau(G)$.

1 **begin**
2 $\quad$ $L_0 \leftarrow \emptyset$;
3 $\quad$ $L_{t+1} \leftarrow \emptyset$;
4 $\quad$ **for** $w \leftarrow 0$ *to* $t+1$ **do**
5 $\quad\quad$ compute $l(v) = \min\{w : v \in L_w\}$ and $r(v) = \max\{k : v \in L_w\}$ for every $v \in V$, and then compute all minimal cutsets $C_s = L_s \cap L_{s-1}$, $s \in \{1, 2, \cdots, t-1\}$. For all l, r $(1 \le l \le r \le t)$ compute the vertex set $\mathfrak{P}(l, r)$;
6 $\quad\quad$ **if** $\mathfrak{P}(l, r) = \emptyset$ **then**
7 $\quad\quad\quad$ mark (l, r) 'empty';
8 $\quad\quad$ **end**
9 $\quad\quad$ **if** $\mathfrak{P}(l, r) \ne \emptyset$ *and* $G[\mathfrak{P}(l, r)]$ *is a complete induced subgraph* **then**
10 $\quad\quad\quad$ mark (l, r) 'complete'.
11 $\quad\quad$ **end**
12 $\quad\quad$ For all nonmarked tuples (l, r), check whether $G[\mathfrak{P}(l, r)]$ is connected;
13 $\quad\quad$ **if** $G[\mathfrak{P}(l, r)]$ *is connected* **then**
14 $\quad\quad\quad$ mark (l, r) 'incomplete', and for every $s \in \{l, l+1, \ldots, r-1\}$, compute the components $Q_j = \mathfrak{P}(l_j, r_j)$ of $G[\mathfrak{P}(l, r) - C_s]$, $1 \le j_t \le k$. Check whether $C_s \cap \mathfrak{P}(l, r)$ is a minimal cutset of $G[\mathfrak{P}(l, r)]$, and if so mark (s, i, j) 'minimal', store $(l_1, r_1), (l_2, r_2), \ldots, (l_k, r_k)$ in a linked list with a pointer from (s, l, r) to the head of this list, and compute $\kappa(G[\mathfrak{P}(l, r)]) = \min\{|C_s \cap \mathfrak{P}(l, r)|\}$ for (s, l, r) marked 'minimal'.
15 $\quad\quad$ **end**
16 $\quad\quad$ **if** $G[\mathfrak{P}(l, r)]$ *is disconnected* **then**
17 $\quad\quad\quad$ compute the components $Q_j = \mathfrak{P}(l_j, r_j)$, $1 \le j \le q$, of $G[\mathfrak{P}(l, r)]$ and store $(l_1, r_1), (l_2, r_2), \ldots, (l_q, r_q)$ in a linked list with a pointer from (l, r) to the head of this list.
18 $\quad\quad$ **end**
19 $\quad\quad$ For every pair (l, r) marked 'complete' compute $i_k(G[\mathfrak{P}(l, r)])$, $k \in \{0, 1, \ldots, |\mathfrak{P}(l, r)|\}$, according to equation (1);
20 $\quad\quad$ **for** $d \leftarrow 1$ *to* t *and for* $l \leftarrow 1$ *to* $t-d$ **do**
21 $\quad\quad\quad$ **if** $(l, l+d)$ is marked 'incomplete', compute $i_k(G[\mathfrak{P}(l, r)])$ for every $k \in \{\kappa(G[\mathfrak{P}(l, l+d)]), \ldots, |\mathfrak{P}(l, l+d)| - 2\}$ according to equation (2). Set $i_k(G[\mathfrak{P}(l, r)]) = 0$ for $k = |\mathfrak{P}(l, l+d)|$ or $k < \kappa(G[\mathfrak{P}(l, l+d)])$, and let $i_k(G[\mathfrak{P}(l, r)]) = 1$ for $k = |\mathfrak{P}(l, l+d)| - 1$.
22 $\quad\quad$ **end**
23 $\quad$ **end**
24 **end**

$(i_0(Q_j), i_1(Q_j), \ldots, i_{|Q_j|}(Q_j), j \in \{1, 2, \ldots, t\}$, for each component $Q_j = (l_j, r_j)$ of $G[\mathfrak{P}(l, l+d) - \mathcal{C}(s)]$ can be determined in constant time $O(n^2)$ by table look-up, since these lists of smaller pieces are already known. Thus

$$
\max_{0 \leq r_{h+1}, r_{h+2}, \ldots, r_p \leq n} \left\{ \sum_{j=h+1}^{p} i_{r_j}(G[Q_j]) + h : \sum_{j=h+1}^{p} r_j = k - |\mathfrak{P}(l, l+d) \cap \mathcal{C}(s)| \right\}
$$

can be evaluated in time $O(n^3)$ for a given minimal cutset $\mathfrak{P}(l, l+d) \cap \mathcal{C}(s)$ and for every k with $|\mathfrak{P}(l, l+d) \cap \mathcal{C}(s)| \leq k$ in time $O(n^3)$.

Consequently, from the above analysis we know that the running time of isolated toughness algorithm is $O(n^6)$.

This completes the proof.   ∎

Acknowledgements. This work was supported by NSFC (No.11871280), Natural Science Foundation of Zhejiang Province(China) (No. LY17A010017) and Qing Lan Project. Especially, the authors are very thankful to anonymous referees for their constructive suggestions and critical comments, which led to this improved version.

References

1. Bondy, J., Murty, U.: Graph Theory with Applications. Macmillan, London and Elsevier, New york (1976)
2. Booth, K., Lueker, G.: Testing for the consecutive ones property, interval graphs, and graph planarity using PQ-tree algorithms. J. Comput. System Sci. **13**(3), 335–379 (1976)
3. Broersma, H., Fiala, J., Golovach, P., Kaiser, T., Paulusma, D., Proskurowski, A.: Linear-time algorithms for scattering number and hamilton-connectivity of interval graphs. J. Graph Theor. **79**(4), 282–299 (2015)
4. Carlisle, M.C., Lloyd, E.L.: On the k-coloring of intervals. In: Dehne, F., Fiala, F., Koczkodaj, W.W. (eds.) ICCI 1991. LNCS, vol. 497, pp. 90–101. Springer, Heidelberg (1991). https://doi.org/10.1007/3-540-54029-6_157
5. Chvátal, V.: Tough graphs and hamiltonian circuits. Discrete Math. **5**, 215–228 (1973)
6. Enomoto, H., Jackson, B., Katerinis, P., Saito, A.: Toughness and the existence of k-factors. J. Graph Theor. **9**, 87–95 (1985)
7. Fabri, J.: Automatic Storage Optimization. UMI Press Ann Arbor, MI (1982)
8. Gilmore, P., Hoffman, A.: A characterization of comparability graphs and of interval graphs. Can. J. Math. **16**(99), 539–548 (1964)
9. Golumbic, M.C.: Algorithmic Graph Theory and Perfect Graphs. Academic Press (1980)
10. Jungck, J., Dick, O., Dick, A.: Computer assisted sequencing, interval graphs and molecular evolution. Biosystem **15**, 259–273 (1982)
11. Kloks, T., Kratschz, D.: Listing all minimal separators of a graph. SIAM J. Comput. **27**(3), 605–613 (1998)
12. Kloks, A.J.J., Kratsch, D., Spinrad, J.P.: Treewidth and pathwidth of co comparability graphs of bounded dimension. Computing Science Note. Eindhoven University of Technology, Eindhoven, The Netherlands **9346** (1993)

13. Kratsch, D., Klocks, T., Müller, H.: Computing the toughness and the scattering number for interval and other graphs. IRISA resarch report. France (1994)
14. Li, F., LI, X.: Neighbor-scattering number can be computed in polynomial time for interval graphs. Comput. Math. Appl. **54**(5), 679–686 (2007)
15. Ma, Y., Liu, G.: Isolated toughness and the existence of fractional factors in graphs. Acta Appl. Math. Sinica (in Chinese) **62**, 133–140 (2003)
16. Ma, Y., Liu, G.: Fractional factors and isolated toughness of graphs. Mathematica Applicata **19**(1), 188–194 (2006)
17. Ma, Y., Wang, A., Li, J.: Isolated toughness and fractional (g, f)-factors of graphs. Ars Comb. **93**, 153–160 (2009)
18. Ma, Y., Yu, Q.: Isolated toughness and existence of f-factors. In: Akiyama, J., Chen, W.Y.C., Kano, M., Li, X., Yu, Q. (eds.) CJCDGCGT 2005. LNCS, vol. 4381, pp. 120–129. Springer, Heidelberg (2007). https://doi.org/10.1007/978-3-540-70666-3_13
19. Ma, Y., Yu, Q.: Isolated toughness and existence of $[a, b]$-factors in graphs. J. Combin. Math. Combin. Comput. **62**, 1–12 (2007)
20. Ohtsuki, T., Mori, H., Khu, E., Kashiwabara, T., Fujisawa, T.: One dimensional logic gate assignment and interval graph. IEEE Trans. Circ. Syst. **26**, 675–684 (1979)
21. Scheinerman, E.R., Ullman, D.H.: Fractional Graph Theory. John Wiley and Son Inc., New York (1997)
22. Yang, J., Ma, Y., Liu, G.: fractional (g, f)-factor in graphs. Acta Mathematica Scientia **16**(4), 385–390 (2001)

Graceful Performance Degradation in Apache Storm

Mohammad Reza HoseinyFarahabady[1]([⊠]), Javid Taheri[2], Albert Y. Zomaya[1], and Zahir Tari[3]

[1] School of Computer Science, Center for Distributed and High Performance Computing, The University of Sydney, Sydney, NSW, Australia
{reza.hoseiny,albert.zomaya}@sydney.edu.au
[2] Department of Mathematics and Computer Science, Karlstad University, Karlstad, Sweden
javid.taheri@kau.se
[3] School of Science, RMIT University, Melbourne, VIC, Australia
zahir.tari@rmit.edu.au

Abstract. The concept of stream data processing is becoming challenging in most business sectors where try to improve their operational efficiency by deriving valuable information from unstructured, yet, contentiously generated high volume raw data in an expected time spans. A modern streamlined data processing platform is required to execute analytical pipelines over a continues flow of data-items that might arrive in a high rate. In most cases, the platform is also expected to dynamically adapt to dynamic characteristics of the incoming traffic rates and the ever-changing condition of underlying computational resources while fulfill the tight latency constraints imposed by the end-users. *Apache Storm* has emerged as an important open source technology for performing stream processing with very tight latency constraints over a cluster of computing nodes. To increase the overall resource utilization, however, the service provider might be tempted to use a consolidation strategy to pack as many applications as possible in a (cloud-centric) cluster with limited number of working nodes. However, collocated applications can negatively compete with each other, for obtaining the resource capacity in a shared platform that, in turn, the result may lead to a severe performance degradation among all running applications.

The main objective of this work is to develop an elastic solution in a modern stream processing ecosystem, for addressing the shared resource contention problem among collocated applications. We propose a mechanism, based on design principles of Model Predictive Control theory, for coping with the extreme conditions in which the collocated analytical applications have different quality of service (QoS) levels while the shared-resource interference is considered as a key performance limiting parameter. Experimental results confirm that the proposed controller can successfully enhance the p-99 latency of high priority applications by 67%, compared to the default round robin resource allocation strategy in Storm, during the high traffic load, while maintaining the requested quality of service levels.

© Springer Nature Switzerland AG 2021
Y. Zhang et al. (Eds.): PDCAT 2020, LNCS 12606, pp. 389–400, 2021.
https://doi.org/10.1007/978-3-030-69244-5_35

Keywords: Apache storm streaming processing platform · Elastic resource controller · Performance modeling of computer system · Quality of Services (QoS)

1 Introduction

Modern analytical applications require performing data analytic computations over an endless flow of data, known as stream data processing, which is different from the conventional workloads which are commonly found in well-established relational database management systems (RDBMS) and non-relational (NoSQL) data-storage systems. With the advent of social networking sits and internet of things (IoT) platforms, enterprises have to look for new technologies to replace traditional methods, which is built upon a core concept around a central storage system. Such traditional approaches are proved to be inefficient and expensive to deal with the complexity that stream data processing applications bring into content, particularly in terms of the data variety and velocity aspects. With the advent of social networking sits and internet of things (IoT) platforms, the need to satisfy the *latency* requirements for processing huge volumes of data in a short time makes it extremely challenging in traditional storage-centric systems, where all data need to be first stored on disk before further processing [17]. *Real-time analytic* relies on a collection of tools and frameworks for performing analytic over streaming, yet distributed data, which can be generated from various sources, with the main aim of capturing the value of data before it reaches a specified time threshold [2]. Example of such usages can be found in multiple domains like financial markets, healthcare, traffic and monitoring systems, and radio astronomy.

Apache Storm has emerged as a stream processing platform to reliably build distributed data processing applications for performing *continuous computations* over large amount of data in real time in a highly scalable and highly available manner. The Storm platform can execute an intense of a given query in a parallel fashion across many distributed working nodes, through ZooKeeper coordination cluster, which, in turn, can increase the processing capacity of the application while achieves higher availability and exhibits a linear scalability. The Storm engine can also provide a strong consistency guarantee model, in which each message is processed at least once, even in the presence of node failures [9]. If there is a burst of incoming data tuples, Storm is equipped with a simple mechanism to prevent becoming overwhelmed due to the increased load. To cope with the a burst of incoming data items, Storm relies on a topology-wide configuration setting that adjusts the maximum number of not-yet-processed tuples emitted from every data sources, known as spout. Once such a threshold is reached, the corresponding data source is asked to stop emitting further data-tuples until a new state for pending data-tuples (*e.g.*, being acknowledged or timed out) is detected.

However, the experimental results (such as those run in our local cluster) confirms that when a Storm platform is operating under a significant traffic load, the end-to-end processing time of data-tuples belonging to different analytical applicants is determined by the total waiting delays in the corresponding buffers instead of the actual processing time. Such an observation motivates us to propose an elastic resource (*i.e.*, CPU and memory) allocation strategy for Storm platform to scale such computing resources up or down whenever the rate of incoming data-tuple traffic fluctuates, mainly in order to comply with the strict requirements on processing latency defined by the end-users. To this ends, we develop a performance model based on queuing theory to estimate the required processing budget for each stream application, particularly, when the underlying resources are suffering from a high traffic load. The performance model is used along with a low-overhead prediction module for predicting the end-to-end response time of each application withing the forthcoming controlling intervals.

This paper is organized as follows. Section 2 motivates the research problem by showing the performance gap exists in performing multiple concurrent high-throughput stream processing applications in a shared platform when sudden workload surges occur. Section 2 formalizes the research question and the system model targeted in this paper. Section 3 gives insights into the proposed controller. Section 4 summarizes the experimental evaluation results, followed by our conclusion in Sect. 6.

2 Motivation

We conducted an experiment to show how a static resource allocation strategy can poorly result in a performance degradation among 1000 stream processing applications which are concurrently running within a cluster with eight nodes. Nevertheless, such inefficiency is more significant when there is an unexpected surge in the incoming traffic. We divided the entire application set into three different quality of service (QoS) classes, denoted by $Q_{\{1,2,3\}}$, where each class represents a different priority level in terms of requested performance level. In the run-time, however, we deliberately increase the incoming traffic rate of applications belonging to the lowest priority class, *i.e.*, Q_3, while at the same time we measure the impact of such load surges on the performance level of applications belonging to other two QoS classes.

The aim is to understand how well can internal mechanism in the Storm platform *isolate* the obtained performance among collocated applications in a platform with shared resources, particularly when there are sudden surges in the incoming traffic of some applications (low priority classes in our example). The results confirm that a poor level of performance isolation among collocated applications can be achieved by applying the default Storm resource allocation strategy (which is majorly based on the round-robin policy). We observe that during such congested intervals, the default policy starts giving service to Q_1 and Q_2 requests only after all requests from Q_3 are being served. An optimal strategy, on the other hand, would avoid the unsatisfactory responsiveness during the congested periods, most likely by dynamically reducing the rate of QoS violation incidents in the run-time while improving the overall users satisfaction.

Problem Statement

We showed the inflexibility of a round-robin policy when responding to the temporal performance variations in a shared platform. Particularly, as a round-robin policy depends upon the arrival time of data-items for prioritizing of the incoming requests, it can lead to a poor performance when fulfilling the complex quality of service (QoS) enforcement which are often imposed by end-users with different priority levels. We explain such a performance isolation issue in this section in more details.

Shared Resource Interference. To enhance the resource utilization level (*i.e.*, reaching a better scaling at a lower cost) a service provider is often tempted to employ a "consolidation" technique to host as many applications as possible into a fixed set of computing devices. However, devising an effective consolidation plan to be applied across a platform with shared resources is a challenging problem. The main reason lies on this fact that each application could have its own characteristic (in terms of resource usage or sensitiveness to the available shared resources like the last level cache size and memory bandwidth) which may also depend on the incoming traffic rate. The problem is exacerbated by considering the fact that such an incoming traffic rate can unexpectedly fluctuate over time of execution (at frequent, yet short intervals). In most practical situation, the contention among consolidated applications to obtain the capacity of shared resources is an important factor that makes it difficult to design a resource allocation and consolidation policy that fully supports an isolated performance among collocated workloads. A well-known scenario in this context occurs when a query that belongs to a low-priority application evicts the data of other high-priority applications from the last level cache when a operating system context switch occurs. Such unpredictable behavior can cause an undesirable performance degradation to consolidation applications, as they need to continuously fetch their own data from the main memory to the last level cache (which is shared among multiple applications) in the next CPU cycles. Such incident has been reported across other platforms, such as [10, 12–15, 18–20, 22], but to the best of our knowledge, no systematic research exists to prevent the negative impact of shared resource contention among consolidated stream processing applications.

QoS Violation Semantic. Our approach in this paper is to dynamically allocate the CPU and memory capacity to each Storm application based on the continuously monitored run-time conditions and QoS violation incidents. In particular, we assume that there are exactly Q different QoS classes that an application owner can select from, and is billed accordingly. We identify a QoS class as a value pair, denoted by $\langle r^*_{q,m}, \mathcal{V}_{q,\Delta t} \rangle$, where $r^*_{q,m}$ is the maximum response time that an application in class q can tolerate before collecting the result. If a delay takes longer than r^*, it is identified as a *QoS violation incident* in the target platform. In the above-mentioned formula, $\mathcal{V}_{q,\Delta t}$ denotes an upper bound for the percentage of QoS violation incidents, to be tolerated by applications in class q, during any interval of length Δt.

3 Design Principals

This section provides details of the core modules we used to design a resource allocation mechanism for an Apache Storm platform. Having such approach not only does facilitate the implementation of the controlling layer, but also provides a more elastic solution as its scalability limit becomes independent from the Storm scalability barrier.

3.1 System Design Consideration

The main design considerations in our approach can be summarized as follows:

- The controller must obey a non-centralized architecture, in which the controlling actions are made solely based on gathering the local information in each working node. The non-centralized architecture enables the controller to not only receive data streams from various data generators, but also to dispatch the outputs to different data sinks (*e.g.*, databases, HDFS files, among others). The controller must also impose a low overhead to the system (*e.g.*, for monitoring activities and calculating the near-optimal solution).
- The controller must be implemented as a separate proxy layer to the underlying platform, in which it does not interfere with the internal operational strategy of the Storm platform or the CPU/memory management units embedded in the JVM layer.

3.2 Solution Overview

The proposed solution follows the well-established principles of a modern controlling theory know as "model predictive control (MPC)" design that applies run-time throttling over the utilization of CPU and memory resources for every running application. The controller is developed based on a simple idea of deferring the CPU share allocation to the run-time in which the controller can monitor the performance level of each application in addition to the arrival rate to each streaming application for making a more accurate resource allocation decision and to prevent each working node in the entire cluster from becoming a bottleneck. One of the advantages of using MPC approach is its performance robustness against system modeling or prediction errors [16].

Figure 1 presents an architectural overview and the core components of the proposed solution. To comply with the above-mentioned design consideration points, each application runs within its own isolated Linux container across the distributed Storm platform. The Linux kernel provides a lightweight operating-system-level virtualization method for running multiple applications (*i.e.*, the spout execution engine in Storm terminology) on a host. The proposed controller uses the Linux container APIs to limit and prioritize the amount of CPU,

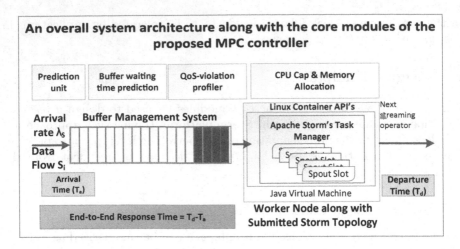

Fig. 1. The core components of the proposed solution

and memory resources per container (it can be further extended for controlling block I/O bandwidth). We employ the Linux control group (`cgroups`) to limit the resource usage for a group of isolated spout execution processes by adjusting *cpu.share* and *mem.share* parameters for the corresponding *cgroup*. The spout tasks can share a single operating system kernel and are isolated from other applications running in working machines. At every epoch of $\tau \in \{t_1, t_1 + \Delta t, t_1 + 2\Delta t, \cdots\}$, each local controlling module running at each working node compares the monitored performance metric of each submitted processing application to the desired absolute end-to-end delay. It aims to reduce the measured error to zero in the forthcoming controlling steps by adjusting the amount of processing/memory budget for each spout execution process.

The proposed architecture comprised of the following components: (1) a queuing model, (2) an incoming rate forecaster module, (3) a shared resource anti-saturation module, and (4) a CPU and memory cap optimization module. The controller goal is to adjust configurable system parameters to achieve an accurate result without causing any significant over- or under-shooting around the target set-point performance metric.

3.3   Performance Model

We present a performance model to predict the response time for analytical processing of data items, which is required by the optimization module, to allocate enough CPU and memory size to each application. We use Allen-Cunneen (A-C) approximation of $G/G/N$ queues [1] to estimate an upper-bound for the mean response time (*i.e.*, the sum of the waiting time in the main buffer and the required processing time) as experienced by each data item, as follows.

$$W_M = \frac{P_{cb,N}}{\mu N(1 - \rho)} \left(\frac{C_S^2 + C_D^2}{2} \right) \tag{1}$$

where W_M denotes the average waiting time of customers in a $G/G/N$ queue consists of N working nodes with a general distribution of both arrival time and service time, ρ is the service traffic intensity (server utilization), $C_D = \sigma_D/E_D$ is the coefficient of variation for inter-arrival time, $C_S = \sigma_S/E_S$ is the coefficient of variation for service time, $\frac{C_S^2 + C_D^2}{2}$ is the stochastic variability of the queue, and $P_{cb,N}$ represents the probability that all workers in the queuing system are busy (*i.e.*, the waiting time of a new customer is greater than zero).

3.4 Prediction Module

We assume that the probability distribution of the arrival rate of data tuples toward each streaming application is not known in advance. Hence, we employ the famous auto regressive integrated moving average (ARIMA) model [4] to estimate the arrival tuple rate as a non-controllable parameter during forthcoming sampling intervals $\tau + 1, \tau + 2 \cdots$. We estimate the future values of such a parameter by applying a stochastic analysis over a series of recent h observations.

3.5 Anti-saturation

To cope with the performance degradation bottleneck issue caused by contention among consolidated workloads to obtain the shared CPU cache or the memory bandwidth, we pursue an effective method firstly proposed in [20] to appropriately quantify the performance slowdown due to micro-architecture interference. This method identifies the workloads contention by detecting any abnormal increase in the *memory bandwidth utilization* of each node, which is calculated by monitoring two standard hardware events of *UNC_QMC_READS* (an indicator of memory reads) and *UNC_QMC_WRITES* (an indicator of memory writes).

3.6 Optimization Process

To determine the possibility of fulfilling total CPU cap and buffer size requested from all applications, the controller first checks if the entire demand is *higher* than the available resource capacity of each working node. If the available capacity is not adequate to fulfill all requests, then the controller performs a cost-benefit analysis to prioritize requests issued by applications in the high QoS class. Let $\mathcal{C}_{S,\tau}^*$ denote the amount of resource demanded by a stream processing application S, and $\mathcal{R}$ denote the total resource capacity which is available in that working node. We define a *contribution* function for a partial fulfillment of the requested resource to every stream application S as

$$\mathcal{C}(r_S) = \mathcal{I}_{q_S} \times (r_S - \mathcal{C}_S^*), \tag{2}$$

where r_S is the amount of resource share to be allocated to application S at the next epoch, q_S represents the QoS class that application S belongs to, and $\mathcal{I}$ is a predefined weight which is set by system administrators to represent the importance of each q_S compared to other QoS classes. The optimization module will formulate the total contribution received by the service provider by solving the following optimization problem.

$$\max_r \sum_{S \in \mathcal{S}} \mathcal{C}(r_S), \tag{3}$$

subject to the obvious constraints of $r_S \leq \mathcal{C}^*_{S,\tau}$, $\forall S \in \mathcal{S}$, and $\sum_{S \in \mathcal{S}} r_S = \mathcal{R}$, where $\mathcal{S}$ represents the set of all running applications. We transfer the continuous variable into a set of several discrete counterparts. By recursively solving the following Bellman's equation, the controller can find an optimal solution for Eq. (3) as follows.

$$V_\omega(\mathcal{R}_\omega) = \max_{0 \leq r_\omega \leq \mathcal{R}_\omega} V_{\omega+1}(\mathcal{R}_\omega - r_\omega) + \mathcal{C}(r_\omega), \tag{4}$$

where $V_\omega(.)$ denotes the optimal reward of allocating R_ω resources among all non-allocated applications.

4   Experimental Evaluation

The proposed controlling system is implemented using C++ programming language as a proxy-tier layer and buffer management tool while it has integrated with Apache Storm based on the methodologies we described in Sect. 3.

4.1   Experimental Setup and Benchmark Suite

The hardware platform we used consists of eight worker nodes with a total of 32 CPU cores. Each computing node runs a 64-bit Linux 5.0-generic kernel and is equipped with four Intel Xeon E-5620 cores working at 2.4 GHz clock rate, 8 GB RAM, and 64 GB SCSI-v3 storage drive.

We develop different scenarios by choosing a variety of stream analytical applications chosen from the following benchmark suits.

- A data pipeline benchmark (namely DPIP) that emulates a distributed pub-sub messaging system for consuming large text streams generated from a synthetic data load generator.
- A computational benchmark chosen from BigDataBench 4.0 that mainly comprises of Grep, Sort, and MD5 operations running on randomly generated streaming data.

Each application runs inside its own Linux container and is initialized with a series of bench-marking workloads to exhibit different run-time workload intensiveness. We run three scenarios for each application by setting the corresponding incoming data rate to $\lambda \in \{500, 1500, 3000\}$ data-items per second as the incoming rate. We referred to such scenarios as *low*, *middle* and *high* traffic intervals, respectively. Each analytical application is randomly assigned to one of the three predefined QoS classes, denoted by $Q_{i=1\cdots3}$, where Q_1 has the most stringent latency requirements (highest priority).

4.2   Result Summary

Latency and Throughput Figure 2 depicts the performance degradation of applications belonging to different QoS classes with regards to the respective p-99 latency. Under *low* traffic intervals, both policies can meet the target QoS requirements for each class. Such results show more dramatic effect under *high* workload conditions. Particularly, the improvements in p-99 latency during high workloads for q_1 and q_2 classes are 67.8% and 41%, respectively. Moreover, during the high traffic, the variation of latency achieved by the proposed controller shows 37% improvement comparing to the default policy. Such responsiveness for high priority applications is the main aim of this project, even though our approach only uses a proxy-tier architecture that does not change the internal structure of Storm.

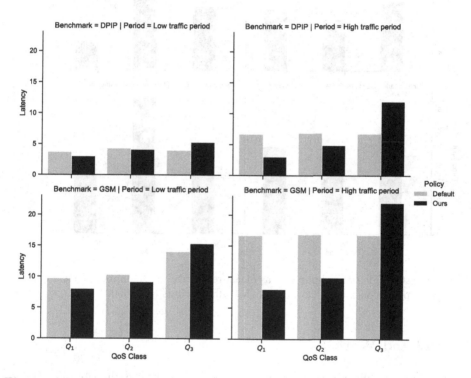

Fig. 2. The p-99 response time of a set of streaming analytical applications chosen from two different bench-marking tools with different QoS requirements, denoted by $Q_{\{1,2,3\}}$, when the default resource allocation policy of Storm engine versus the proposed controller is being used.

Figure 3 represents the average cluster-wide throughput of working nodes in terms of number of streaming query operations accomplished per second as achieved by applying the proposed controller versus the default scheduler. The experimental results confirm that the overall throughput remains almost the same by using either of the mentioned resource allocation strategies. However, there is a significant difference of how each strategy assigns the shared throughput among requests belonging to different QoS classes. During the burst period, the default policy assigns 45% of the total resource utilization to applications in the lowest priority class, q_3, while this number is only 16% for the proposed controller.

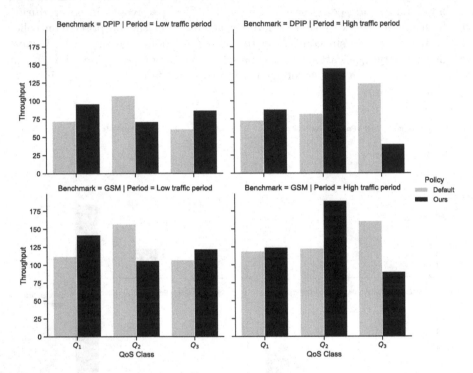

Fig. 3. Cluster wide throughput, $|\bar{O}|$, as measured by the total number of accomplished operations per second, under low and high traffic periods ($n = 6000$).

5   Related Work

The design of efficient scheduling schemes for real-time distributed stream processing received significant attention in recent years. Most of the studies addressed the limitation of RR (Round Robin) default scheduler of Apache Storm. For example, the offline algorithm proposed in [3] could successfully reduce the processing delays of streams when compared to the Storm default scheduler. However, such offline scheduling decisions require executing before an event is triggered in lieu of making schedule decisions during execution. Hence, its limitation is quickly revealed as it fails to adapt in run-time to varying traffic conditions.

The work in [8] presented an online mechanism to automatically explore the parallel level of a given topology based on the measured congestion status and the throughput of the system. It provided a solution regarding the stateful migration if rescheduling happens. The issue of stateful migration has not been covered in our work, and thus we plan to consider it for future works. The online approach in T-Storm [21] is concerned with the run-time traffic patterns. T-storm enables dynamic adjustment of schedule parameters to support running fewer worker nodes while speeding up the overall time for data processing. The evaluation showed that T-storm provides 84% and 27% speedup on lightly and heavily loaded topology, respectively, while it achieves 30% less utilization of worker nodes compared the default scheduler. However, T-storm does not support any mechanism to guarantee QoS enforcement.

MPC-based RA solutions have been successfully exploited by different researchers in the past for devising capacity provisioning in cloud computing [5,11], and elastic scaling mechanism in stream data processing [6,7]. The key idea is to exploit Model Predictive Control that takes into account the behavior of the platform in the time horizon in future to find out the best configuration.

6 Conclusion

High-throughput and low-latency stream processing algorithms are of increasing importance due to their wide applicability in large-scale distributed processing systems, which often require running continuous pipeline queries over streamed high-volume data. Measuring run-time system behavior can be of great practical importance to design a QoS-aware resource allocation mechanism for a low-latency stream processing engine. This paper presented a distributed resource controller for Apache Storm engine to achieve such goals without the need for exchanging run-time states to any centralized system. The proposed solution enhances the QoS satisfaction level up to 90%, while reducing the p-99 query response time by 67% during the high-rate workloads for analytical applications belonging to the QoS classes with the highest priority.

Acknowledgments. Professor Albert Y. Zomaya would like to acknowledge the support of the Australian Research Council Discovery scheme (grant DP200103494). Professor Zahir Tari would like to acknowledge the support of the Australian Research Council Discovery scheme (grant DP200100005). Professor Javid Taheri would like to acknowledge the support of the Knowledge Foundation of Sweden through the AIDA project. Dr. M.Reza HoseinyFarahabady would like to acknowledge continued support of *The Centre for Distributed and High Performance Computing* in *The University of Sydney* for providing access to advanced high-performance computing and cloud facilities, digital platforms and tools.

References

1. Allen, A.O.: Probability, Statistics, Queueing Theory. Academic Press (1990)

2. Andrade, H.C.M., Gedik, B., Turaga, D.S.: Fundamentals of Stream Processing. Cambridge University, New York, NY, USA (2014)
3. Aniello, L., Baldoni, R., Querzoni, L.: Adaptive online scheduling in storm. In: Proceedings of the 7th ACM International Conference on Distributed Event-Based Systems, pp. 207–218. ACM (2013)
4. Box, G., et al.: Time Series: Forecasting & Control. Wiley (2008)
5. Casalicchio, E., et al.: Autonomic resource provisioning in cloud systems with availability goals. In: Proceedings of the 2013 ACM Cloud and Autonomic Computing Conference, pp. 1–10. ACM (2013)
6. De Matteis, T., Mencagli, G.: Keep calm and react with foresight: strategies for low-latency and energy-efficient elastic data stream processing. In: Proceedings of the 21st ACM SIGPLAN Symposium on Principles and Practice of Parallel Programming, PPoPP 2016, ACM, New York, NY, USA, pp. 13:1–13:12 (2016)
7. DeMatteis, T., et al.: Proactive elasticity and energy awareness in data stream processing. J. Syst. Softw. **127**, 302–319 (2016)
8. Gedik, B., et al.: Elastic scaling for data stream processing. Trans. Parallel Distrib. Syst. **25**(6), 1447–1463 (2014)
9. Jain, A.: Mastering Apache Storm. Packt Publishing (2017)
10. Kim, Y., et al.: Scalable & high-performance scheduling algorithm for multiple memory controllers. In: HPCA-16 2010 The Sixteenth International Symposium on High-Performance Computer Architecture, pp. 1–12. IEEE (2010)
11. Kusic, D., et al.: Power and performance management of virtualized computing environments via lookahead control. In: International Conference on Autonomic Computing. ICAC 2008, IEEE, Washington, DC, pp. 3–12 (2008)
12. Moraveji, R., Taheri, J., HoseinyF., M., et al.: Data-intensive workload consolidation for the hdfs systems. In: ACM/IEEE 13th International Conference on Grid Computing, pp. 95–103. IEEE (2012)
13. Mutlu, O., Moscibroda, T.: Stall-time fair memory access scheduling for chip multiproc. In: 40th Annual IEEE/ACM International Symposium on Microarchitecture (MICRO 2007), pp. 146–160. IEEE (2007)
14. Nathuji, R., et al.: Q-clouds: managing performance interference effects for QoS-aware clouds. In: Proceedings of the 5th European conference on Computer systems, pp. 237–250. ACM (2010)
15. Onur, M., Thomas, M.: Parallelism-aware batch scheduling: enhancing both performance & fairness of shared dram systems. In: International Symposium on Computer Architecture, pp. 63–74. IEEE (2008)
16. Rawlings, J., et al.: Model predictive control: theory & design. NobHill (2009)
17. Stonebraker, M., et al.: The 8 requirements of real-time stream processing. ACM Sigmod Rec. **34**(4), 42–47 (2005)
18. Subramanian, L., et al.: MISE: Providing performance predictability & improving fairness in shared memory systems. In: High Performance Computer Architecture, pp. 639–650. IEEE (2013)
19. Tembey, P., et al.: Application & platform-aware RA in consolidated servers. In: SOCC, pp. 1–14 (2014)
20. Wang, H., et al.: A-DRM: architecture-aware distributed RA of virtualized clusters. In: ACM SIGPLAN/SIGOPS on Virtual Execution Env. pp. 93–106 (2015)
21. Xu, J., et al.: T-storm: traffic-aware online scheduling in storm. In: IEEE 34th International Conference on Distributed Computing Systems, pp. 535–544. IEEE (2014)
22. Yang, H., et al.: Precise online QoS man. for increased util. in warehouse computer, pp. 607–618 (2013)

Author Index

Printed in the United States
By Bookmasters